MW01674364

Hybrid Power Cycle Arrangements for Lower Emissions

Science, Technology, and Management Series

Series Editor: J. Paulo Davim, *Professor, Department of Mechanical Engineering, University of Aveiro, Portugal*

This book series focuses on special volumes from conferences, workshops, and symposiums, as well as volumes on topics of current interested in all aspects of science, technology, and management. The series will discuss topics such as, mathematics, chemistry, physics, materials science, nanosciences, sustainability science, computational sciences, mechanical engineering, industrial engineering, manufacturing engineering, mechatronics engineering, electrical engineering, systems engineering, biomedical engineering, management sciences, economical science, human resource management, social sciences, engineering education, etc. The books will present principles, models techniques, methodologies, and applications of science, technology and management.

Manufacturing and Industrial Engineering
Theoretical and Advanced Technologies
Edited by Pakaj Agarwal, Lokesh Bajpai, Chandra Pal Singh, Kapil Gupta, and J. Paulo Davim

Multi-Criteria Decision Modelling
Applicational Techniques and Case Studies
Edited by Rahul Sindhwani, Punj Lata Singh, Bhawna Kumar, Varinder Kumar Mittal, and J. Paulo Davim

High-k Materials in Multi-Gate FET Devices
Edited by Shubham Tayal, Parveen Singla, and J. Paulo Davim

Advanced Materials and Manufacturing Processes
Edited by Amar Patnaik, Malay Kumar, Ernst Kozeschnik, Albano Cavaleiro, J. Paulo Davim, and Vikas Kukshal

Computational Technologies in Materials Science
Edited by Shubham Tayal, Parveen Singla, Ashutosh Nandi, and J. Paulo Davim

Hybrid Power Cycle Arrangements for Lower Emissions
Edited by Anoop Kumar Shukla, Onkar Singh, Meeta Sharma, Rakesh Kumar Phanden, and J. Paulo Davim

For more information about this series, please visit: www.routledge.com/Science-Technology-and-Management/book-series/CRCSCITECMAN

Hybrid Power Cycle Arrangements for Lower Emissions

Edited by

Anoop Kumar Shukla

Department of Mechanical Engineering, Amity School of Engineering & Technology, Amity University, Noida, India

Onkar Singh

Department of Mechanical Engineering, Harcourt Butler Technical University, Kanpur, India

Meeta Sharma

Department of Mechanical Engineering, Amity School of Engineering & Technology, Amity University, Noida, India

Rakesh Kumar Phanden

Department of Mechanical Engineering, Amity School of Engineering & Technology, Amity University, Noida, India

J. Paulo Davim

Department of Mechanical Engineering, University of Aveiro, Aveiro, Portugal

CRC Press
Taylor & Francis Group
Boca Raton London New York

CRC Press is an imprint of the
Taylor & Francis Group, an **informa** business
A BALKEMA BOOK

First published 2022
by CRC Press/Balkema
Schipholweg 107C, 2316 XC Leiden, The Netherlands

e-mail: enquiries@taylorandfrancis.com

www.routledge.com – www.taylorandfrancis.com

CRC Press/Balkema is an imprint of the Taylor & Francis Group, an informa business

© 2022 selection and editorial matter, Anoop Kumar Shukla, Onkar Singh, Meeta Sharma, Rakesh Kumar Phanden and J. Paulo Davim; individual chapters, the contributors

The right of Anoop Kumar Shukla, Onkar Singh, Meeta Sharma, Rakesh Kumar Phanden and J. Paulo Davim to be identified as the authors of the editorial material, and of the authors for their individual chapters, has been asserted in accordance with sections 77 and 78 of the Copyright, Designs and Patents Act 1988.

All rights reserved. No part of this book may be reprinted or reproduced or utilised in any form or by any electronic, mechanical, or other means, now known or hereafter invented, including photocopying and recording, or in any information storage or retrieval system, without permission in writing from the publishers.

Although all care is taken to ensure integrity and the quality of this publication and the information herein, no responsibility is assumed by the publishers nor the author for any damage to the property or persons as a result of operation or use of this publication and/ or the information contained herein.

Library of Congress Cataloging-in-Publication Data
Names: Shukla, Anoop Kumar, editor. | Singh, Onkar, 1968– editor. |
Sharma, Meeta, editor. | Phanden, Rakesh Kumar, editor. | Davim, J. Paulo, editor.
Title: Hybrid power cycle arrangements for lower emissions / edited by
Anoop Kumar Shukla, Onkar Singh, Meeta Sharma, Rakesh Kumar Phanden, J. Paulo Davim.
Description: Leiden, The Netherlands : CRC Press/Balkema, 2022. |
Includes bibliographical references and index. |
Identifiers: LCCN 2021050836 (print) | LCCN 2021050837 (ebook) |
ISBN 9781032072531 (hardback) | ISBN 9781032101293 (paperback) |
ISBN 9781003213741 (ebook)
Subjects: LCSH: Hybrid power systems. | Hybrid power.
Classification: LCC TK1041 .H94 2022 (print) |
LCC TK1041 (ebook) | DDC 621.31/21–dc23/eng/20220111
LC record available at https://lccn.loc.gov/2021050836
LC ebook record available at https://lccn.loc.gov/2021050837

ISBN: 978-1-032-07253-1 (hbk)
ISBN: 978-1-032-10129-3 (pbk)
ISBN: 978-1-003-21374-1 (ebk)

DOI: 10.1201/9781003213741

Typeset in Times New Roman
by Newgen Publishing UK

Contents

List of contributors vii

1 Hybrid power cycle: an introduction 1

ONKAR SINGH, ANOOP KUMAR SHUKLA, AND MEETA SHARMA

2 Geothermal-based power system integrated with Kalina and organic Rankine cycle 11

AMIRMOHAMMAD BEHZADI AND AHMAD ARABKOOHSAR

3 Integrated gasification combined cycle with co-gasification 35

T. SRINIVAS, B. V. REDDY, AND A. V. S. S. K. S. GUPTA

4 Supercritical CO_2 cycle powered by solar thermal energy 45

RAMNEEK SINGH, RUPINDER PAL SINGH, AND DIBAKAR RAKSHIT

5 Integrated fuel cell hybrid technology 73

ABHINAV ANAND SINHA, TUSHAR CHOUDHARY, AND MOHD. ZAHID ANSARI

6 CHP coupled with a SOFC plant 93

NIMA NOROUZI, YASHAR PEYDAYESH, SAEED TALEBI, AND MARYAM FANI

7 Fuel cell hybrid power system 117

KRITI SRIVASTAVA AND ABHINAV ANAND SINHA

8 Solid oxide fuel cell integrated blade cooled gas turbine hybrid power cycle 135

TUSHAR CHOUDHARY, SANJAY, MITHILESH KUMAR SAHU, AND TIKENDRA NATH VERMA

9 Municipal solid waste-fueled plants 167

AHMAD ARABKOOHSAR AND AMIRMOHAMMAD BEHZADI

10 4E-analysis of sustainable hybrid tri-generation system 191

MEISAM SADI AND AHMAD ARABKOOHSAR

11 Trigeneration system: exergoeconomic and environmental analysis 219

JOY NONDY AND TAPAN KUMAR GOGOI

12 Organic Rankine cycle integrated hybrid arrangement for power generation 239

MOHAMMAD BAHRAMI, FATHOLLAH POURFAYAZ, AND ALI GHEIBI

13 Power-to-fuel: a new energy storage technique 269

ALPER CAN INCE, CAN OZGUR COLPAN, AND MUSTAFA FAZIL SERINCAN

Index 297

Contributors

Mohd. Zahid Ansari, PDPM Indian Institute of Information Technology, Design and Manufacturing, Jabalpur, India

Ahmad Arabkoohsar, Department of Energy Technology, Aalborg University, Aalborg, Denmark

Mohammad Bahrami, Department of Renewable Energies and Environment, Faculty of New Sciences and Technologies, University of Tehran, Tehran, Iran

Amirmohammad Behzadi, Department of Energy Technology, Aalborg University, Aalborg, Denmark

Tushar Choudhary, PDPM Indian Institute of Information Technology, Design and Manufacturing, Jabalpur, India

Can Ozgur Colpan, Dokuz Eylul University, Faculty of Engineering, Mechanical Engineering Department, Buca, Izmir, Turkey

Maryam Fani, Department of Energy Engineering and Physics, Amirkabir University of Technology (Tehran Polytechnic), Iran

Ali Gheibi, Department of Mechanical Engineering, University of Kashan, Kashan, Iran

Tapan Kumar Gogoi, Department of Mechanical Engineering, Tezpur University, Napaam, Tezpur, India

A. V. S. S. K. S. Gupta, Department of Mechanical Engineering, JNTU College of Engineering, Kukatpally, Hyderabad, India

Alper Can Ince, Gebze Technical University, Faculty of Engineering, Mechanical Engineering Department, Gebze, Kocaeli, Turkey

Joy Nondy, Department of Mechanical Engineering, Tezpur University, Napaam, Tezpur, India

Nima Norouzi, Department of Energy Engineering and Physics, Amirkabir University of Technology (Tehran Polytechnic), Iran

Yashar Peydayesh, School of Industrial Engineering, Tabriz University, Tabriz, Iran

Fathollah Pourfayaz, Department of Renewable Energies and Environment, Faculty of New Sciences and Technologies, University of Tehran, Tehran, Iran

Dibakar Rakshit, Centre for Energy Studies, Indian Institute of Technology Delhi, Hauz Khas New Delhi, India

B. V. Reddy, Faculty of Engineering and Applied Sciences, University of Ontario Tech University, Oshawa, ON, Canada

Meisam Sadi, Department of Engineering, Shahrood Branch, Islamic Azad University, Shahrood, Iran

Mithilesh Kumar Sahu, Gayatri Vidya Parishad College of Engineering (A), Visakhapatnam, India

Sanjay, National Institute of Technology Jamshedpur, India

Mustafa Fazıl Serincan, Gebze Technical University, Faculty of Engineering, Mechanical Engineering Department, Gebze, Kocaeli, Turkey

Meeta Sharma, Amity University, Noida, Uttar Pradesh, India

Anoop Kumar Shukla, Amity University, Noida, Uttar Pradesh, India

Onkar Singh, Harcourt Butler Technical University, Kanpur, India

Ramneek Singh, Department of Mechanical Engineering, Punjab Agricultural University, Ludhiana, India

Rupinder Pal Singh, Department of Mechanical Engineering, Punjab Agricultural University, Ludhiana, India

Abhinav Anand Sinha, PDPM Indian Institute of Information Technology, Design and Manufacturing, Jabalpur, India

T. Srinivas, Department of Mechanical Engineering, Dr. B. R. Ambedkar National Institute of Technology, Jalandhar, Punjab, India

Kriti Srivastava, Mechanical Engineering Department, IET Dr. R. M. L. Avadh University, Ayodhya, India

Saeed Talebi, Department of Energy Engineering and Physics, Amirkabir University of Technology (Tehran Polytechnic), Iran

Tikendra Nath Verma, Maulana Azad National Institute of Technology, Bhopal, India

Chapter 1

Hybrid power cycle

An introduction

Onkar Singh
Harcourt Butler Technical University, Kanpur, India

Anoop Kumar Shukla and Meeta Sharma
Amity University, Noida, India

1.1 INTRODUCTION

Present-day civilization depends on energy availability and its utilization. Most of the activities around us are realized by the use of electricity and all the ongoing technological advancements are electricity-centric. This has led to the world electricity consumption crossing 24,000 billion terawatt hours. Also, the constant increase in electricity requirement is evident from the civilization that depends heavily upon it. The power sector is under stress to enhance its generation potential. As a result, efforts have been made for capacity addition by setting up new power plants, improving the performance of the existing ones, and evolving newer arrangements by mitigating the losses so that these are more efficient. Simultaneously, concerns about depleting fuel reserves and emissions causing global warming has resulted in the increasing use of renewable sources of energy for power generation and direct energy conversion systems along with emission control in the conventional power plants.

Looking at the spectrum of power generation, it is evident that the conventional approaches of electricity generation like coal-based, hydroelectric, nuclear, renewable energy-based, etc. contribute to the total electricity generation across the world. Therefore, for meeting the growing demand, the increase in output from the existing power plants by plugging the losses and/or utilizing the energy effectively for augmenting power generation appears a reasonable proposition apart from setting up of new power plants. In this regard there are evidences of integrating different power generation propositions in synergy and operating the different power plants in hybrid mode for realizing;

- energy sustainability
- optimal utilization of energy in power cycle
- better efficiency
- better output
- combined cooling, heating, and power (CCHP)
- shifting from depleting fossil fuel reserves to renewable energy resources

To introduce the subject, a brief description of some popular arrangements is detailed below for understanding the basics involved in them.

DOI: 10.1201/9781003213741-1

1.2 COMBINED CYCLE

The synergetic combination of Brayton cycle (gas turbine) with Rankine cycle (steam turbine) resulting in the combined gas/steam cycle is a popular example of the integration of two thermodynamic cycles that are capable of operating in isolation with the sole aim of utilizing the gas turbine exhaust which would have been lost otherwise. Thus, the high temperature cycle, such as GT cycle, and lower temperature cycle, i.e., steam turbine cycle, combine together to give enhanced work output without additional fuel supply, increased thermal efficiency, low temperature exhaust with minimum emission impact, etc.

1.3 HYBRID SOLAR ASSISTED COMBINED COOLING, HEATING, AND POWER

The hybrid solar-assisted CCHP arrangement is an arrangement that is capable of offering power, heating, and cooling. It is worth mentioning that along with power the industrial processes do require heating and/or cooling. The proper usage of solar energy has huge potential of contributing to the cooling, heating, and power demands with negligible CO_2 emission (Sharma et al. 2021). The solar energy is harnessed through photovoltaic modules, heat collectors, and their combinations for CCHP systems. Figure 1.1 demonstrates different possible combinations for accommodating hybrid CCHP systems (Wang, Han, & Guan 2020). Different possible arrangements can be set up by integrating the respective locks as shown in the diagram. The solar energy utilization being at the core helps in mitigating the emissions and makes the complex arrangement environment friendly.

The hybrid solar gas turbine shown in Figure 1.2 has the air flowing through the solar receiver for harnessing the high temperatures before passing into the combustion chamber with provision of fuel addition for receiving the expanding gases at desired turbine inlet

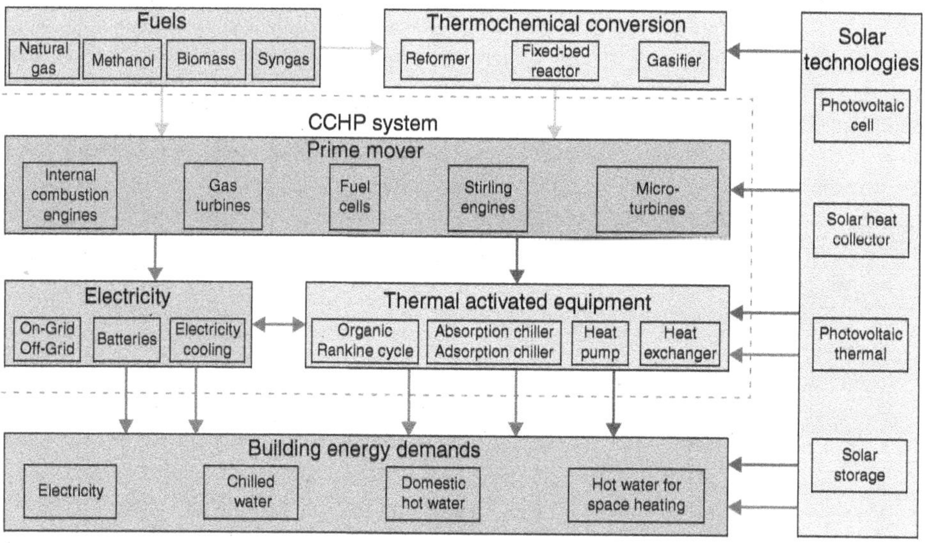

Figure 1.1 Hybrid solar assisted CCHP systems. (Source: Wang, Han, & Guan 2020.)

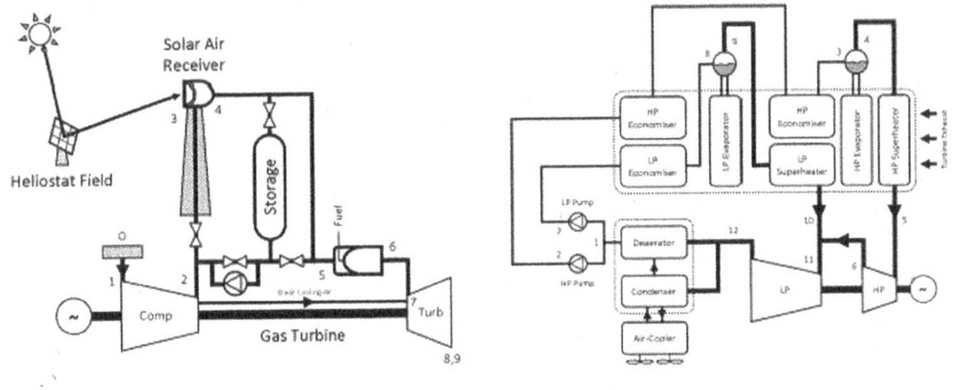

Figure 1.2 Hybrid solar gas turbine with steam turbine using heat recovery steam generator. (Source: Spelling, Laumert, & Fransson 2014.)

temperature of gas turbine (Spelling, Laumert, & Fransson 2014). Here additional thermal energy storage (TES) is deployed for storing the excess solar energy in day time and releasing the same during night or cloudy weather conditions. The heat recovery steam generator integrates the two cycles through steam generation by the gas turbine exhaust. The steam thus generated is used in the steam turbine run on the Rankine cycle.

The hybrid power systems can also be run on solar heat and methanol, which is an alternative fuel for vehicles and power generation. The methanol being produced by chemical synthesis of the syngas created by reforming natural gas or from coal gasification is better when compared to natural gas as it is easy to transport and store at atmospheric state. The coal-based polygeneration system having a combined cycle integrated with coal gasification and methanol synthesis using solar energy for decomposition and reforming makes such hybrid power systems efficient and less polluting (Li, Zhang, & Lior 2015).

Solar energy has the potential to be a viable alternative to fossil fuels with global availability, no pollution, and sustainability. The hybrid systems of concentrated solar power (CSP) with renewable energy resources are environmentally friendly with minimum impact on surroundings, while the environmental impact of hybrid CSP-non-renewable energy systems will be determined by the solar energy share in the hybrid system. Hybrid power cycles of CSP with coal, natural gas, biofuels, geothermal, photovoltaic, wind, etc. offer differing impacts on the surroundings, reliability, and flexibility as per (Mohammadi, Ellingwood, & Powell 2020). Pramanik and Ravikrishna (2017) classified the hybrid systems as

- high hybrid power plants of CSP with wind energy, biomass energy, geothermal energy, etc. with least impact on global warming and climate change
- medium hybrid power plants of CSP with supplementary firing using fossil fuels which have an impact on global warming and climate change more than high hybrid power plants
- low hybrid power plants having conventional fossil fuel-fired power plants integrated with CSP-based plants which have an impact on global warming and climate change more than medium hybrid power plants. These can be solar-Brayton cycle, solar-aided coal power plants, integrated solar combined power cycles, etc.

Also, hybrid biomass and solar energy systems have been tried in the form of a triple combined cycle to ensure better utilization of energy in the power cycle (Cao et al. 2021). Figure 1.3 shows the schematic arrangement of using solar energy for hydrogen generation followed by biomass-powered gas turbine in topping cycle operating together with bottoming cycle consisting of Brayton cycle and Rankine cycle, which yields reduced carbon dioxide emissions and enhanced power generation.

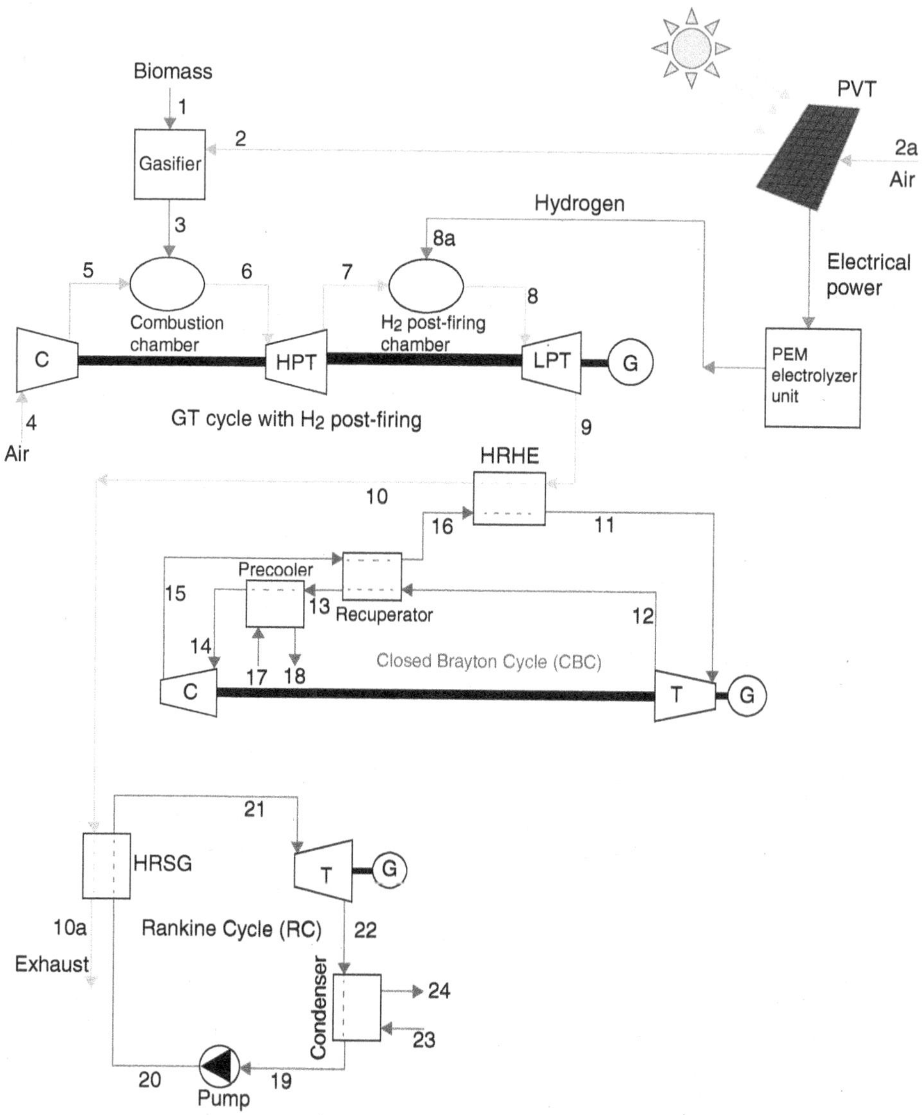

Figure 1.3 Hybrid biomass and solar energy system in triple combined cycle. (Source: Cao et al. 2021.)

Notes: C – compressor; G – generator; GT – gas turbine; HPT – high pressure turbine; HRHE – heat recovery heat exchanger; HRSG – heat recovery steam generator; LPT – low pressure turbine; PEM – proton exchange membrane; T – turbine.

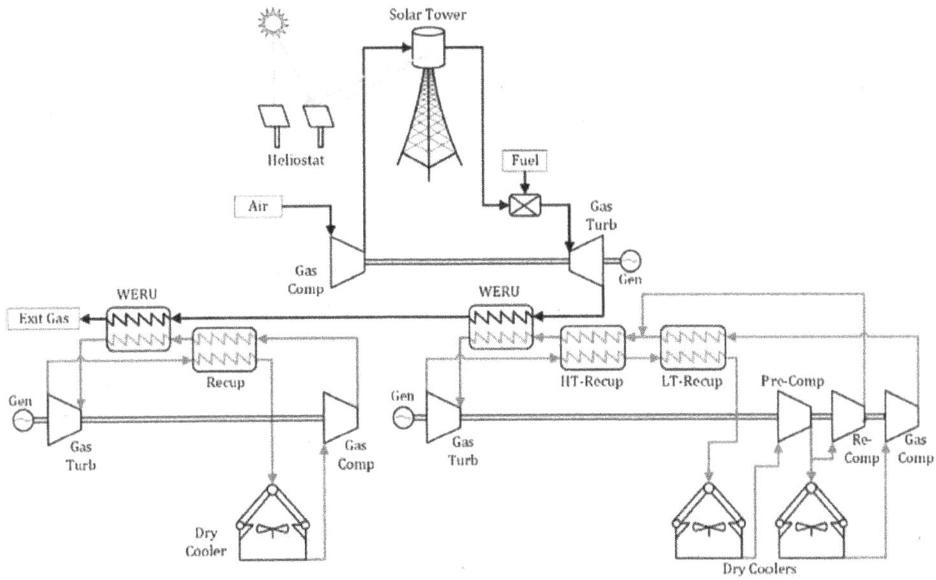

Figure 1.4 Hybrid solar GT combined cycle having supercritical carbon dioxide-based bottoming cycles as recuperative and recompression power cycles. (Source: Scaccabarozzi et al. 2021.)

Notes: Gas Comp – gas compressor; Gas Turb – gas turbine, Gen – generator; HT – high temperature; LT – low temperature; Pre-Comp – pre-compressor; Re-comp – recompressor; Recup – recuperator; WERU – waste energy recovery unit.

The integration of solar energy-operated gas turbine cycles with supercritical carbon dioxide-run power cycles utilize the waste energy from gas turbines and also the solar energy captured using solar towers (Mohammadi, Ellingwood, & Powell 2020). The bottoming cycle can have supercritical carbon dioxide cycles operating as a recompression power cycle trailed by a recuperative power cycle as shown in Figure 1.4.

The bottoming cycle can also be a supercritical carbon dioxide-based partial cooling cycle coupled with a recuperative cycle as shown in Figure 1.5. These arrangements of hybrid mode of operation help in better energy utilization and lesser emissions.

1.4 SOLID OXIDE FUEL CELL-BASED HYBRID SYSTEMS

The direct energy conversion systems working on solid oxide fuel cells (SOFC) have the high temperature discharge whose energy is used to power a gas turbine works on Brayton cycle or other power/cooling cycles to yield SOFC-based hybrid systems. Such hybrid systems consisting of SOFC and gas turbines run on SOFC waste energy can be further integrated with steam turbine-based Rankine cycle/organic Rankine cycle which are run on the exhaust waste energy available from the gas turbine. The integration of carbon capture systems to SOFC-based hybrid systems make them an environmentally friendly option. Figure 1.6 shows a simple block diagram for an SOFC-based hybrid power cycle.

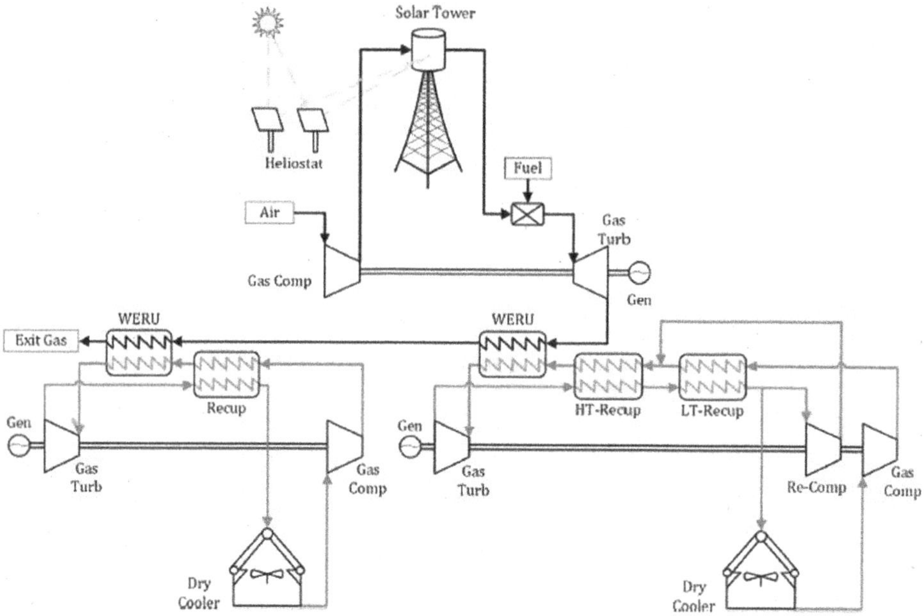

Figure 1.5 Hybrid solar GT combined cycle having bottoming cycles with supercritical carbon dioxide-based limited cooling and recuperative power cycles. (Source: Scaccabarozzi et al. 2021.)

Notes: Gas Comp – gas compressor; Gas Turb – gas turbine, Gen – generator; HT – high temperature; LT – low temperature; Pre-Comp – pre-compressor; Re-comp – recompressor; Recup – recuperator; WERU – waste energy recovery unit.

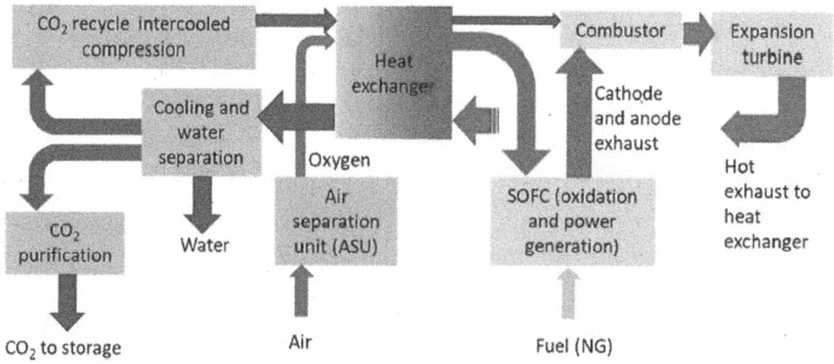

Figure 1.6 SOFC-based hybrid power cycle. (Source: Scaccabarozzi et al. 2021.)

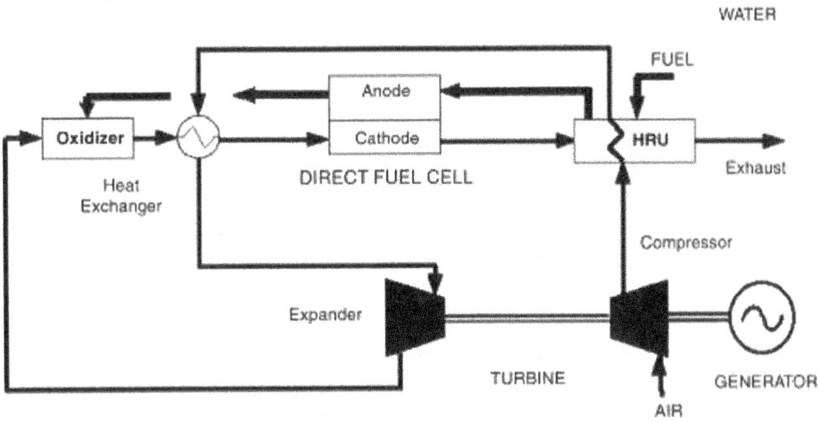

Figure 1.7 Hybrid power cycle involving molten carbonate fuel cell and microturbine. (Source: Jurado 2002.)

Note: HRU: Heat Recovery Unit

1.4.1 Molten carbonate fuel cell-based microturbine hybrid power cycle

A hybrid power cycle involving a molten carbonate fuel cell operating with an indirectly fired turbine has the fuel cell positioned at turbine's exhaust (Jurado 2002). In Figure 1.7, air and residuals leaving the fuel cell are sent to the atmospheric combustor which heats the air coming from compressor in the heat recovery unit and sends the same to the turbine whose exhaust goes to the fuel cell anode exhaust oxidizer and then to the heat exchanger to heat the compressed air. Exhaust from the heat exchanger passes to the fuel cell cathode to provide the oxygen and carbon dioxide required in the carbonate fuel cell.

1.5 GEOTHERMAL ENERGY-BASED HYBRID POWER SYSTEMS

Geothermal energy, being a sustainable and carbon-free resource, can be combined with other sources of energy to evolve a power system. The hybrid operation of a solar thermal electric power plant with a geothermal energy-based power plant offers good performance. Figure 1.8 shows the self-explanatory layout of the hybrid operation of solar and geothermal energy-based power systems that is environmentally friendly (Bonyadi, Johnson, & Baker 2018).

In view of the perennial requirement of improving energy efficiency through better utilization of energy and minimizing global warming impact to control climate change, the hybrid cycles have been tried with a series of combinations some of which have been introduced above. The hybrid power cycles based on different arrangements demonstrate hope for energy efficient and sustainable operation of power plants based on them.

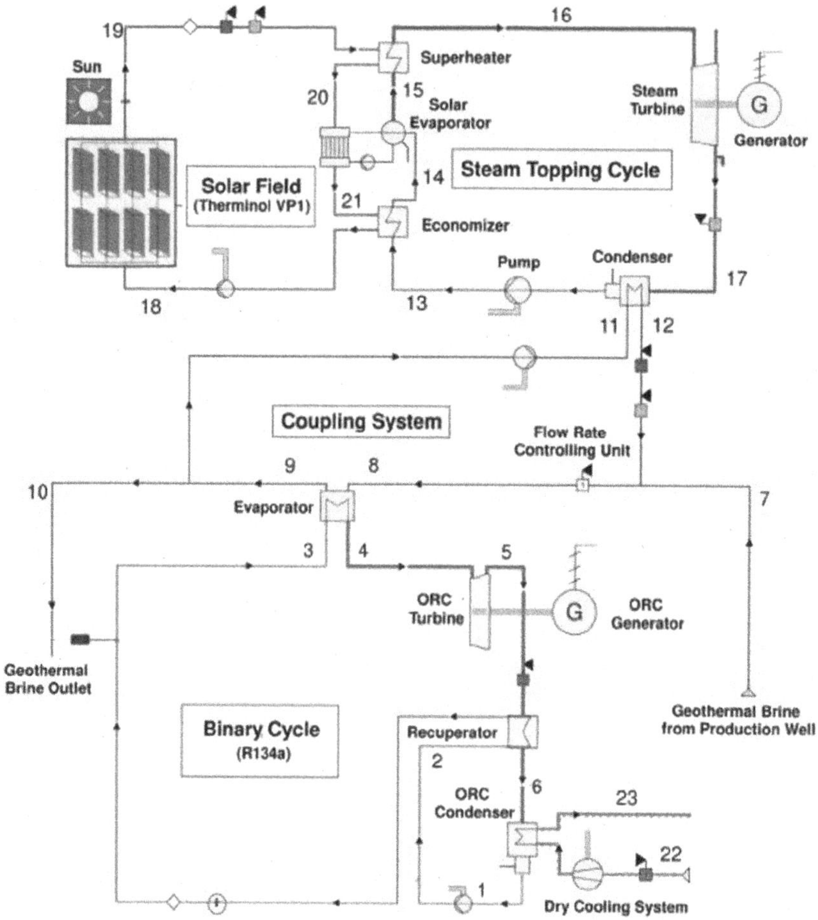

Figure 1.8 Hybrid solar and geothermal energy-based power system. (Source: Bonyadi, Johnson, & Baker 2018.)

Notes: ORC – organic Rankine cycle; Therminol VP1 – synthetic oil that acts as heat transfer fluid; R134a – organic fluid in ORC.

REFERENCES

Bonyadi, Nima, Evan Johnson, & Derek Baker. 2018. "Technoeconomic and exergy analysis of a solar geothermal hybrid electric power plant using a novel combined cycle." *Energy Conversion and Management* 156: 542–554. doi:10.1016/J.ENCONMAN.2017.11.052.

Cao, Yan, Hayder A. Dhahad, Hussein Togun, Ali E. Anqi, Naeim Farouk, & Babak Farhang. 2021. "A novel hybrid biomass-solar driven triple combined power cycle integrated with hydrogen production: Multi-objective optimization based on power cost and CO_2 emission." *Energy Conversion and Management* 234: 113910. doi:10.1016/J.ENCONMAN.2021.113910.

Jurado, Francisco. 2002. "Study of molten carbonate fuel cell: Microturbine hybrid power cycles." *Journal of Power Sources* 111(1): 121–129. doi:10.1016/S0378-7753(02)00340-3.

Li, Yuanyuan, Na Zhang, & Noam Lior. 2015. "Performance comparison of two low-CO_2 emission solar/methanol hybrid combined cycle power systems." *Applied Energy* 155: 740–752. doi:10.1016/J.APENERGY.2015.06.052.

Mohammadi, Kasra, Kevin Ellingwood, & Kody Powell. 2020. "Novel hybrid solar tower-gas turbine combined power cycles using supercritical carbon dioxide bottoming cycles." *Applied Thermal Engineering* 178: 115588. doi:10.1016/J.APPLTHERMALENG.2020.115588.

Pramanik, Santanu, & R. V. Ravikrishna. 2017. "A review of concentrated solar power hybrid technologies." *Applied Thermal Engineering* 127: 602–637. doi:10.1016/J.APPLTHERMALENG.2017.08.038.

Scaccabarozzi, Roberto, Manuele Gatti, Stefano Campanari, & Emanuele Martelli. 2021. "Solid oxide semi-closed CO_2 cycle: A hybrid power cycle with 75% net efficiency and zero emissions." *Applied Energy* 290: 116711. doi:10.1016/J.APENERGY.2021.116711.

Sharma, Achintya, Anoop Kumar Shukla, Onkar Singh, & Meeta Sharma. 2021. "Recent advances in gas/steam power cycles for concentrating solar power." *International Journal of Ambient Energy*. doi:10.1080/01430750.2021.1919552.

Spelling, James, Björn Laumert, & Torsten Fransson. 2014. "Advanced hybrid solar tower combined-cycle power plants." *Energy Procedia* 49: 1207–1217. doi:10.1016/J.EGYPRO.2014.03.130.

Wang, Jiangjiang, Zepeng Han, & Zhimin Guan. 2020. "Hybrid solar-assisted combined cooling, heating, and power systems: A review." *Renewable and Sustainable Energy Reviews* 133: 110256. doi:10.1016/J.RSER.2020.110256.

Chapter 2

Geothermal-based power system integrated with Kalina and organic Rankine cycle

Amirmohammad Behzadi and Ahmad Arabkoohsar
Department of Energy Technology, Aalborg University, Denmark

2.1 INTRODUCTION

In recent years, an increase in power demand and environmental pollutions has been observed unprecedentedly (Sadi & Arabkoohsar 2019). Renewable energy expansion and appropriate performance improvement approaches such as waste heat recovery are essential areas of researchers' attention in power generation systems (Behzadi, Arabkoohsar, & Perić 2021; Lorzadeh et al. 2021).

2.1.1 The worldwide availability and potential of geothermal energy sources

Of all kinds of renewable resources, low-temperature geothermal is a secured and high potential alternative resource to fossil fuels due to the consistency, reliability, and minimal environmental pollution (Wang et al. 2021). Moreover, compared to other renewable sources, geothermal energy resources have the highest capacity factor because they are available 24 hours per day, 365 days a year and are only shut down for maintenance (Spittler et al. 2020). Also, it has an intrinsic storage capability and can be run continuously to supply needs during regular use periods and at peak demand when there is surplus demand (Mahzari et al. 2021). According to its specifications, geothermal energy can be applied to generate power or for heat and cooling production. While power production from the geothermal resource is used in about 30 countries, geothermal heating is currently extracted in more than 70 countries (El Bassam, Maegaard, & Schlichting 2013). Based on the latest report of the geothermal energy association, the worldwide geothermal power capacity is anticipated to rise about 18.4 GW by the end of 2021 (El Bassam, Maegaard, & Schlichting 2013). While power generation with geothermal is rather spread between countries in diverse economic categories, direct use of geothermal source (heating and cooling production) is performed in the top 14 high-income countries (Lund & Toth 2021). According to the latest reports, Iceland and El Salvador have the highest percentage share of geothermal energy in their national electricity production (Lund & Toth 2021).

2.1.2 Techno-economic-environmental comparison

Organic Rankine cycle (ORC) and Kalina cycle system (KCS) as two competitors are suitable choices for utilizing the low-temperature heat source. The main distinction between the two cycles is that while ORC uses a pure working fluid, KCS utilizes an ammonia-water

DOI: 10.1201/9781003213741-2

mixture leading to different techno-economic metrics (Gholamian & Zare 2016). Many researchers have investigated the thermodynamic and economic analysis of ORC and KCS driven by various sources like solar, geothermal, and exhaust gases. The comparison of KCS 34 and ORC integrated with internal heat exchanger driven by high- and low-temperature geothermal heat source was studied by Fiaschi et al. (2017). They concluded that ORC using R1233zd(E) as the working fluid is superior to the Kalina from the exergoeconomic facet because of the 3% lower levelized cost of electricity. In another study, Shokati, Ranjbar, and Yari (2015) compared and optimized the exergoeconomic indicators of a dual-fluid ORC and Kalina 34. They revealed that at the optimal condition, the maximum value of produced power corresponds to the dual-pressure ORC, and the turbine is the most crucial section from the technoeconomic aspect. Rodríguez et al. (2013) investigated and compared the exergetic and economic metrics of simple ORC using 15 different working fluids against KCS 34 with three unlike compositions for the case of Brazil. They showed that the best techno-economic performance is achieved at the ammonia-water mass fraction of (84%–16%) and using R-290 as the ORC working fluid.

Techno-economic comparison of transcritical CO_2 cycles and Kalina cycle was assessed by Meng et al. (2019), concluding that while transcritical CO_2 cycle integrated with internal heat exchanger has a higher produced power, Kalina cycle is economically superior. Wang et al. (2017) studied replacing ORC with the Kalina cycle for multi-stream low-temperature waste heat recovery. They concluded that ORC and Kalina cycle are appropriate with straight and convex heat, respectively. A thermodynamic assessment to compare the pros and cons of simple ORC and KCS 11 for recovering the gas turbine cycle's waste heat was investigated by Nemati et al. (2017), reporting that ORC is superior due to the higher power production. Bombarda et al. (2010) evaluated the thermodynamic comparison of KCS 34 against simple ORC for the case of heat recovery from diesel engines and demonstrated that the Kalina cycle is not a proper option for low power level applications. Lately, Wang, Yang, and Xu (2021) studied and compared the use of ORC against implementing KCS 34 to exploit the waste heat of the flue gas of a roller skin from an environmental viewpoint. They revealed that while the Kalina cycle is a proper option from an emission mitigation aspect, ORC results in a lower environmental impact. Gogoi and Hazarika (2020) studied four solar-driven systems combined with different absorption chillers, ORC, and Kalina cycles. They showed that a lower irreversibility and exergy destruction rate is obtained using Kalina. Wang et al. (2017) concluded that the ratio of latent to the sensible heat source and waste heat temperature are two significant parameters affecting the operation of ORC and Kalina cycles. Based on their results, when the ratio is small, the Kalina cycle operates better than the ORC. However, due to the more power generation, ORC is superior to the Kalina cycle when the waste heat temperature is higher than 190°C. Techno-economic-environmental comparison of different configurations of ORC and KCS 34 integrated with gas turbine system was proposed and compared by Köse, Koç, and Yağlı (2021), resulting in a payback period of 3.48, 3.22, and 3.39 years for ORC, Kalina, and ORC-Kalina arrangements. Ghaebi and Rostamzadeh (2020) investigated and compared the effect of main operational parameters on energy and exergy indicators of a solar-based cogeneration system combined with ORC and KCS 11. They showed that the integration with ORC using R600a leads to a better performance compared to the KCS. The influence of turbine inlet pressure and NH_3 mass fraction on a power system driven by liquefied natural gas integrated with KCS 34 and ORC was carried out by Pan et al. (2017) to obtain a lower exergy loss and higher power production.

2.2 MULTI-CRITERIA OPTIMIZATION

Although numerous studies have been carried out on evaluating and comparing the ORC and KCS systems from thermodynamic and thermoeconomic perspectives, less work has been focused on optimizing the proposed systems and comparing them at the optimal condition. Lately, Musharavati et al. (2021) applied multi-objective optimization based on the genetic approach to a combined cooling, heating, and power system driven by geothermal. According to their results, considerably higher exergy efficiency and lower levelized cost of electricity are achieved at the optimum evaporation pressure of 8 bars. A micro combined cooling, heating, and power system integrated with KCS 11 and simple ORC was compared and optimized by Rostamzadeh et al. (2019), concluding that the Kalina-based system has a higher optimum energy efficiency of 0.78% and lower optimum exergy efficiency of 17.17% than the ORC-based configuration. Behzadi et al. (2018) optimized a waste-fueled power system integrated with ORC using the genetic algorithm method in MATLAB. They reported that the highest optimum exergy efficiency and electricity cost of 19.61% and 24.65 $/GJ is attained when R123 is selected as the ORC fluid. A power system equipped with gas turbine and Kalina cycles was examined and augmented by Parikhani et al. (2019). They resulted in the maximum exergy efficiency of 53.2% at the optimal ammonia concentration and evaporation temperature of 37.39% and 402 K, respectively. Xie et al. (2020) implemented multi-objective optimization to an integrated system comprising Kalina cycle to enhance the system's performance from energy, economic, and environmental standpoints. Various configurations of low-temperature power systems driven by a gas engine were optimized and compared by Milani, Saray, and Majafi (2020). They demonstrated that for the best configuration (ORC with the internal heat exchanger), at the optimum point, higher exergy efficiency of 14.31% and a lower investment cost of 8.2% is achieved compared to the design condition. In another study, Ebrahimi-Moghadam et al. (2021) optimized an innovative tri-generation system integrated with the Kalina cycle using a genetic algorithm and artificial neural network. They revealed that the system's techno-economic indicators improve remarkably at the best optimum condition.

2.2.1 Contributions of this chapter

Despite the intensive interest and extensive work in the Kalina cycle and ORC, the literature suffers from a thorough techno-economic-sustainability comparison between all the possible configurations to obtain an efficient, cost-effective, and clean option for waste heat recovery at low-temperature. According to the literature, most of the research works investigate only a few scenarios at the design condition, focusing on the comparison at the best optimal condition obtained by multi-criteria optimization.

In this chapter, five feasible waste heat recovery technologies, including simple ORC, ORC integrated with internal heat exchanger, ORC integrated with open feed heater, KCS 11, and KCS 34, driven by geothermal source, are analyzed and compared. The comparison is performed by assessing the performance efficiencies, sustainability index, levelized cost of power, and total cost rate of each configuration via parametric study. Each component's share on the configurations' total exergy destruction is evaluated and compared through a comprehensive exergy analysis. Finally, multi-criteria genetic algorithm optimization is applied to each system to find the most favorable condition from techno-economic facets.

2.3 SELECTION AND DESCRIPTION OF PROPOSED SYSTEM CONFIGURATIONS

Figure 2.1 presents each geothermal-driven power system's schematic comprising evaporator, turbine, condenser, pump, open feed heater, internal heat exchanger, separator, absorber, regenerator, and valve. According to the working fluid and the configuration, they are categorized into five scenarios, including simple ORC (configuration (a)), ORC integrated with an internal heat exchanger (configuration (b)), ORC integrated with open feed heater (configuration (c)), Kalina cycle system 11 (configuration (d)), and Kalina cycle system 34 (configuration (e)). According to the figure, the first system (configuration (a)) is a simple ORC using geothermal water as the heat source. As depicted, the geothermal water coming from the production well (state 1) enters the evaporator to heat the ORC working fluid. The superheated fluid goes

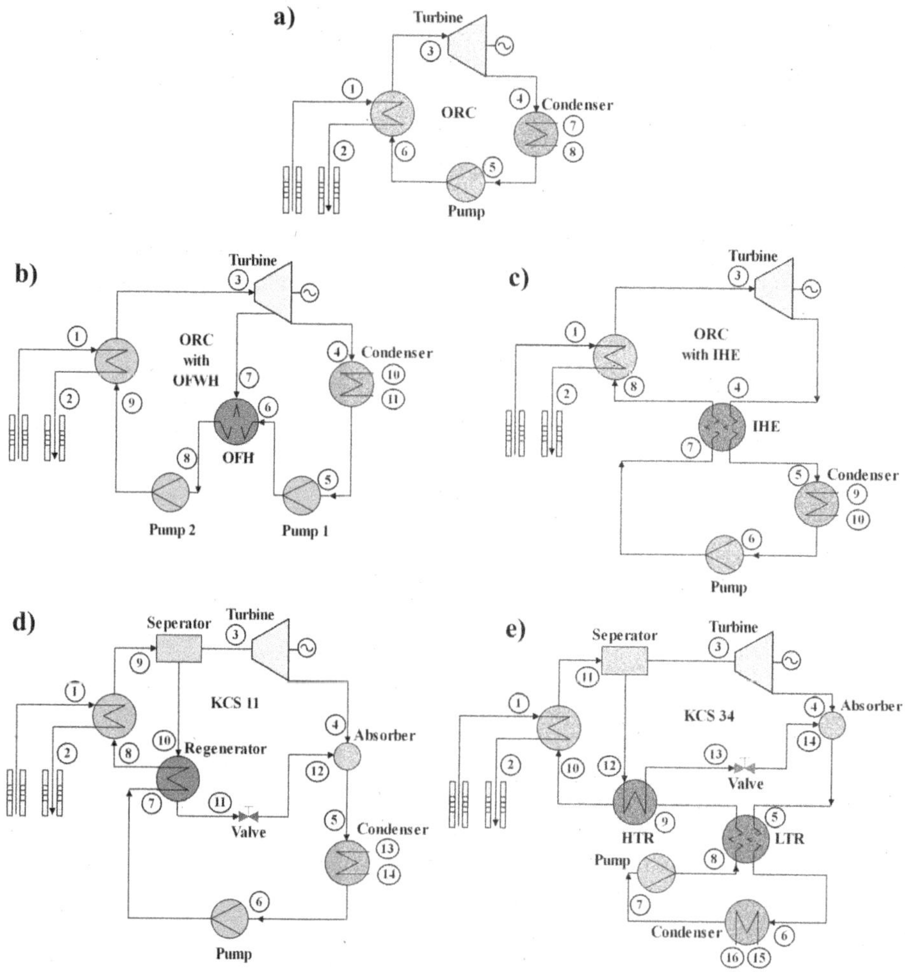

Figure 2.1 Schematic diagram of each configuration.

into the turbine (state 3) generating power, and after being condensed in the condenser heat exchanger (state 4), it is pumped into the evaporator (state 6).

Configuration (b) is equipped with an open feed heater (OFH) as a mixing chamber, where the steam extracted from the turbine (state 6) mixes with the fluid exiting the pump (state 6) to generate a saturated liquid at the heater pressure (state 9). Configuration (c) is another modified version of ORC integrated with an internal heat exchanger improving the system efficiency by preheating the pump outlet fluid before entering the evaporator.

The Kalina cycle is an innovative low-temperature power cycle using a zeotropic mixture of two fluids (ammonia and water), leading to a lower temperature mismatching in the evaporator, improving the quality of the WHR process. Among various Kalina cycles proposed in the literature, the newest and advanced ones (KCS 11 and KCS 34) are investigated and compared in the present study. According to Figure 2.1(d), the evaporated ammonia-water mixture (state 9) with a mass fraction of 0.8 leaves the evaporator toward the separator. In the separator, the mixture is divided into the ammonia-rich vapor (state 3) entering the turbine to produce power and the weak saturated solution (state 10) going into the regenerator to heat the cold stream from the pump (state 7). In the absorber, the turbine outlet mixture (state 4) is mixed with the fluid passing through the valve (state 12). Afterward, the mixed solution (state 5) passes through the condenser to become a saturated liquid (state 6). As depicted in Figure 2.1(e), the KCS 34 is a modified version of KCS 11. A low-temperature regenerator is added to preheat the pump outlet mixture before entering the high-temperature regenerator.

2.3.1 Methodology

The thermo-economic, and sustainability equations of every component are developed in engineering equation software (EES) to examine and compare each configuration from various perspectives. Then, the proposed configurations are optimized by linking EES to MATLAB based on a genetic algorithm approach.

2.3.2 Thermodynamic analysis

Thermodynamic evaluation is carried out by comparing the energy, exergy, and sustainability indicators of each configuration. For this, the conservation laws of mass and energy, as well as the exergy balance equation for every component, are written as below (Behzadi & Arabkoohsar 2020; Arabkoohsar, Behzadi, & Alsagri 2021):

$$\sum \dot{m}_{in} = \sum \dot{m}_{out} \tag{2.1}$$

$$\dot{Q} - \dot{W} = \sum \dot{m}_{out} h_{out} - \sum \dot{m}_{in} h_{in} \tag{2.2}$$

$$\dot{E}x_Q - \dot{E}x_W = \sum \dot{m}_{out} e_{out} - \sum \dot{m}_{in} e_{in} + \dot{E}x_D \tag{2.3}$$

In Equation (2.3), $\dot{E}x_D$ is the rate of exergy destruction, which is a critical thermodynamic parameter indicating each component's irreversibility. Also, Q and W indices correspond to the exergy associated with heat and work. The following assumptions are made to fulfill the thermodynamic modeling of each configuration:

- Potential and kinetic energy changes are zero.
- Heat loss in the components and pressure reduction within the pipes are neglected.
- The weather temperature and pressure are, respectively, 25°C and 1 bar.
- The geothermal water enters the evaporator with a temperature and mass flow rate of 160°C and 100 kg/s.
- The ammonia-water mass fraction is assumed to be (80%–20%), and isobutane is selected as the ORC working fluid.
- The turbines and pump isentropic efficiencies are assumed to be 85% and 90%.
- The condensers and evaporator's pinch point temperature difference are 5°C, and the superheat temperature difference is 10°C.
- The turbine inlet pressure is 35 bar, and the condenser cooling water goes into the condenser with the ambient temperature and pressure.

The detailed mathematical equations of energy and exergy analysis of each configuration are listed in Table 2.1.

After applying first- and second-law analysis to each configuration and solving the set of equations, the net produced and consumed power and each component's exergy destruction rate are determined. The energy and exergy efficiencies as the ratio of net power to the input heat and exergy are, respectively, calculated as:

$$\eta_I = \frac{\dot{W}_{turbine} - \dot{W}_{pump}}{\dot{Q}_{geothermal}} \times 100 \tag{2.4}$$

$$\dot{Q}_{geothermal} = \dot{m}_? \left(h_? - h_?\right) \tag{2.5}$$

$$\eta_{II} = \frac{\dot{W}_{turbine} - \dot{W}_{pump}}{\dot{E}_{geothermal}} \times 100 \tag{2.6}$$

$$\dot{E}_{geothermal} = \dot{m}_? \left(e_? - e_?\right) \tag{2.7}$$

Defining the depletion factor (D_P) as the ratio of destructed exergy to the input exergy, the sustainability index (*SI*), which evaluate the impact of exergy destruction decline on the enhancement of the system's environmental friendliness, is calculated as (Nami, Arabkoohsar, & Anvari-Moghaddam 2019):

$$SI = \frac{1}{D_P} \tag{2.8}$$

Table 2.1 The energy and exergy balance for each component

Configuration	*Component*	*Energy balance*	*Exergy balance*
Simple ORC (a)	**Evaporator**	$\dot{m}_1(h_1 - h_2) = \dot{m}_3(h_3 - h_6)$	$\dot{E}x_{D,evap} = \dot{E}x_1 + \dot{E}x_6 - \dot{E}x_2 - \dot{E}x_3$
	Turbine	$\dot{W}_T = \dot{m}_3(h_3 - h_4)$	$\dot{E}x_{D,T} = \dot{E}x_3 - \dot{W}_T - \dot{E}x_4$
	Condenser	$\dot{m}_4(h_4 - h_5) = \dot{m}_7(h_8 - h_7)$	$\dot{E}x_{D,cond} = \dot{E}x_4 + \dot{E}x_7 - \dot{E}x_5 - \dot{E}x_8$
	Pump	$\dot{W}_{Pu} = \dot{m}_5(h_6 - h_5)$	$\dot{E}x_{D,Pu} = \dot{E}x_5 + \dot{W}_{Pu} - \dot{E}x_6$
ORC with open feed heater (b)	**Evaporator**	$\dot{m}_1(h_1 - h_2) = \dot{m}_3(h_3 - h_9)$	$\dot{E}x_{D,evap} = \dot{E}x_1 + \dot{E}x_9 - \dot{E}x_2 - \dot{E}x_3$
	Turbine	$\dot{W}_T = \dot{m}_3(h_3 - h_4)$	$\dot{E}x_{D,T} = \dot{E}x_3 - \dot{W}_T - \dot{E}x_4$
	OFH	$\dot{m}_6h_6 + \dot{m}_7h_7 = \dot{m}_8h_8$	$\dot{E}x_{D,OFH} = \dot{E}x_6 + \dot{E}x_7 - \dot{E}x_8$
	Condenser	$\dot{m}_4(h_4 - h_5) = \dot{m}_{10}(h_{11} - h_{10})$	$\dot{E}x_{D,cond} = \dot{E}x_4 + \dot{E}x_{10} - \dot{E}x_5 - \dot{E}x_{11}$
	Pump 1	$\dot{W}_{Pu1} = \dot{m}_5(h_6 - h_5)$	$\dot{E}x_{D,Pu1} = \dot{E}x_5 + \dot{W}_{Pu} - \dot{E}x_6$
	Pump 2	$\dot{W}_{Pu2} = \dot{m}_5(h_9 - h_{58})$	$\dot{E}x_{D,Pu2} = \dot{E}x_8 + \dot{W}_{Pu} - \dot{E}x_9$
ORC with internal heat exchanger (IHE) (c)	**Evaporator**	$\dot{m}_1(h_1 - h_2) = \dot{m}_3(h_3 - h_8)$	$\dot{E}x_{D,evap} = \dot{E}x_1 + \dot{E}x_8 - \dot{E}x_2 - \dot{E}x_3$
	Turbine	$\dot{W}_T = \dot{m}_3(h_3 - h_4)$	$\dot{E}x_{D,T} = \dot{E}x_3 - \dot{W}_T - \dot{E}x_4$
	IHE	$\dot{m}_4(h_4 - h_5) = \dot{m}_7(h_8 - h_7)$	$\dot{E}x_{D,IHE} = \dot{E}x_4 + \dot{E}x_7 - \dot{E}x_5 - \dot{E}x_8$
	Condenser	$\dot{m}_5(h_5 - h_6) = \dot{m}_9(h_{10} - h_9)$	$\dot{E}x_{D,cond} = \dot{E}x_5 + \dot{E}x_9 - \dot{E}x_6 - \dot{E}x_{10}$
	Pump	$\dot{W}_{Pu} = \dot{m}_6(h_7 - h_6)$	$\dot{E}x_{D,Pu} = \dot{E}x_6 + \dot{W}_{Pu} - \dot{E}x_7$
KCS 11	**Evaporator**	$\dot{m}_1(h_1 - h_2) = \dot{m}_9(h_9 - h_8)$	$\dot{E}x_{D,evap} = \dot{E}x_1 + \dot{E}x_8 - \dot{E}x_2 - \dot{E}x_9$
	Separator	$\dot{m}_9h_9 = \dot{m}_{10}h_{10} + \dot{m}_3h_3$	$\dot{E}x_{D,sep} = \dot{E}x_9 - \dot{E}x_3 - \dot{E}x_{10}$
	Turbine	$\dot{W}_T = \dot{m}_3(h_3 - h_4)$	$\dot{E}x_{D,T} = \dot{E}x_3 - \dot{W}_T - \dot{E}x_4$
	Absorber	$\dot{m}_4h_4 + \dot{m}_{12}h_{12} = \dot{m}_5h_5$	$\dot{E}x_{D,abs} = \dot{E}x_4 + \dot{E}x_{12} - \dot{E}x_5$
	Condenser	$\dot{m}_5(h_5 - h_6) = \dot{m}_{13}(h_{14} - h_{13})$	$\dot{E}x_{D,cond} = \dot{E}x_5 + \dot{E}x_{13} - \dot{E}x_6 - \dot{E}x_{14}$
	Valve	$h_{11} = h_{12}$	$\dot{E}x_{D,valve} = \dot{E}x_{11} - \dot{E}x_{12}$
	Pump	$\dot{W}_{Pu} = \dot{m}_6(h_7 - h_6)$	$\dot{E}x_{D,Pu} = \dot{E}x_6 + \dot{W}_{Pu} - \dot{E}x_7$
	Regenerator	$\dot{m}_{10}(h_{10} - h_{11}) = \dot{m}_7(h_8 - h_7)$	$\dot{E}x_{D,Reg} = \dot{E}x_7 + \dot{E}x_{10} - \dot{E}x_8 - \dot{E}x_{11}$
KCS 34	**Evaporator**	$\dot{m}_1(h_1 - h_2) = \dot{m}_{10}(h_{11} - h_{10})$	$\dot{E}x_{D,evap} = \dot{E}x_1 + \dot{E}x_{10} - \dot{E}x_2 - \dot{E}x_{11}$
	Separator	$\dot{m}_{11}h_{11} = \dot{m}_{12}h_{12} + \dot{m}_3h_3$	$\dot{E}x_{D,sep} = \dot{E}x_{11} - \dot{E}x_3 - \dot{E}x_{12}$
	Turbine	$\dot{W}_T = \dot{m}_3(h_3 - h_4)$	$\dot{E}x_{D,T} = \dot{E}x_3 - \dot{W}_T - \dot{E}x_4$
	Absorber	$\dot{m}_4h_4 + \dot{m}_{14}h_{14} = \dot{m}_5h_5$	$\dot{E}x_{D,abs} = \dot{E}x_4 + \dot{E}x_{14} - \dot{E}x_5$
	LTR	$\dot{m}_5(h_5 - h_6) = \dot{m}_8(h_9 - h_8)$	$\dot{E}x_{D,LTR} = \dot{E}x_5 + \dot{E}x_8 - \dot{E}x_6 - \dot{E}x_9$
	HTR	$\dot{m}_{12}(h_{12} - h_{13}) = \dot{m}_9(h_{10} - h_9)$	$\dot{E}x_{D,LTR} = \dot{E}x_5 + \dot{E}x_8 - \dot{E}x_6 - \dot{E}x_9$
	Condenser	$\dot{m}_6(h_6 - h_7) = \dot{m}_{15}(h_{16} - h_{15})$	$\dot{E}x_{D,cond} = \dot{E}x_6 + \dot{E}x_{15} - \dot{E}x_7 - \dot{E}x_{16}$
	Pump	$\dot{W}_{Pu} = \dot{m}_7(h_8 - h_7)$	$\dot{E}x_{D,Pu} = \dot{E}x_7 + \dot{W}_{Pu} - \dot{E}x_8$
	Valve	$h_{13} = h_{14}$	$\dot{E}x_{D,valve} = \dot{E}x_{13} - \dot{E}x_{14}$

$$D_P = \frac{\sum_{k=1}^{n_k} \dot{E}_{D,k}}{\dot{E}_{geothermal}} \tag{2.9}$$

2.3.3 Exergoeconomic analysis

Exergoeconomic assessment is a robust method to evaluate energy systems' economics based on the second law standpoint. According to this approach, the inlet ($\dot{C}_{in}$) and outlet ($\dot{C}_{out}$) stream cost rates, the total component cost rate ($\dot{Z}$), power cost ($\dot{C}_w$), and heat cost ($\dot{C}_q$) are calculated based on the specific cost theory to find the total cost rate (TCR), and levelized cost of power (LCOP) as the essential economic indicators. In conducting this evaluation, the cost balance equation for the ith component is calculated and written as (Wang et al. 2020):

$$\sum \dot{C}_{out.k} + \dot{C}_{w.k} = \sum \dot{C}_{in.k} + \dot{C}_{q.k} + \dot{Z}_k \tag{2.10}$$

Here the total component expenses and stream costs are defined as following (Fakhari et al. 2021):

$$\dot{Z}_k = \dot{Z}_k^{CI} + \dot{Z}_k^{OM} \tag{2.11}$$

$$\dot{C}_{in} = c_{in} \dot{E}x_{in} \tag{2.12}$$

$$\dot{C}_{out} = c_{out} \dot{E}x_{out} \tag{2.13}$$

Where c is the unit exergy cost and the CI and OM superscripts correspond to the capital asset and operational and maintenance expenses. Defining CRF, τ, i, and n as the capital recovery factor, operating hours over the year, interest rate, and the working years' number, $\dot{Z}_k^{CI}$ and $\dot{Z}_k^{OM}$ are obtained from (Mojaver et al. 2020):

$$\dot{Z}_k^{CI} = \left(\frac{CRF}{\tau}\right) Z_k \tag{2.14}$$

$$\dot{Z}_k^{OM} = \left(\frac{\gamma_k}{\tau}\right) Z_k \tag{2.15}$$

$$CRF = \frac{i(1+i)^n}{(1+i)^n - 1} \tag{2.16}$$

Detailed information about the components cost (Z_k) and their corresponding formulas are tabulated in Appendix A. After writing the cost balance equations and finding the unknown economic parameters, the exergoeconomic evaluation is fulfilled by calculating the unit product and fuel costs and exergoeconomic factor (Mehrabadi & Boyaghchi 2019):

$$c_{F.k} = \frac{\dot{C}_{F.k}}{\dot{E}x_{F.k}} \tag{2.17}$$

$$c_{P.k} = \frac{\dot{C}_{P.k}}{\dot{E}x_{P.k}} \tag{2.18}$$

$$\dot{C}_{D.k} = c_{F.k}\dot{E}x_{D.k} \tag{2.19}$$

$$\dot{C}_{L.k} = c_{F.k}\dot{E}x_{L.k} \tag{2.20}$$

$$f_k = \frac{\dot{Z}_k}{\dot{Z}_k + \dot{C}_{D.k+}\dot{C}_{L.k}} \tag{2.21}$$

The subscripts D and L correspond to the exergy destruction and loss, respectively. Eventually, TCR and LCOP are calculated for each configuration to assess and compare the economic aspect as:

$$TCR = \sum_{k=1}^{n_k} \dot{Z}_k + \sum_{i=1}^{n_F} \dot{C}_{F_i} \tag{2.22}$$

$$LCOP = \frac{TCR}{\dot{W}_{turbine} - \dot{W}_{pump}} \tag{2.23}$$

2.3.4 Optimization procedure

Optimization is one of the most powerful tools in engineering problems to obtain the best design relative to a set of prioritized criteria or constraints. It comprises either maximizing the favorable factors such as produced power, sustainability index, and performance efficiency or minimizing undesirable objectives like total cost rate and environmental contamination. While single-objective optimization aims to find the best solution for a particular indicator, the aim of multi-criteria optimization (MCO) is to ascertain the best operating condition considering various conflictive metrics simultaneously. A genetic algorithm approach is an excellent option for different optimization methods because of higher calculation speed and fitness value. It can also optimize complex problems which other approaches cannot solve. While other methods require information or even a complete understanding of the problem structure and variables, the genetic algorithm is more flexible and does not need such specific information. The basic concept of a genetic algorithm is presented in Figure 2.2 as an iterative loop.

In the present study, MCO using the genetic algorithm method in MATLAB is applied to find each configuration's best optimal condition from exergy and economic aspects. The optimum significant variables are found to minimize the total cost rate (Equation (2.2)) while maximizing the exergy efficiency (Equation (2.6)).

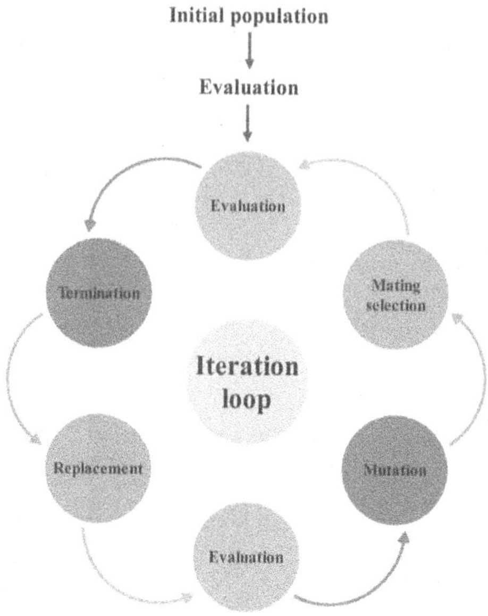

Figure 2.2 The concept of genetic algorithm.

2.4 RESULTS AND DISCUSSION

A parametric study is performed to assess the influence of main decision variables on techno-economic indicators. For this, the influence of geothermal temperature, turbine inlet pressure, ambient temperature, and turbine isentropic efficiency on total cost rate, exergy efficiency, sustainability index, and levelized cost of power of each configuration is evaluated and compared. Then, the exergy destruction rate is assessed to find the significant sources of each configuration's irreversibility. Eventually, each configuration's best optimum condition is found considering the total cost rate to be minimized and exergy efficiency to be maximized simultaneously.

Figure 2.3 illustrates the variation of the total cost rate, exergy efficiency, sustainability index, and levelized cost of power with the geothermal temperature. As depicted, each configuration's exergy efficiency and total cost rate increase by increasing the geothermal temperature. Therefore, while a worse economic condition is obtained, the quality of energy conversion will improve, indicating the significance of multi-criteria optimization. The figure further presents that the levelized cost of power decreases as the geothermal temperature increases from 160°C to 175°C. This is reasonable since at constant turbine inlet temperature and pressure, by increasing the geothermal temperature, the mass flow rate of the power cycle increases (according to the energy balance equation for evaporator), which leads to a higher turbine produced power. Figure 2.3 depicts that when the geothermal temperature increases, the sustainability index decreases up to a certain value and then increases. According to the figure, ORC with an internal heat exchanger (configuration (c)) has the highest exergy efficiency and sustainability index in the whole geothermal temperature range.

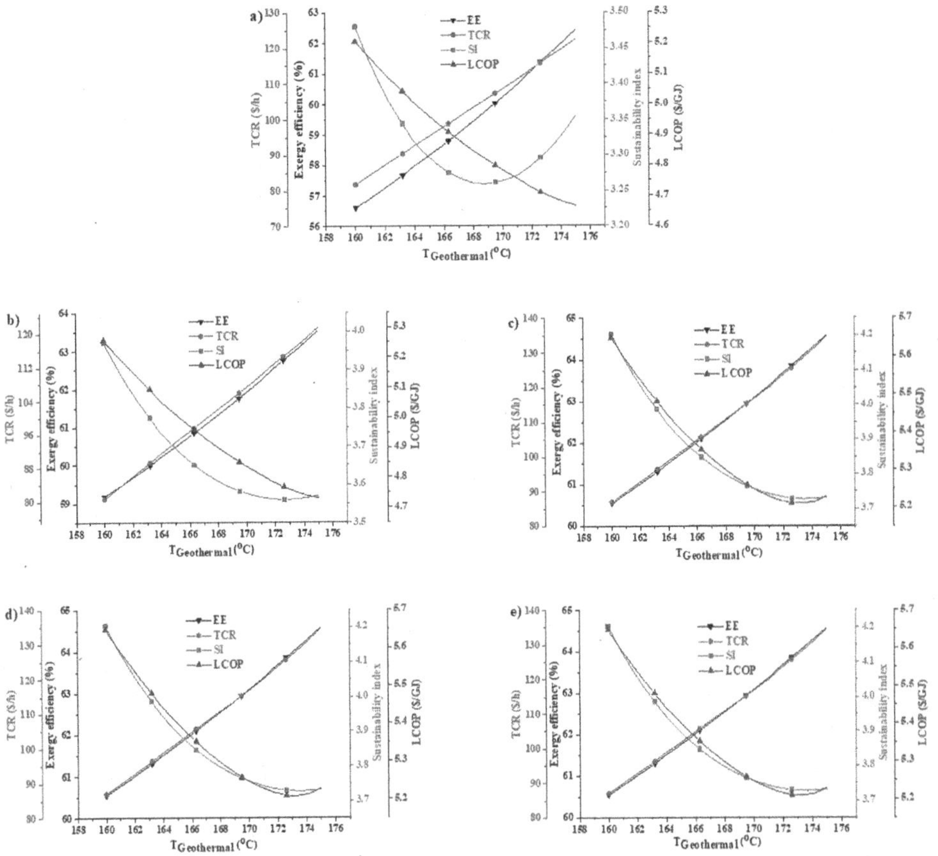

Figure 2.3 The effect of geothermal temperature on techno-economic-sustainability indicators of each configuration.

Moreover, the figure shows that KCS 34 (configuration (e)) is superior to KCS 11 (configuration (d)) from the irreversibility standpoint because of greater exergy efficiency and sustainability index. In contrast, the figure reveals that the lowest total cost rate corresponds to the KCS 11 due to the lowest sum of components costs. It can be further concluded that configuration (c) is the worst economic model among ORC options because of a higher total cost rate and levelized cost of power.

Since each configuration's energy, exergy, and economic aspects are highly influenced by the turbine inlet pressure, its effect on the total cost rate, levelized cost of power, exergy efficiency, and sustainability index is demonstrated in Figure 2.4. According to the figure, the variation of techno-economic indicators with the turbine inlet pressure highly depends on the system configuration. As shown, while the exergy efficiency and sustainability index of configuration (a), (b), (c), and (d) increase by increasing the turbine inlet pressure, they will decrease in the case of KCS 34 (configuration (e)). The figure also indicates that while the total cost rate of ORC-based systems increases by increasing the turbine inlet pressure from 25

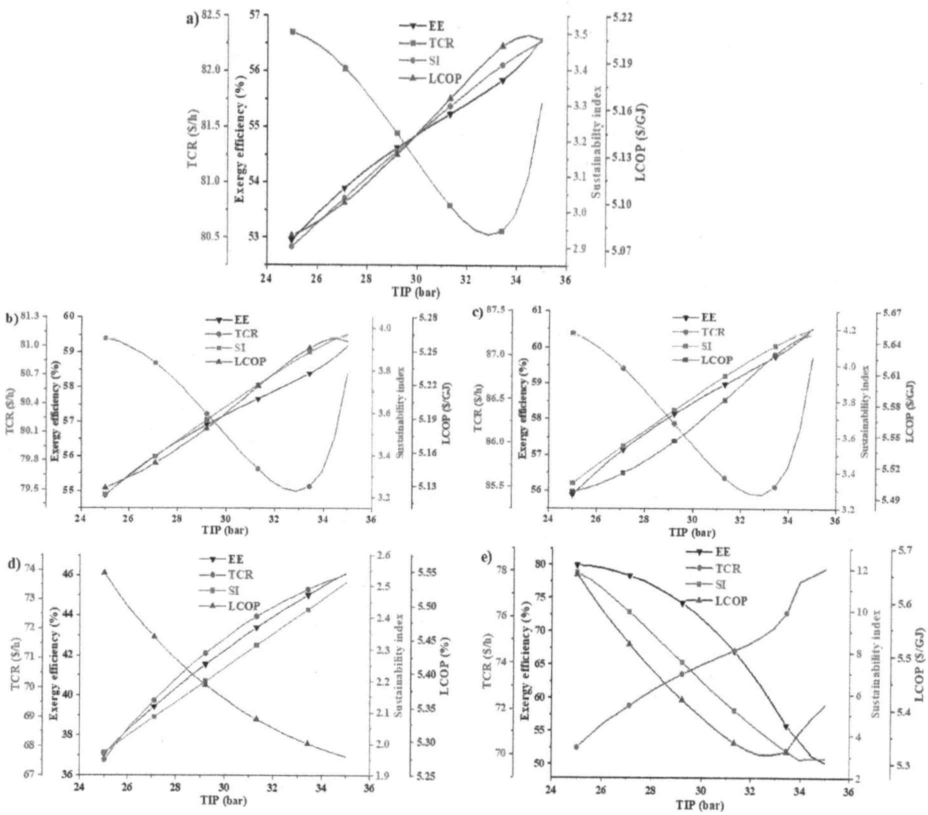

Figure 2.4 The effect of turbine inlet pressure on techno-economic-sustainability indicators of each configuration.

bar to 35 bar, it decreases for the Kalina cycle systems (configuration (d) and (e)). According to Figure 2.4(a), (b), and (c), when the turbine inlet pressure increases from 25 bar to 33 bar, the levelized cost of power decreases to a specific value and then increases by increasing the pressure up to 35 bar. Eventually, Figure 2.4(d) and (e) indicate that as the turbine inlet temperature increases, the total cost rate of KCS 11 and KCS 34 increases from 67.5 $/h and 70.3 $/h to 73.8 $/h and 78.1 $/h, respectively. This is justified since the turbine purchase cost increases as the inlet pressure increases.

The effect of ambient temperature on each configuration's techno-economic facets is depicted in Figure 2.4. According to the figure, the increase of ambient temperature has a negative effect on the irreversibility of each configuration (except KCS 34) due to the reduction of exergy efficiency and sustainability index. This is rational since the sum of exergy destruction rate of components increases in a higher environmental temperature. It can be further concluded that when the ambient temperature increases from 0°C to 40°C, each configuration's total cost rate decreases, which is economically appropriate. Furthermore, the figure illustrates that for each configuration, since the reduction of net produced power is higher than the total cost rate decrement, the levelized cost of power declines as the ambient temperature increases. Besides, Figure 2.4(e) depicts that for KCS 34, the ambient temperature increase would be either desired

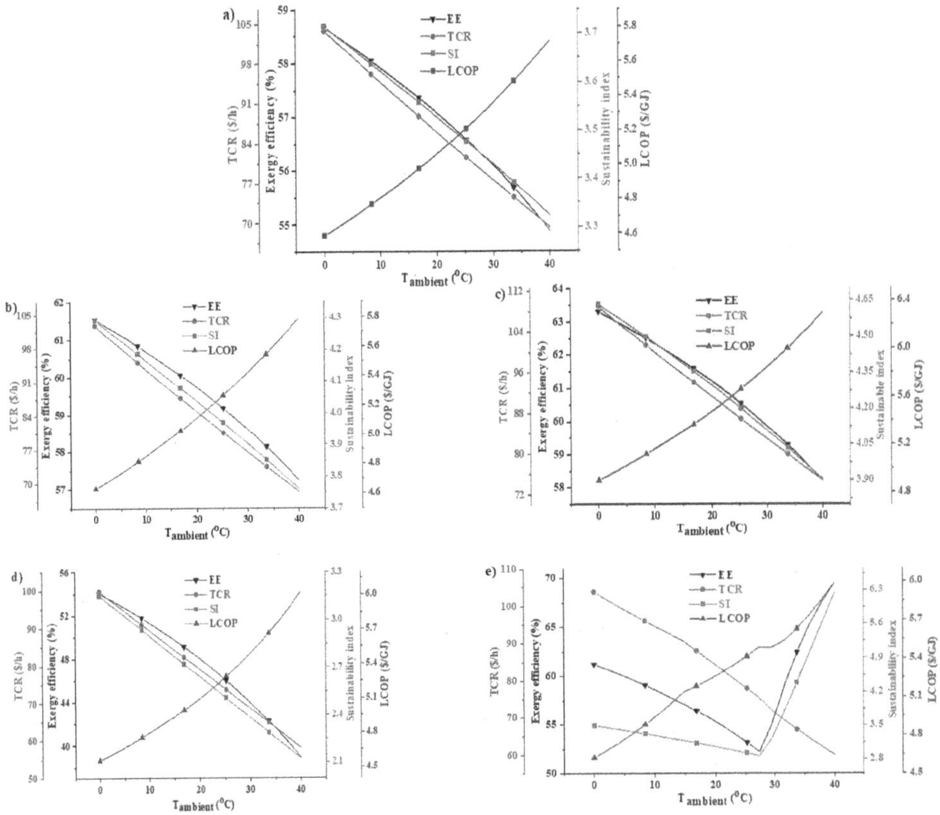

Figure 2.5 The effect of ambient temperature on techno-economic-sustainability indicators of each configuration.

or adverse from the irreversibility and sustainability points of view. As indicated, while a lower exergy efficiency and irreversibility are obtained by increasing the ambient temperature from 0°C to 28°C, they will increase in a higher temperature.

The more isentropic efficiency results in a higher enthalpy difference in the turbine; therefore, the net produced power and exergy efficiency will increase for each configuration, as shown in Figure 2.5. The figure also presents that when the isentropic efficiency increases, the total cost rate will increase. This is rational because the turbine cost is proportional to the turbine-produced power. Conversely, the figure demonstrates that a lower levelized cost of power is attained when the isentropic efficiency increases since the increase in total cost rate is lower than the produced power increment. Finally, it can be concluded that the sustainability index of each configuration increases by increasing the isentropic efficiency from 0.7% to 0.95%.

The exergy destruction rate of each configuration component is shown in Figure 2.6 to find the main source of irreversibility and compare each component's contribution to the overall systems' destructions. Based on the second law of thermodynamics, mixing the streams, the high temperature difference between the hot and cold flows, and chemical reaction result in higher irreversibility. According to Figure 2.6(a), the ORC evaporator with an exergy destruction rate of 1,198 kW (41.3% of destruction) is the key source of deviation from the

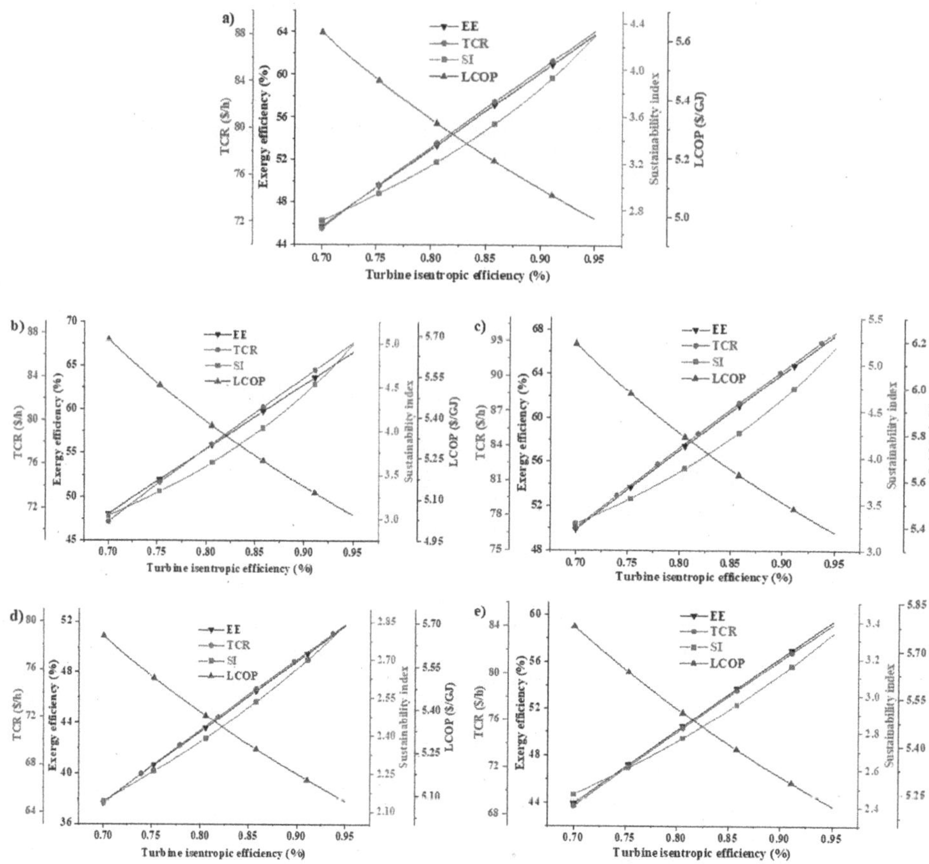

Figure 2.6 The effect of turbine isentropic efficiency on techno-economic-sustainability indicators of each configuration.

ideal state in the simple ORC system. From Figure 2.6(a) and 2.6(b), it can be concluded that the evaporator and condenser have the highest destruction due to the high temperature difference between hot and cold sides. The comparison of exergy destruction rate of configuration (b) and (c) illustrates that the feedwater heater has a higher destruction rate of 5.9 kW than the internal heat exchanger due to the mixing of the inlet streams. Figure 2.6(d) indicates that in KCS 11, the highest exergy destruction rate of 1,760 kW corresponds to the condenser due to the higher temperature difference between hot and cold streams compared to the evaporator. According to Figure 2.6 (e), the low-temperature generator has a higher destruction rate of 235.2 kW than a high-temperature generator. As depicted in Figure 2.7, in each configuration, the best component from the quality of energy conversion is the pump because of the lowest (near zero 1% of destruction) irreversibility.

After performing a comparative parametric study, each configuration's techno-economic indicators are compared at the optimum condition. For this, a multi-criteria optimization based on a genetic algorithm approach is applied to each system to find the most appropriate condition considering exergy efficiency and TCR as objectives. Figure 2.8 depicts the Pareto frontier

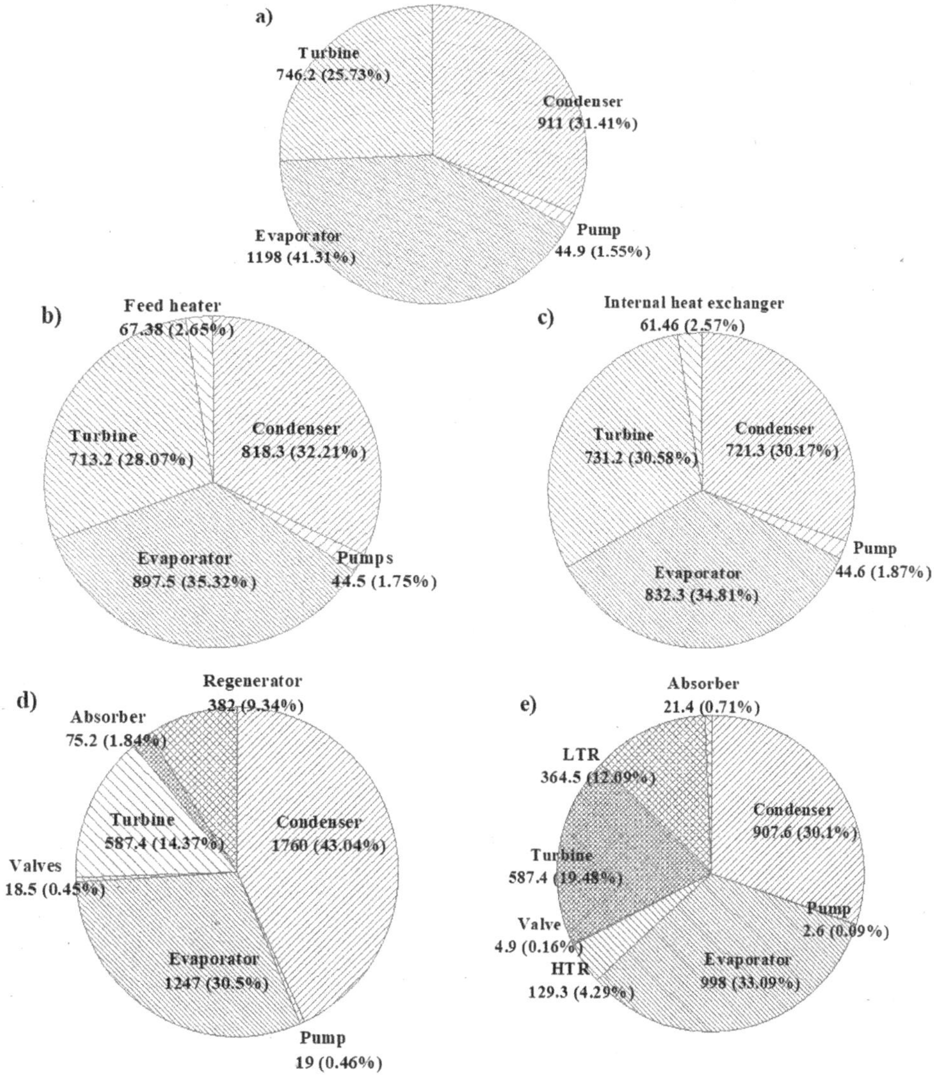

Figure 2.7 The exergy destruction rate of every component of each configuration.

diagram of objective functions for each scenario. According to the figure, each curve consists of many Pareto points prioritized based on the policy of decision-makers. As depicted, since the ideal point, having the maximum exergy efficiency and minimum TCR simultaneously, is not on the curve, according to the dimensionless Pareto diagram (Appendix B), the closest point is selected as the best point.

The detailed information about the most significant Pareto diagram points for each configuration is listed in Table 2.2. The table indicates the optimum values of decision parameters, including the geothermal temperature, turbine inlet pressure, ambient temperature, and isentropic efficiency, and the optimal objectives at the multi- and single-criteria optimization

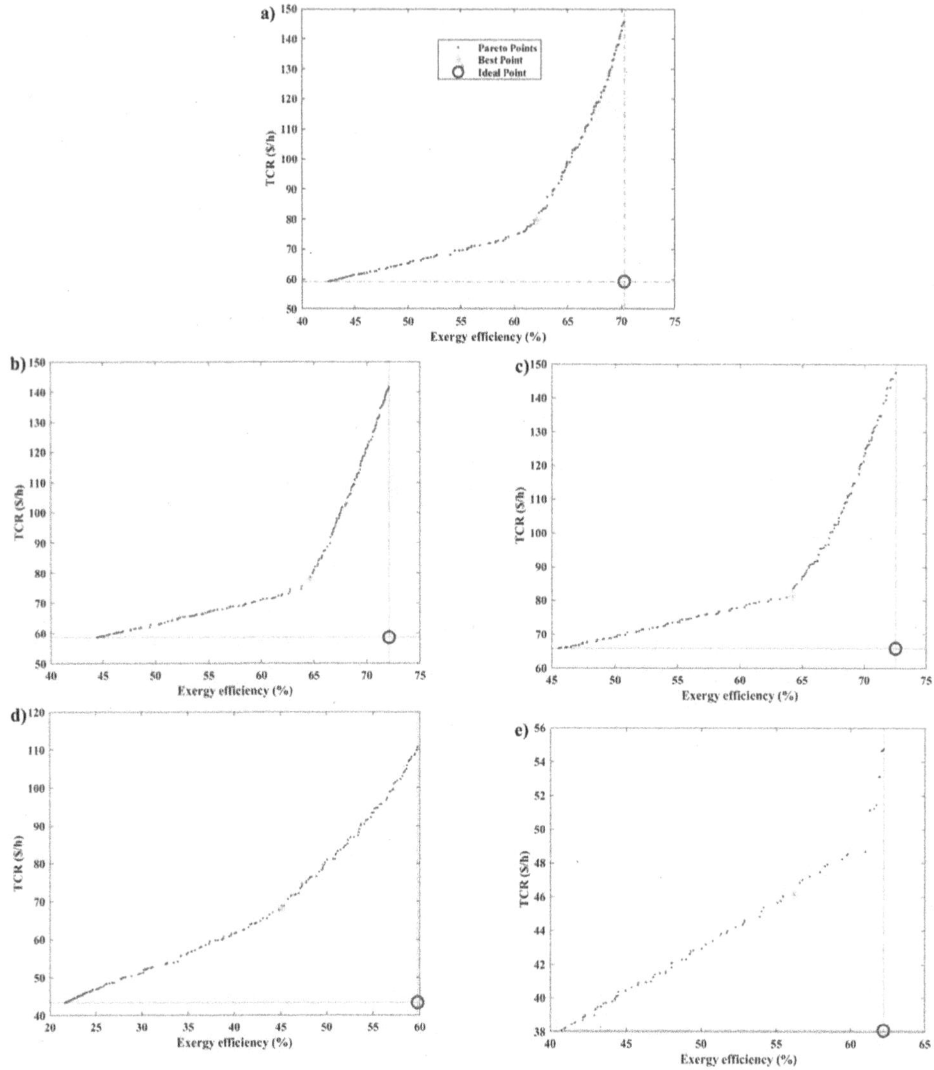

Figure 2.8 The Pareto frontier diagram of TCR and exergy efficiency for each configuration.

points. According to the table, at the MCO point, the exergy efficiency of configurations (a), (b), (c), (d), and (e) increases about 5.55%, 5.42%, 3.66%, 1.02%, and 2.95%, respectively, compared to the operating condition. Also, the table indicates that TCR decreases by about 1.9 $/h, 3.04 $/h, 5.68 $/h, 5.86 $/h, and 6.21 $/h. The comparison of optimization results demonstrates that although at the operating condition, ORC with an internal heat exchanger is the most excellent option from the performance facet, ORC with an open feed heater has the highest exergy efficiency of 64.61% at the MCO point.

The scatter distribution of significant decision variables for each configuration is shown in Figure 2.9 to demonstrate the distribution of optimum points across the population size.

Table 2.2 The detailed information of the Pareto frontier diagram

Condition	Decision parameters				Objective functions	
	$T_{geothermal}$ (°C)	TIP (bar)	$T_{ambient}$ (°C)	$\eta i_{sentropic}$ (%)	η_{exergy} (%)	TCR ($/h)
For configuration (a): Simple ORC						
Maximum efficiency	169.9	34.9	2.3	0.94	70.24	145.61
Minimum TCR	160	30.9	39.9	0.7	42.56	59.31
MCO point	160.6	34.1	34.9	0.94	62.16	79.82
For configuration (b): ORC with open feed heater						
Maximum efficiency	169.9	34.9	3.9	0.95	72.07	141.54
Minimum TCR	160	30.8	39.9	0.7	44.39	58.59
MCO point	160.4	34.6	35.8	0.94	64.61	77.67
For configuration (c): ORC with internal heat exchanger						
Maximum efficiency	168.9	34.75	2.7	0.95	72.52	147.62
Minimum TCR	160.5	30.25	39.1	0.7	45.55	65.81
MCO point	160.8	33.4	38.9	0.94	64.21	81.31
For configuration (d): KCS 11						
Maximum efficiency	163.6	34.9	1.1	0.95	59.74	110.65
Minimum TCR	161.6	25.1	39.9	0.71	21.63	43.32
MCO point	162.1	34.6	35.9	0.94	47.07	67.99
For configuration (e): KCS 34						
Maximum efficiency	160.1	25.9	32.1	0.94	62.21	54.76
Minimum TCR	160.7	29.6	39.9	0.7	40.757	38.08
MCO point	160.1	29.2	39.7	0.88	56.2	71.89

According to the figure, while some parameters are sensitive and should be kept at a specific value to reach the best operating condition, the others are ineffective and dispersed in the whole domain. According to the figure, since the optimum points of ambient temperature and turbine isentropic efficiency are distributed in the whole range, they are ineffective, and their variation will not change the configurations' techno-economic condition. In contrast, it can be concluded that the geothermal temperature and turbine inlet pressure are sensitive variables. As depicted, the appropriate range for the geothermal temperature and turbine inlet pressure in ORC-based systems (configuration (a), (b), and (c)) are, respectively, $160°C<T_{geothermal}<170°C$ and $31\ bar<TIP<35\ bar$. Besides, the geothermal temperature for KCS 11 and KCS 34 should be kept lower than 164°C.

2.5 SUMMARY

In this chapter, various configurations of geothermal-driven power cycles comprising simple ORC, ORC with internal heat exchanger, ORC with open feedwater heater, KCS 11, and KCS 34 are analyzed and compared from energy, exergy, sustainability, and economic aspects. For this, the variation of exergy efficiency, levelized cost of power, sustainability index, and total cost rate with geothermal temperature, turbine inlet pressure, and isentropic efficiency, and ambient temperature are investigated and compared for each arrangement through a parametric investigation. Besides, the exergy destruction rate is evaluated to assess every components'

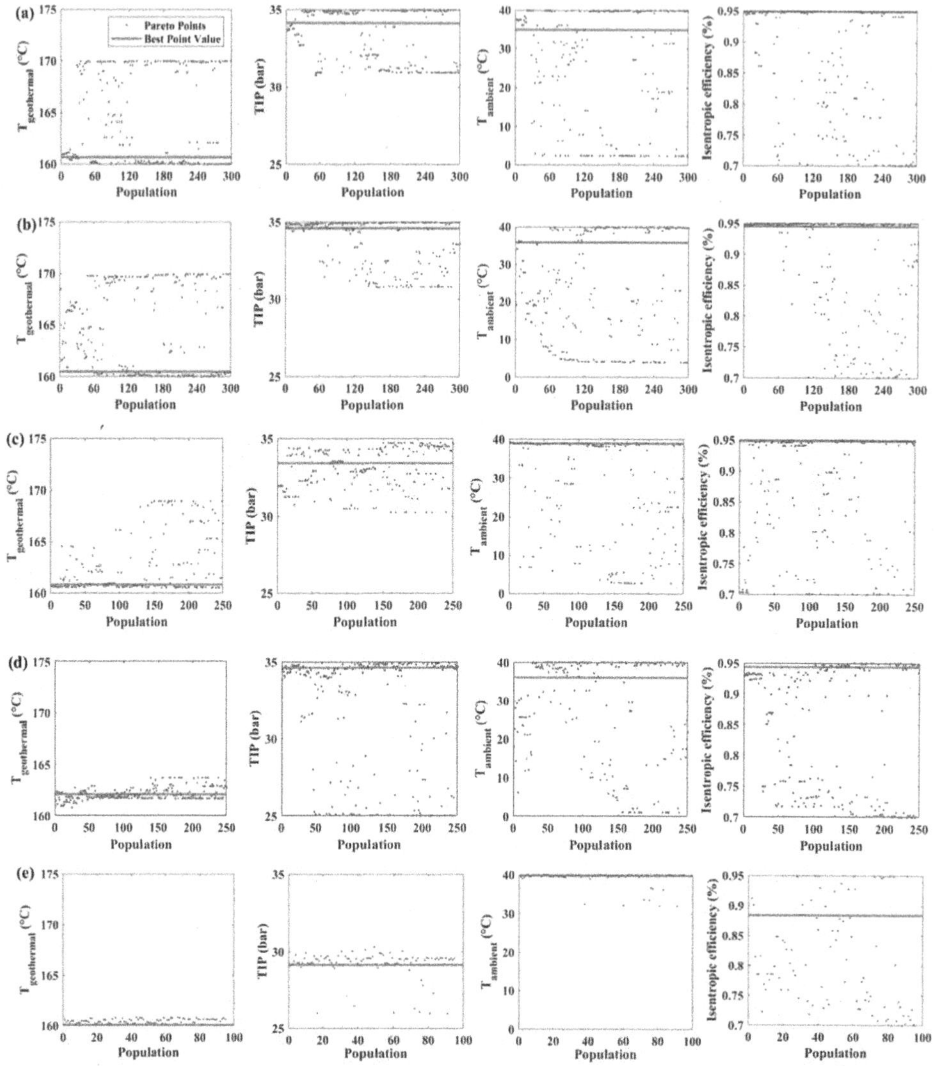

Figure 2.9 The scatter distribution of significant operational variables of each configuration.

deviation from the ideal performance from the irreversibility facet. Eventually, multi-criteria optimization employing genetic algorithm approach based on developed MATLAB code is applied to each configuration to determine the most suitable condition from techno-economic standpoints. The results indicate that while the best performance is obtained using ORC integrated with an internal heat exchanger, it is not economically a favorable option due to the highest total cost rate and levelized cost of power. The results further show that KCS 11 is the excellent option from an economic viewpoint because of the lowest total cost rate. According to the performance comparison results, KCS 34 is superior to the KCS 11 because of a higher exergy efficiency and sustainability index of 7.16% and 0.44. Based on the parametric study results, techno-economic indicators are noticeably changed with turbine inlet pressure and geothermal temperature variation. What stands out from the optimization results is that at the optimum condition while the second law efficiency of basic ORC, ORC with internal heat exchanger, ORC with open feed heater, KCS 11, and KCS 34 increases about 5.55%, 5.42%, 3.66%, 1.02%, and 2.95%, in comparison to the design state, the total cost rate decreases about 1.9 $/h, 3.04 $/h, 5.68 $/h, 5.86 $/h, and 6.21 $/h, respectively. Finally, the scatter distribution of the significant decision variables illustrates that the exergetic and economic outputs are not sensitive to the turbine isentropic efficiency and ambient temperature.

APPENDIX A

Marshall and Swift cost equation is implemented to convert the component cost from the reference to the present year as following (Fakhari et al. 2020):

$$\dot{Z}_k^{PY} = \frac{Cost\ index\ at\ the\ present\ year}{Cost\ index\ at\ the\ reference\ year} \times \dot{Z}_k^{RY} \tag{A1}$$

The purchased costs and the cost index at reference year for each component are tabulated in Table 2.3.

The heat exchanger area needed to calculate the purchased cost are evaluated by:

$$\dot{Q} = UA\Delta T_{LMTD} \tag{A2}$$

Moreover, the values of the heat exchangers coefficient for each component are tabulated in Table 2.4.

Table 2.3 The purchased cost and the reference year for each component

Component	*Purchased cost*	*Reference year*	*Cost index*
Pump	$Z_{Pu} = 200\dot{W}_{Pu}^{0.65}$	2010	522.8
Turbine	$Z_T = 4750\dot{W}_T^{0.75}$	2010	522.8
Evaporator	$Z_{Evap} = 309.14A_{Evap}^{0.85}$	2003	402.3
Condenser	$Z_{Cond} = 516.62A_{Cond}^{0.6}$	2005	468.2
IHE	$Z_{IHE} = 130\left(\frac{A_{IHE}}{0.093}\right)^{0.78}$	2005	468.2
OFH	$Z_{OFH} = 66\dot{Q}\left(\frac{1}{T_{TTD}+4}\right)^{0.1}$	2010	522.8
HTR	$Z_{HTR} = 2143A_{HTR}^{0.514}$	2010	522.8
LTR	$Z_{LTR} = 2143A_{LTR}^{0.514}$	2010	522.8
Absorber	$Z_{Abs} = 130\left(\frac{A_{Abs}}{0.093}\right)^{0.78}$	2005	468.2

Sources: Behzadi, Arabkoohsar, and Gholamian (2020); Habibollahzade, Mehrabadi, and Markides (2021); Zare and Palideh (2018).

Table 2.4 The heat exchangers coefficients

Evaporator	*Condenser*	*Heat exchanger*	*Absorber*	*Ammonia-water heat exchanger*
0.9	1.1	1	0.6	1.1

APPENDIX B

The Pareto frontier of TCR and exergy efficiency for each configuration are presented in Figure 2B.1 to have a vivid vision into the interaction of best and ideal points. According to the figure, the best point is the closest one to the ideal point after normalizing the objectives.

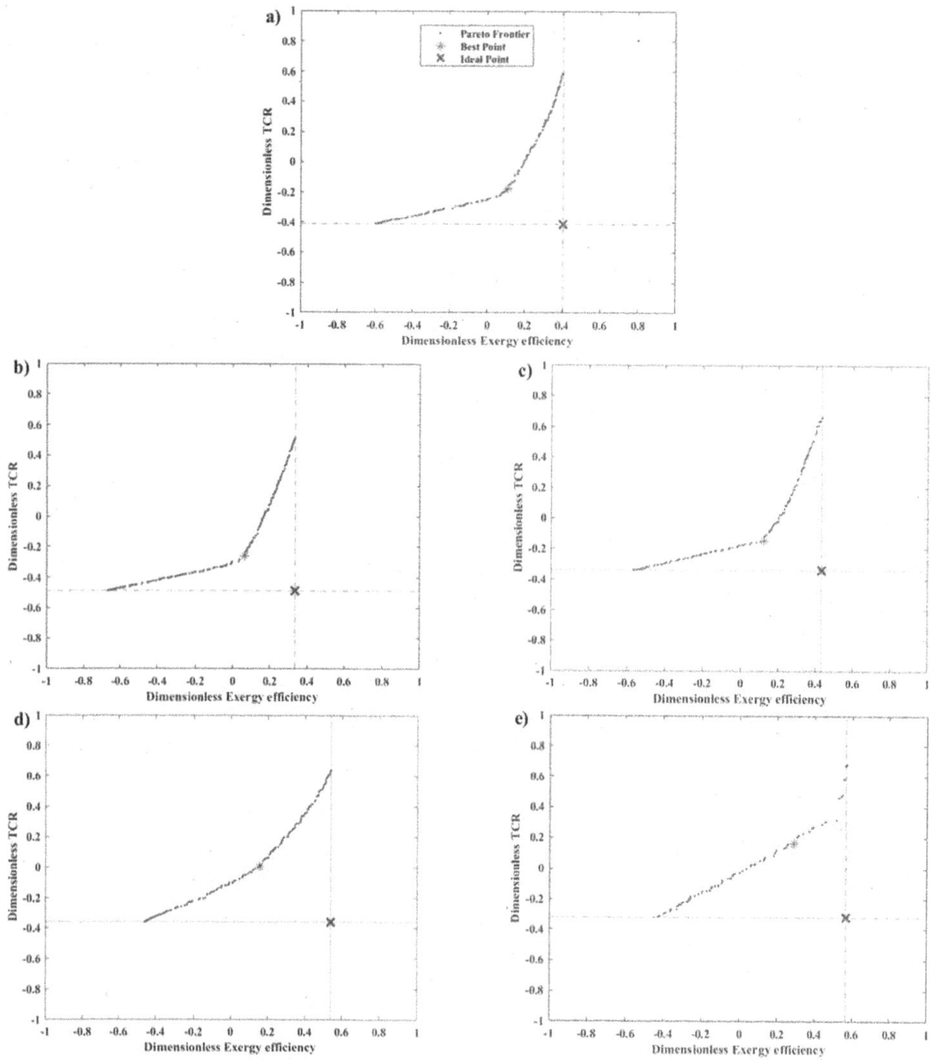

Figure 2B.1 The normalized Pareto frontier diagram of TCR and exergy efficiency.

Nomenclature and abbreviations

Nomenclature

c	Exergy cost (\$/GJ)
D_P	Depletion factor
$\dot{E}$	Exergy rate (kW)
f	Exergoeconomic factor
h	Specific enthalpy (kJ/kg)
$LCOP$	Levelized cost of power (\$/GJ)
$\dot{m}$	Mass flow rate (kg/s)
PP	Payback period
$\dot{Q}$	Heat (kW)
SI	Sustainability index
T	Temperature (°C)
TCR	Total cost rate (\$/h)
TIP	Turbine inlet pressure (kPa)
Z	The investment cost of components (\$)
$\dot{Z}$	Investment cost rate of component (\$/hr)

Subscript and abbreviations

Ch	Chemical
CI	Capital investment
D	Destruction
EES	Engineering equation solver
HTR	High-temperature regenerator
i	Inlet
IHE	Internal heat exchanger
KCS	Kalian cycle system
L	Loss
LTR	Low-temperature regenerator
MCO	Multi-criteria optimization
OFH	Open feed heater
OM	Operation and maintenance
P	Product
Ph	Physical
Pu	Pump
T	Turbine
tot	Total

Greek symbols

η_{II}	Exergy efficiency
η_{I}	Energy efficiency
η_{is}	Isentropic efficiency

REFERENCES

Arabkoohsar, Ahmad, Amirmohammad Behzadi, & Ali Sulaiman Alsagri. 2021. "Techno-economic analysis and multi-objective optimization of a novel solar-based building energy system: An effort to reach the true meaning of zero-energy buildings." *Energy Conversion and Management* 232: 113858. doi:10.1016/j.enconman.2021.113858.

Behzadi, Amirmohammad, & Ahmad Arabkoohsar. 2020. "Comparative performance assessment of a novel cogeneration solar-driven building energy system integrating with various district heating designs." *Energy Conversion and Management* 220: 113101. https://doi.org/10.1016/j.enconman.2020.113101.

Behzadi, Amirmohammad, Ahmad Arabkoohsar, & Ehsan Gholamian. 2020. "Multi-Criteria optimization of a biomass-fired proton exchange membrane fuel cell integrated with organic Rankine cycle/thermoelectric generator using different gasification agents." *Energy* 201: 117640. doi:10.1016/j.energy.2020.117640.

Behzadi, Amirmohammad, Ahmad Arabkoohsar, & Vedran S. Perić. 2021. "Innovative hybrid solar-waste designs for cogeneration of heat and power, an effort for achieving maximum efficiency and renewable integration." *Applied Thermal Engineering* 190 (February). doi:10.1016/j.applthermaleng.2021.116824.

Behzadi, Amirmohammad, Ehsan Gholamian, Ehsan Houshfar, & Ali Habibollahzade. 2018. "Multi-objective optimization and exergoeconomic analysis of waste heat recovery from Tehran's waste-to-energy plant integrated with an ORC unit." *Energy* 160: 1055–1068. https://doi.org/10.1016/j.energy.2018.07.074.

Bombarda, Paola, Costante M. Invernizzi, & Claudio Pietra. 2010. "Heat recovery from diesel engines: A thermodynamic comparison between Kalina and ORC Cycles." *Applied Thermal Engineering* 30(2–3): 212–219. doi:10.1016/j.applthermaleng.2009.08.006.

Ebrahimi-Moghadam, Amir, Mahmood Farzaneh-Gord, Ali Jabari Moghadam, Nidal H. Abu-Hamdeh, Mohammad Ali Lasemi, Ahmad Arabkoohsar, & Ashkan Alimoradi. 2021. "Design and multi-criteria optimisation of a trigeneration district energy system based on gas turbine, Kalina, and ejector cycles: Exergoeconomic and exergoenvironmental evaluation." *Energy Conversion and Management* 227: 113581. https://doi.org/10.1016/j.enconman.2020.113581.

El Bassam, Nasir, Preben Maegaard, & Marcia Lawton Schlichting. 2013. "Geothermal energy." In Nasir El Bassam, Preben Maegaard, & Marcia Lawton Schlichting (eds.), *Distributed Renewable Energies for Off-Grid Communities*, pp. 185–192. New York: Elsevier. https://doi.org/10.1016/B978-0-12-397178-4.00012-8.

Fakhari, Iman, Amirmohammad Behzadi, Ehsan Gholamian, Pouria Ahmadi, & Ahmad Arabkoohsar. 2020. "Design and tri-objective optimization of a hybrid efficient energy system for tri-generation of power, heat, and potable water." *Journal of Cleaner Production*, 290: 125205. https://doi.org/10.1016/j.jclepro.2020.125205.

Fakhari, Iman, Amirmohammad Behzadi, Ehsan Gholamian, Pouria Ahmadi, & Ahmad Arabkoohsar. 2021. "Comparative double and integer optimization of low-grade heat recovery from PEM fuel cells employing an organic Rankine cycle with zeotropic mixtures." *Energy Conversion and Management* 228: 113695. doi:10.1016/j.enconman.2020.113695.

Fiaschi, Daniele, Giampaolo Manfrida, E. Rogai, & Lorenzo Talluri. 2017. "Exergoeconomic analysis and comparison between ORC and Kalina cycles to exploit low and medium-high temperature heat from two different geothermal sites." *Energy Conversion and Management* 154: 503–516. doi:10.1016/j.enconman.2017.11.034.

Ghaebi, Hadi, & Hadi Rostamzadeh. 2020. "Performance comparison of two new cogeneration systems for freshwater and power production based on organic Rankine and Kalina cycles driven by salinity-gradient solar pond." *Renewable Energy* 156: 748–767. doi:10.1016/j.renene.2020.04.043.

Gholamian, Ehsan, & Vahid Zare. 2016. "A comparative thermodynamic investigation with environmental analysis of SOFC waste heat to power conversion employing Kalina and organic Rankine cycles." *Energy Conversion and Management* 117: 150–161. doi:10.1016/j.enconman.2016.03.011.

Gogoi, Tapan Kumar, & Prarthana Hazarika. 2020. "Comparative assessment of four novel solar based triple effect absorption refrigeration systems integrated with organic Rankine and Kalina cycles." *Energy Conversion and Management* 226: 113561. doi:10.1016/j.enconman.2020.113561.

Habibollahzade, Ali, Zahra Kazemi Mehrabadi, & Christos N. Markides. 2021. "Comparative thermoeconomic analyses and multi-objective particle swarm optimization of geothermal combined cooling and power systems." *Energy Conversion and Management* 234: 113921. https://doi.org/10.1016/j.enconman.2021.113921.

Köse, Özkan, Yıldız Koç, & Hüseyin Yağlı. 2021. "Energy, exergy, economy and environmental (4E) analysis and optimization of single, dual and triple configurations of the power systems: Rankine cycle/Kalina cycle, driven by a gas turbine." *Energy Conversion and Management* 227. doi:10.1016/j.enconman.2020.113604.

Lorzadeh, Omid, Iman Lorzadeh, Mohsen Nourbakhsh Soltani, & Amin Hajizadeh. 2021. "Source-side virtual RC damper-based stabilization technique for cascaded systems in DC microgrids." *IEEE Transactions on Energy Conversion*, 1. doi:10.1109/TEC.2021.3055897.

Lund, John W., & Aniko N. Toth. 2021. "Direct utilization of geothermal energy 2020 worldwide review." *Geothermics* 90: 101915. https://doi.org/10.1016/j.geothermics.2020.101915.

Mahzari, Pedram, Ashley Stanton-Yonge, Catalina Sanchez-Roa, Giuseppe Saldi, Thomas Mitchell, Eric H. Oelkers, Vala Hjorleifsdottir, et al. 2021. "Characterizing fluid flow paths in the Hellisheidi geothermal field using detailed fault mapping and stress-dependent permeability." *Geothermics* 94: 102127. https://doi.org/10.1016/j.geothermics.2021.102127.

Mehrabadi, Zahra Kazemi, & Fateme Ahmadi Boyaghchi. 2019. "Thermodynamic, economic and environmental impact studies on various distillation units integrated with gasification-based multi-generation system: Comparative study and optimization." *Journal of Cleaner Production* 241: 118333. doi:10.1016/j.jclepro.2019.118333.

Meng, Fanxiao, Enhua Wang, Bo Zhang, Fujun Zhang, & Changlu Zhao. 2019. "Thermo-economic analysis of transcritical CO_2 power cycle and comparison with Kalina cycle and ORC for a low-temperature heat source." *Energy Conversion and Management* 195: 1295–1308. doi:10.1016/j.enconman.2019.05.091.

Milani, Samira Marami, Rahim Khoshbakhti Saray, & Mohammad Najafi. 2020. "Comparison of different optimized heat driven power-cycle configurations of a gas engine." *Applied Thermal Engineering* 179. doi:10.1016/j.applthermaleng.2020.115768.

Mojaver, Parisa, Ata Chitsaz, Mohsen Sadeghi, & Shahram Khalilarya. 2020. "Comprehensive comparison of SOFCs with proton-conducting electrolyte and oxygen ion-conducting electrolyte: Thermoeconomic analysis and multi-objective optimization." *Energy Conversion and Management* 205: 112455. doi:10.1016/j.enconman.2019.112455.

Musharavati, Farayi, Shoaib Khanmohammadi, Amir Hossein Pakseresht, & Saber Khanmohammadi. 2021. "Enhancing the performance of an integrated CCHP system including ORC, Kalina, and refrigeration cycles through employing TEG: 3E analysis and multi-criteria optimization." *Geothermics* 89: 101973. doi:10.1016/j.geothermics.2020.101973.

Nami, Hossein, Ahmad Arabkoohsar, & Amjad Anvari-Moghaddam. 2019. "Thermodynamic and sustainability analysis of a municipal waste-driven combined cooling, heating and power (CCHP) plant." *Energy Conversion and Management* 201: 112158. https://doi.org/10.1016/j.enconman.2019.112158.

Nemati, Arash, Hossein Nami, Faramarz Ranjbar, & Mortaza Yari. 2017. "A comparative thermodynamic analysis of ORC and Kalina cycles for waste heat recovery: A case study for CGAM cogeneration system." *Case Studies in Thermal Engineering* 9: 1–13. doi:10.1016/j.csite.2016.11.003.

Pan, Zhen, Li Zhang, Zhien Zhang, Liyan Shang, & Shujun Chen. 2017. "Thermodynamic analysis of KCS/ORC integrated power generation system with LNG cold energy exploitation and CO_2 capture." *Journal of Natural Gas Science and Engineering* 46: 188–198. doi:10.1016/j.jngse.2017.07.018.

Parikhani, Towhid, Javad Jannatkhah, Afshar Shokri, & Hadi Ghaebi. 2019. "Thermodynamic analysis and optimization of a novel power generation system based on modified Kalina and GT-MHR cycles." *Energy Conversion and Management* 196: 418–429. doi:10.1016/j.enconman.2019.06.018.

Rodríguez, Carlos Eymel Campos, José Carlos Escobar Palacio, Osvaldo J. Venturini, Electo E. Silva Lora, Vladimir Melián Cobas, Daniel Marques Dos Santos, Fábio R. Lofrano Dotto, & Vernei Gialluca. 2013. "Exergetic and economic comparison of ORC and Kalina cycle for low temperature enhanced geothermal system in Brazil." *Applied Thermal Engineering* 52(1): 109–119. doi:10.1016/j.applthermaleng.2012.11.012.

Rostamzadeh, Hadi, Mohammad Ebadollahi, Hadi Ghaebi, & Afshar Shokri. 2019. "Comparative study of two novel micro-CCHP systems based on organic Rankine cycle and Kalina cycle." *Energy Conversion and Management* 183: 210–229. doi:10.1016/j.enconman.2019.01.003.

Sadi, Meisam, & Ahmad Arabkoohsar. 2019. "An efficient and reliable district heating and cooling supplier." In *2019 9th International Conference on Power and Energy Systems (ICPES)*, pp. 1–5. Perth, Australia. doi:10.1109/ICPES47639.2019.9105392.

Shokati, Naser, Faramarz Ranjbar, & Mortaza Yari. 2015. "Exergoeconomic analysis and optimization of basic, dual-pressure and dual-fluid ORCs and Kalina geothermal power plants: A comparative study." *Renewable Energy* 83: 527–542. doi:10.1016/j.renene.2015.04.069.

Spittler, Nathalie, Ehsan Shafiei, Brynhildur Davidsdottir, & Egill Juliusson. 2020. "Modelling geothermal resource utilization by incorporating resource dynamics, capacity expansion, and development costs." *Energy* 190: 116407. https://doi.org/10.1016/j.energy.2019.116407.

Wang, Mengying, Xiao Feng, & Yufei Wang. 2017. "Comparison of energy performance of organic Rankine and Kalina cycles considering combined heat sources at different temperature." In Antonio Espuña, Moisès Graells, and Luis Puigjaner (eds.), *27th European Symposium on Computer Aided Process Engineering*, pp. 2419–2424. New York: Elsevier. https://doi.org/10.1016/B978-0-444-63965-3.50405-0.

Wang, Yali, Haidong Yang, & Kangkang Xu. 2021. "Comparative environmental impacts and emission reductions of introducing the novel organic Rankine and Kalina cycles to recover waste heat for a roller kiln." *Applied Thermal Engineering*, 116821. https://doi.org/10.1016/j.applthermaleng.2021.116821.

Wang, Yazi, Tian Chen, Yingbo Liang, Huaibo Sun, & Yiping Zhu. 2020. "A novel cooling and power cycle based on the absorption power cycle and booster-assisted ejector refrigeration cycle driven by a low-grade heat source: Energy, Exergy and exergoeconomic analysis." *Energy Conversion and Management* 204: 112321. doi:10.1016/j.enconman.2019.112321.

Wang, Yongzhen, Chengjun Li, Jun Zhao, Boyuan Wu, Yanping Du, Jing Zhang, & Yilin Zhu. 2021. "The above-ground strategies to approach the goal of geothermal power generation in China: State of art and future researches." *Renewable and Sustainable Energy Reviews* 138: 110557. https://doi.org/10.1016/j.rser.2020.110557.

Wang, Yufei, Qikui Tang, Mengying Wang, & Xiao Feng. 2017. "Thermodynamic performance comparison between ORC and Kalina cycles for multi-stream waste heat recovery." *Energy Conversion and Management* 143: 482–492. https://doi.org/10.1016/j.enconman.2017.04.026.

Xie, Nan, Zhiqiang Liu, Zhengyi Luo, Jingzheng Ren, Chengwei Deng, & Sheng Yang. 2020. "Multi-objective optimization and life cycle assessment of an integrated system combining LiBr/H_2O absorption chiller and Kalina cycle." *Energy Conversion and Management* 225: 113448. https://doi.org/10.1016/j.enconman.2020.113448.

Zare, Vahid, & Vahid Palideh. 2018. "Employing thermoelectric generator for power generation enhancement in a Kalina cycle driven by low-grade geothermal energy." *Applied Thermal Engineering* 130: 418–428. doi:10.1016/j.applthermaleng.2017.10.160.

Chapter 3

Integrated gasification combined cycle with co-gasification

T. Srinivas

Department of Mechanical Engineering, Dr. B. R. Ambedkar National Institute of Technology, Jalandhar, India

B. V. Reddy

Faculty of Engineering and Applied Sciences, University of Ontario Tech University, Oshawa, Canada

A. V. S. S. K. S. Gupta

Department of Mechanical Engineering, JNTU College of Engineering, Kukatpally, India

3.1 INTRODUCTION

The gas power plant and steam power plant are combined together to address the individual challenges of single systems and gain the combined benefit. Combined cycles are operated from gaseous fuels as the topping cycle is a gas power plant. With the adoption of gasifiers, the combined cycle can be operated with solid fuels. As the cost of the solid fuels is less than the gaseous fuels, combined cycles can be operated on integrated combined cycle (IGCC) power plant mode (Srinivas et al. 2004, 2006; Srinivas, Gupta, & Reddy 2008). Biomass is a renewable energy source that can be used in IGCC in conjunction with the gasifier (Srinivas, Gupta, & Reddy. 2009). The producer gas from the gasifier can be used in a gas turbine combustion chamber (GTCC) with minimum emissions. The existing individual systems of gas power plants and pulverized coal power plants can be shifted to IGCC with a minimum amount of investment.

Coal and biomass have their individual benefits in power plant use. These benefits can be combined with co-gasification. The use of coal in co-gasification augments the calorific value of producer gas. It also addresses the problem of limited availability of biomass. Gasifier sizes are available in wider ranges (Jong, Andries, & Hem 1999). The low cost of solid fuels, easy operation, stability in operation and flexibility are the promising benefits of co-gasification (Chmielniak & Sciazko 2003). A typical gas composition from the biomass gasifier is 9.27%, 9.25%, and 4.21% respectively for hydrogen, carbon monoxide and methane (Cao et al. 2006). In literature, the experimental and simulation results are reported for the gasification process (Pan et al. 2000; Pinto et al. 2002). Kalisz, Pronobis, and Baxter (2008) studied gasification with indirect co-firing of biodegradable wastes in coal-fired power boilers. Mastellone Zaccariello, and Arena (2010) investigated the co-gasification of seven mixtures of coal, plastics, and wood which are palletized and fed into a pre-pilot scale fluidized bed gasifier. Velez et al. (2009) reported the results of an experimental work on co-gasification of biomass/coal blends in a fluidized bed working at atmospheric pressure. They prepared several samples of blends by mixing 6–15 wt% biomass (sawdust, rice, or coffee husk) with coal. Alzate et al. (2009) conducted experiments on co-gasification experiments with two kinds of mixtures. The

DOI: 10.1201/9781003213741-3

first one was composed of granular coal and pellets of 100% wood and the second one was composed of pulverized wood and granular coal pellets. Kumabe et al. (2007) studied the effect of the feedstock with a varying content of woody biomass and coal on the co-gasification behavior by varying the biomass ratio from 0 to 1. Prins, Ptasinski, and Janssen (2007) showed the required equivalence ratio, defined as the amount of oxygen added for gasification relative to the amount of oxygen required for complete combustion, as 0.295. Lee et al. (2010) showed higher syngas heating value and cold gas efficiency with high H_2 and CO composition in syngas. The transport properties of coal/biomass mixtures were greatly improved compared to coal (McLendon et al. 2004). Goyal, Pushpavanam, and Ravi Kumar (2010) developed a mathematical model and simulated a fluidized bed coal gasifier that uses a mixture of coal and petcoke. Valero and Usón (2006) studied the key issue of co-gasification operation for an existing IGCC power plant. Sjostrom et al. (1999) varied the ratio of coal to wood in their investigation at the operation temperatures of 900°C and 700°C, and the pressure of 0.4 MPa. It has been demonstrated that it is advantageous to inject secondary air into a fluidized bed when the gasification temperature reaches values above 830°C (Pan et al. 1999). It is an essential condition that the temperature in the secondary air injection zone should be higher than that in the fluidized bed.

The main aim of the current work is to develop the characteristics of a coal/biomass co-gasification with biomass to coal ratio. This work is focused on the benefits of co-gasification in power plants. The thermodynamic equilibrium model for NOx estimation is not much reported in the literature. The mathematical model is extended to simulate the NOx emissions.

3.2 THERMO-CHEMICAL EVALUATION OF COAL GASIFIER

In the mathematical thermos-chemical model, the reference state is considered at 298.15 K and 1.01325 bar. The pressure drop in the coal gasifier is 5%. The specific heat of ash is taken as 0.85 kJ/kg K. The other heat losses in the gasifier are 10% of lower heating value (LHV) of fuel. It has been assumed that the solid carbon in the fuel is gasified completely. The polytropic state efficiency for air compressor and gas turbine is taken at 87.5% (Saravanamuttoo, Rogers, & Cohen 2003). Lv et al. (2004) suggested the optimum relative air fuel ratio (RAFR) as 0.23 for gasification. In the current study, RAFR is fixed at 0.25. The composition of coal (Hara et al. 2002) and biomass (Jarungthammachote & Dutta 2007) for co-gasification are taken from the literature.

- **Australia coal:** Ultimate analysis: C: 75.8 %, H: 5.1 %, O: 7.8 %, N: 1.7 %, S: 0.46 % (by mass on dry basis).
- **Solid waste biomass:** Ultimate analysis: C: 51.03 %, H: 6.77 %, O: 39.17 %, N: 2.64 %, S: 0.37 % (by mass on dry basis).

$C_{a11} H_{a21} O_{a31} N_{a41} C_{a11} H_{a21} O_{a31} N_{a41}$ and $C_{a12} H_{a22} O_{a32} N_{a42} C_{a12} H_{a22} O_{a32} N_{a42}$ are the C-H-O-N formula for coal and biomass respectively. In co-gasification, they combined to $C_{a12} H_{a22} O_{a32} N_{a42} C_{a12} H_{a22} O_{a32} N_{a42}$. For a single atom of carbon in fuel, the coefficient a_1 becomes one. The coefficients of a_2, a_3 and a_4 are the H/C, O/C and N/C mole ratios, respectively. The moisture content in the solid fuel is neglected. The reactions are solved at thermodynamic equilibrium conditions. The lower heating value of coal is obtained from stoichiometric quantity of air being used for complete combustion of unit mass of fuel.

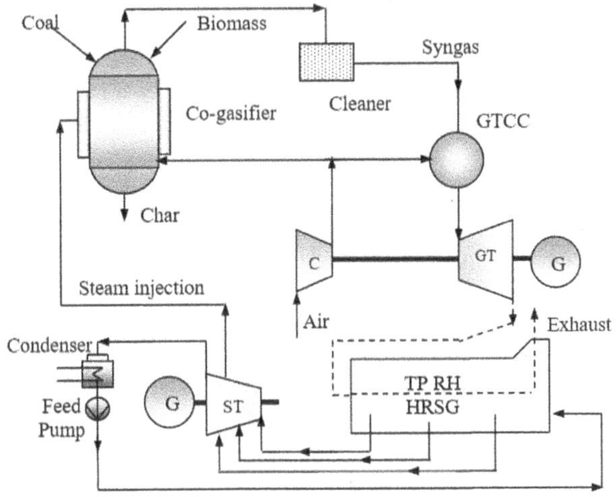

Figure 3.1 Schematic process flow diagram of coal/biomass co-gasification process for IGCC plant.

Figure 3.1 shows the schematic flow diagram of a co-gasification used in an IGCC plant. In co-gasification the two fuels, i.e., coal and biomass, are fed simultaneously. The compressed air and steam enter the gasifier at 12 bars. The syngas, after cleaning in a gas cleaner, goes to the gas turbine combustion chamber (GTCC) where a complete combustion takes place with a supply of compressed air. The exhaust gas from a simple gas power plant generates steam in a triple pressure (TP) reheat (RH) heat recovery steam generator (HRSG) system from the feed water at three pressure levels, i.e., high pressure (HP), intermediate pressure (IP) and low pressure (LP). The required amount of steam from the steam plant is trapped and injected into the co-gasifier.

The gasification products contain CH_4, CO, CO_2, H_2, H_2O and N_2. The following is the chemical reaction in a coal gasifier.

$$C_{a1}H_{a2}O_{a3}N_{a4} + a_5(O_2 + 3.76N_2) + a_6H_2O \Rightarrow b_1CH_4 + b_2CO + b_3CO_2 + b_4H_2 + b_5H_2O + b_6N_2 \tag{3.1}$$

$$C_{a1}H_{a2}O_{a3}N_{a4} + a_5(O_2 + 3.76N_2) + a_6H_2O \Rightarrow b_1CH_4 + b_2CO + b_3CO_2 + b_4H_2 + b_5H_2O + b_6N_2.$$

Thermo-chemical methodology for the above gasification reaction is reported in the literature by the same authors (Srinivas, Gupta, & Reddy 2009). At a typical 80% biomass, 20% coal, 0.25 RAFR, 1 SFR, and 12 bar gasifier pressure, the coefficients of syngas resulted as b_1 (CH_4): 0.068, b_2 (CO): 0.52, b_3 (CO_2): 0.41, b_4 (H_2): 0.88, b_5 (H_2O): 0.90 and b_6 (N_2): 1.07. These are the results obtained at a_1 (C): 1, a_2 (H): 1.436, a_3 (O): 0.476, a_4 (N): 0.0393, a_5 (air): 0.28 and a_6 (steam): 1.2.

Exergy evaluation for energy systems is well developed and published by the same authors (Srinivas et al. 2006) and the irreversibility of coal gasifiers has been determined from the following equation.

Irreversibility of coal gasifier,

$$i_{gasifier} = e_{coal} + e_{air} + e_{steam} - e_{syngas} \tag{3.2}$$

3.3 RESULTS AND DISCUSSIONS

The influence of biomass to coal mass ratio has been tested thermodynamically on the co-gasification process and its plant conditions. The biomass to coal mass ratio varies from 0 to 1, i.e., from complete coal to complete biomass. The responses of gasifier conditions (syngas composition, gasifier temperature, cold gas efficiency, gasifier exergetic loss), plant performance (efficiency and power) and plant emissions (NOx and CO_2) have been investigated with degree of co-gasification. The results are plotted at the gasifier conditions of 0.25 RAFR, 1 SFR and 12 bar pressures.

Figure 3.2 shows the effect of biomass to coal ratio on syngas composition. It varies from pure coal (0) to pure biomass (1). Interestingly, an increased biomass in co-gasification increases the hydrogen portion in syngas. The heating value of syngas increases with methane, hydrogen and carbon monoxide involvement. Hydrogen (11% to 25%) and carbon monoxide (7% to 9%) increase with biomass addition. The methane content is more in coal gasification (14%) and less in biomass gasification (6%). Nitrogen concentration in syngas decreases (47% to 38%) with an increase in biomass-to-coal ratio. There is a slight decrease in carbon dioxide in syngas with a higher biomass-to-coal ratio. Thus, with co-gasification, hydrogen rich syngas with a low amount of carbon dioxide can be generated. It also shows that biomass is a more suggestible fuel for hydrogen generation with low emission power. All the above discussed trends of co-gasification are matched with the Alzate et al. (2009) results.

Figure 3.3 shows the effect of degree of co-gasification on gasification characteristics. The behavior of the gasifier, i.e., cold gas efficiency, exergetic loss, and temperature with biomass to coal mass fraction, is investigated. Cold gas efficiency is defined as the ratio of LHV of

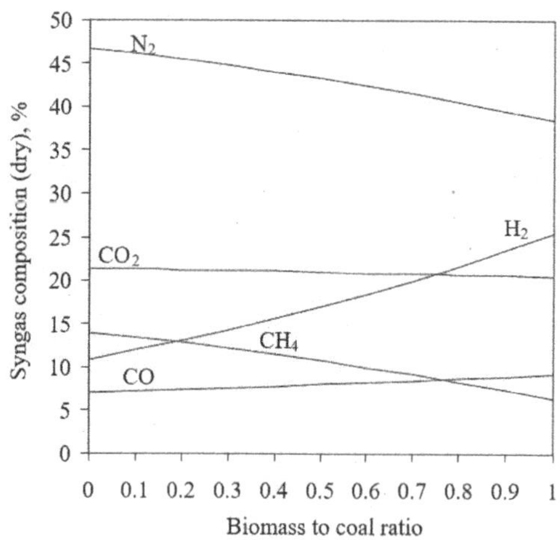

Figure 3.2 Dry syngas species concentrations as a function of biomass to coal ratio.

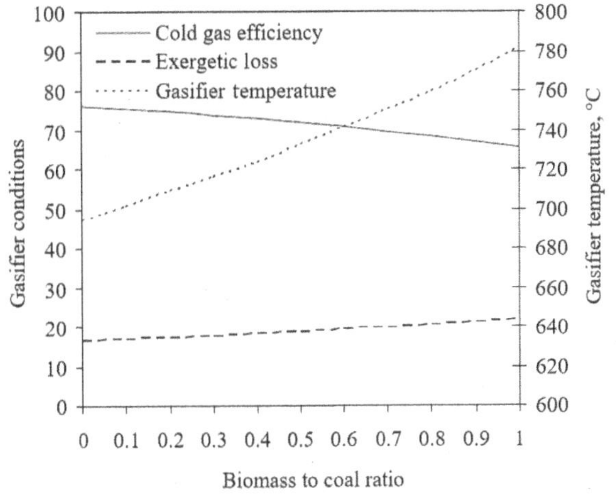

Figure 3.3 Response of gasifier conditions with biomass to coal ratio.

syngas to LHV of fuel. The LHV is determined at the reference temperature and products are cooled to reference temperature (298.15 K) with a theoretically correct air fuel ratio. Though the syngas composition is good with biomass, the gasification conditions are not at a favorable stage at high biomass participation. Cold gas efficiency decreases and exergetic loss increases with an increase in biomass to coal ratio. The gasifier temperature varies between 700°C and 780°C with the variations in biomass ratio from 0 to 1 at the pre-stated conditions. Watanabe and Otaka (2006) resulted in syngas temperature from 850°C to 1,000°C at the gasifier exit.

Figure 3.4 shows the relation between biomass to coal mass ratio with fuel supplies in co-gasifier. For a fixed amount of compressed air (500 kg/s), the plant demands 23 kg/s of only coal or 34 kg/s of only biomass. The fuel demand increases with an increase in biomass ratio because of its lower calorific value compared to coal.

Figure 3.5 shows the effect of biomass to coal mass ratio on plant performance. The performance of the IGCC plant (both specific output and efficiency) decreases with an increase in biomass to coal mass ratio. Owing to the low calorific value of biomass compared to coal, it results in low performance, but it gives a cost-effective solution with low emissions. For fixed air supply to the compressor, biomass gasification demands more fuel, so it generates more power from the plant.

Figure 3.6 depicts the effect of biomass to coal ratio on IGCC plant emissions. Both the CO_2 and NO_X emissions decrease with an increase in biomass participation. The Thermo chemical model for NO*x* emission in the IGCC plant is not well reported in literature. Koornneef et al. (2008) reported 229 mg/kWh of NO_X on average from IGCC plant without CO_2 capture. The current IGCC plant model gives 638 to 705 mg/kWh of NO*x* emission and is above the reported value. The CO_2 emission is expressed in gm/kWh. Amelio et al. (2007) and Ordorica-Garcia et al. (2006) reported 709 gm/kWh and 744 gm/kWh respectively for CO_2 emission without CO_2 capture. In current work, IGCC results are 570 to 630 gm/kWh which is low compared to the reported values.

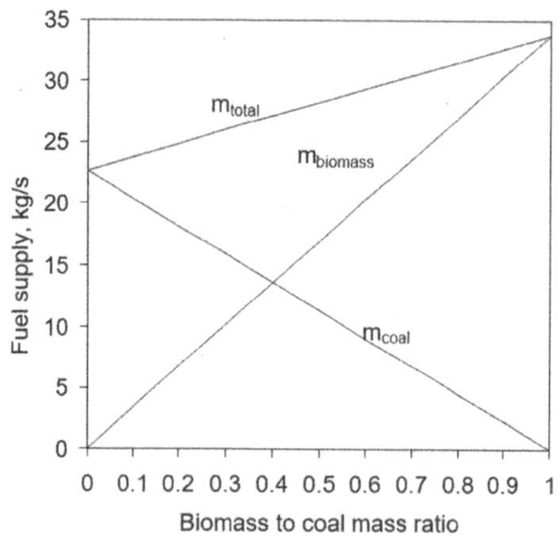

Figure 3.4 Fuel demand variations with biomass to coal mass ratio.

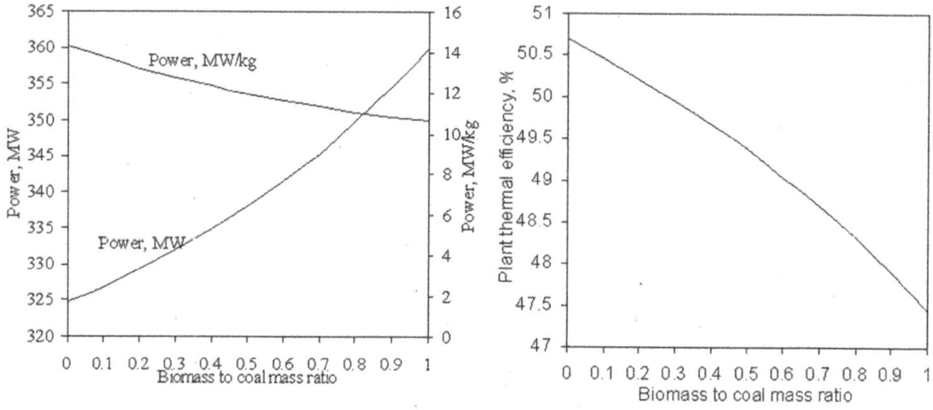

Figure 3.5 IGCC performance variations with biomass to coal mass ratio.

Table 3.1 gives an overview of the IGCC performance with gasifier at the prestated conditions. Even though the RAFR is fixed, stiochiometric air fuel ratio differs in coal and biomass fuels. Therefore, there are differences in actual air fuel ratio and so fuel and air supplies to the gasifier. The change in fuel also affects the steam injection at fixed steam fuel ratio (SFR). More biomass is needed compared to coal at the same RAFR. At the above-stated conditions, the biomass model generates more syngas. CO_2 emission decreases with biomass addition in co-gasification. There is not much variation in NO_X emission with co-gasification. The output increases and efficiency will decrease with biomass increment.

3.4 SUMMARY

The characteristics of coal/biomass co-gasifier have been evaluated with the thermochemical equilibrium model. The results of co-gasification applied to IGCC plants are

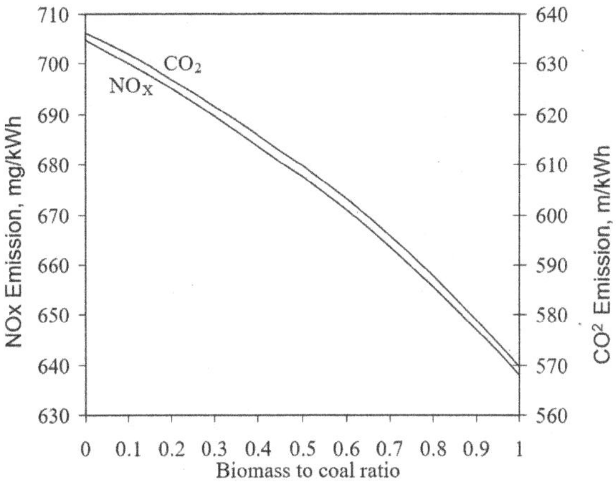

Figure 3.6 IGCC plant emissions as a function of biomass to coal ratio.

Table 3.1 IGCC plant material flow details with gasification/co-gasification at RAFR = 0.25, SFR = 1, gasifier pressure =12 bar, $m_{air,total}$ = 500 kg/s and GTIT = 1,200°C

Particulars	*Coal gasification*	*Biomass gasification*	*20% Coal and 80% Biomass*
Fuel supply, kg/s	22.9	33.8	31.7 (27.2+4.5)
Air supply to gasifier, kg/s	52.5	49.4	50.3
Air supply to GTCC, kg/s	447.5	450.6	449.7
Steam injection, kg/s	22.9	33.8	31.7
Syngas generation, kg/s	96.4	117	112.8
Ash in gasifier, kg/s	1.9	0.13	0.9
Flue gas, kg/s	543.9	567.6	562.5
CH_4, CO, CO_2, & H_2, % (dry)	13.9, 6.4, 21.8 & 11.3	6.3, 9.2, 20.5, & 25.4	8.3, 8.6, 20.8, & 21.8
Gasifier temperature, °C	688.7	780	757
Steam generation (HP, IP, & LP), kg/s	63, 22.4, & 6.8	68.6, 23.8, & 6.9	67, 23.4, & 6.8
NO_X, mg/kWh	692.5	638	651.9
CO_2, gm/kWh	637	569.6	588
CGE, %	76	65.6	68.3
Gasifier exergetic loss, %	16.3	21.6	20.4
Gas plant output, MW	197.6	224.8	216.9
Steam plant output, MW	129.6	135.3	133.7
IGCC output, MW	327.2	360.1	350.6
IGCC plant efficiency, %	50.6	47.4	48.3

compared with the separate coal gasification and biomass gasification. H_2 and CO in syngas increases with an increase in biomass to coal mass ratio. The plant emissions (NO_X and CO_2) are estimated from syngas combustors. An increase in biomass ratio in co-gasification results in high efficiency performance with low emissions. Co-gasification gives a cost-effective solution for power generation. The power is increased from 327.2 MW with coal gasification to 350.6 MW with co-gasification. Similarly, the NO*x* has decreased from 692.5 mg/kWh to 651.9 mg/kWh.

ACKNOWLEDGMENT

Authors T. Srinivas and B. V. Reddy acknowledge the financial support from NSERC, Canada, for the research work.

NOMENCLATURE AND ABBREVIATIONS

Nomenclature

a, b	Coefficients
E	Total exergy in kJ/kg mol
i	Specific irreversibility in kJ/kg mol

Acronyms and abbreviations

AFR	Air fuel ratio
GTCC	Gas turbine combustion chamber
GTIT	Gas turbine inlet temperature
HRSG	Heat recovery steam generator
LHV	Lower heating value
RAFR	Relative air fuel ratio
RH	Reheater
SFR	Steam fuel ratio
ST	Steam turbine
syn	Synthesis gas
TP	Triple pressure

REFERENCES

Alzate, C. A., F. Chejne, C. F. Valdes, A. Berrio, J. D. L. Cruz, & C. A. Londono. 2009. "CO-gasification of pelletized wood residues." *Fuel* 88: 437–445.

Amelio, M., P. Morrone, F. Gallucci, & A. Basile. 2007. "Integrated gasification gas combined cycle plant with membrane reactors: Technological and economical analysis." *Energy Conversion and Management* 48(10): 2680–2693.

Cao, Y., Y. Wang, J. T. Riley, & W. P. Pan. 2006. "A novel biomass air gasification process for producing tar-free higher heating value fuel gas." *Fuel Processing Technology* 87: 343–353.

Chmielniak, T., & M. Sciazko 2003. "Co-gasification of biomass and coal for methanol synthesis." *Applied Energy* 74: 393–403.

Goyal, A., S. Pushpavanam, & V. Ravi Kumar. 2010. "Modeling and simulation of co-gasification of coal and petcoke in a bubbling fluidized bed coal gasifier." *Fuel Processing Technology* 91(10): 1296–1307.

Hara S., J. Inumaru, M., Ashizawa, & K. Ichikawa. 2002. "A study on gasification reactivity of pressurized two-stage entrained flow coal gasifier." *JSME International Journal* 45(3): 518–522.

Jarungthammachote, S., & A. Dutta. 2007. "Thermodynamic equilibrium model and second law analysis of a downdraft waste gasifier." *Energy* 32(9): 1660–1669.

Jong, W., J. Andries, & K. R. G. Hem. 1999. "Coal/biomass co-gasification in a pressurised fluidised bed reactor." *Renewable Energy* 16: 1110–1113.

Kalisz, S., M. Pronobis, & D. Baxter. 2008. "Co-firing of biomass waste-derived syngas in coal power boiler." *Energy* 33: 1770–1778.

Koornneef, J., T. Harmelen, A. Horssen, R. Gijlswijk, A. Ramireza, A. Faaija, & W. Turkenburg. 2008. "The impacts of CO_2 capture on transboundary air pollution in the Netherlands." *Energy Procedia* 1(1): 3787–3794.

Kumabe, K., T. Hanaoka, S. Fujimoto, T. Minowa, & K. Sakanishi, 2007. "Co-gasification of woody biomass and coal with air and steam." *Fuel* 86: 684–689.

Lee, S. H., S. J. Yoon, H. W. Ra, Y. Son, J. C. Hong, & J. G. Lee. 2010. "Gasification characteristics of coke and mixture with coal in an entrained-flow gasifier." *Energy* 35: 3239–3244.

Lv, P. M., Z. H. Xiong, J. Chang, C. Z. Wu, Y. Chen, & J. X. Zhu. 2004. "An experimental study on biomass air–steam gasification in a fluidized bed." *Bioresource Technology* 95(1): 95–101.

Mastellone, M. L., L. Zaccariello, & U. Arena. 2010. "Co-gasification of coal, plastic waste and wood in a bubbling fluidized bed reactor." *Fuel* 89(10): 2991–3000.

McLendon, T. R., A. P. Lui, R. L. Pineault, S. K. Beer, & S. W. Richardson. 2004. "High-pressure co-gasification of coal and biomass in a fluidized bed." *Biomass and Bioenergy* 26: 377–388.

Ordorica-Garcia, G., P. Douglas, E. Croiset, & L. Zheng. 2006. "Technoeconomic evaluation of IGCC power plants for CO_2 avoidance." *Energy Conversion and Management* 47(15–16): 2250–2259.

Pan, Y. G., X. Roca, E. Velo, & L. Puigjaner. 1999. "Removal of tar by secondary air in fluidised bed gasification of residual biomass and coal." *Fuel* 78: 1703–1709.

Pan, Y. G., E. Velo, X. Roca, J. J. Manya, & L. Puigjaner. 2000. "Fluidized-bed co-gasification of residual biomass/poor coal blends for fuel gas production." *Fuel* 79: 1317–1326.

Pinto, F., C. Franco, R. N. Andre, M. Miranda, I. Gulyurtlu, & I. Cabrita. 2002. "Co-gasification study of biomass mixed with plastic waste." *Fuel* 81: 291–297.

Prins, M. J., K. J. Ptasinski, & F. J. J. G., Janssen. 2007. "From coal to biomass gasification: Comparison of thermodynamic efficiency." *Energy* 32: 1248–1259.

Saravanamuttoo, H. I. H., G. F. C. Rogers, & H. Cohen. 2003. *Gas Turbine Theory*, 5nd edn. London: Pearson Education.

Srinivas, T., A. V. Gupta, & B. V. Reddy. 2008. "Parametric exergy analysis of coal gasifier and gas turbine combustion chamber with emission study." *International Energy Journal* 9(1): 33–40.

Srinivas, T., A. V. Gupta, & B. V. Reddy. 2009. "Thermodynamic equilibrium model and exergy analysis of a biomass gasifier." *ASME Journal of Energy Resources Technology* 131(3): 1–7.

Srinivas, T., A. V. Gupta, B. V. Reddy, & P. K. Nag. 2004. "Second law analysis of a coal based combined cycle power plant." In *Proceedings of 17th National Heat and Mass Transfer Conference and 6th ISHMT/ ASME Heat and Mass Transfer Conference*, Indira Gandhi Centre for Atomic Research (IGCAR), Kalpakam, pp. 973–978.

Srinivas T., A. V. Gupta, B. V. Reddy, & P. K. Nag. 2006. "Parametric analysis of a coal based combined cycle power plant." *International Journal of Energy Research* 30(1): 19–36.

Sjostrom, K., G. Chen, Q. Yu, C. Brage, & C. Rosen. 1999. "Promoted reactivity of char in co-gasification of biomass and coal: Synergies in the thermochemical process." *Fuel* 78: 1189–1194.

Valero, A., & S. Usón, 2006. "Oxy-co-gasification of coal and biomass in an integrated gasification combined cycle (IGCC) power plant." *Energy* 31: 1643–1655.

Velez, J. F., F. Chejne, C. F. Valdes, E. J. Emery, & C. A. Londono. 2009. "Co-gasification of Colombian coal and biomass in fluidized bed: An experimental study." *Fuel* 88: 424–430.

Watanabe, H., & M. Otaka. 2006. "Numerical simulation of coal gasification in entrained flow coal gasifier." *Fuel* 85(12–13): 1935–1943.

Chapter 4

Supercritical CO_2 cycle powered by solar thermal energy

Ramneek Singh and Rupinder Pal Singh
Department of Mechanical Engineering, Punjab Agricultural University, Ludhiana, India

Dibakar Rakshit
Centre for Energy Studies, Indian Institute of Technology Delhi, Hauz Khas, New Delhi, India

4.1 INTRODUCTION

4.1.1 Overview of thermodynamic power conversion cycles

The power cycle involves a sequence of certain thermodynamic processes which converts heat energy into useful work by exploiting a working fluid, while varying parameters like pressure, temperature, eventually returning the system to its initial state, henceforth forming a cycle (Cengel, Boles, and Kanoglu 2011). Power conversion cycle is often signified as a thermodynamic cycle or power cycle. Work done may be performed by the system, acting as a heat engine on its surroundings over this cycle. Thermodynamic cycles may be categorized on the basis of critical point as sub critical, Trans and supercritical cycles, which are shown in Figure 4.1.

4.1.2 Subcritical thermodynamic cycle

Thermodynamic cycle in which heat addition and heat rejection occurs below critical point such as the subcritical Rankine cycle. Working fluid from the liquid phase is converted into gas by the boiling process (e.g., organic Rankine cycle)

4.1.3 Transcritical thermodynamic cycle

Thermodynamic cycle in which heat addition occurs at supercritical level and heat rejection occurs below critical point. Working fluid from the liquid phase is converted into gas by boiling then the gas will be superheated to attain a supercritical phase (e.g., transcritical CO_2 cycle).

4.1.4 Supercritical thermodynamic cycle

Thermodynamic power cycle in which heat addition and heat rejection occurs above critical point. The state of working fluid promptly converts into supercritical fluid from liquid by avoiding the boiling process and bypassing two phase regions which results in less loss of exergy (e.g., s-CO_2 cycle).

The thermodynamic cycles are also classified as wet, isentropic, and dry fluid, based on their curve of the vapor saturation of particular working fluid as shown in Figure 4.2.

DOI: 10.1201/9781003213741-4

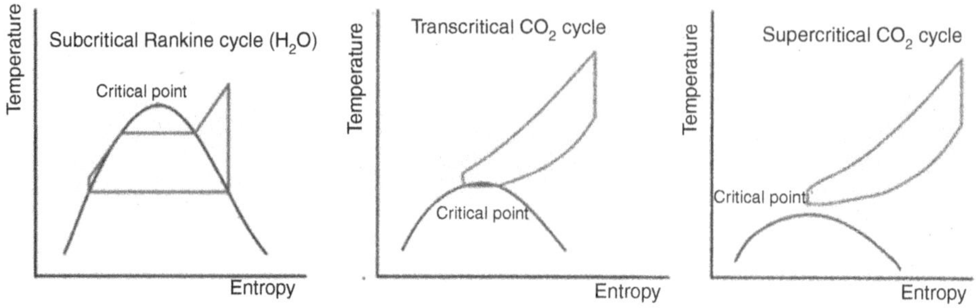

Figure 4.1 Classification of thermodynamic cycle based on critical point.

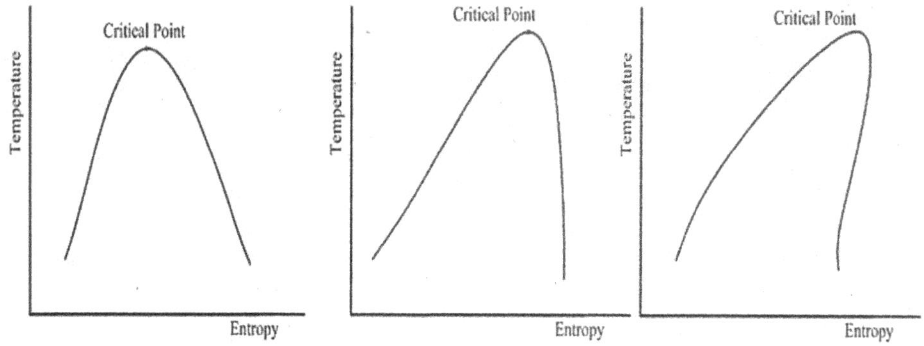

Figure 4.2 Wet fluid, isentropic fluid, dry fluid, respectively.

Wet working fluid: The entropy of wet fluid increases with decreasing saturation temperature. Hence, the slope of the vapor saturation curve is negative (ds/dt < 0). The isentropic expansion of wet working fluid always ends in the two-phase region. Superheating of wet fluid is necessary to avoid blade damage and reduction in isentropic efficiency of the turbine caused by saturated liquid. Therefore, a minimum quality factor of 85% must be required at the turbine outlet (Bao & Zhao 2013) (e.g., H_2O, ammonia, R707, R502).

Dry working fluid: The entropy of dry fluid declines with a fall in saturation temperature. The slope of the vapor saturation curve is positive in dry fluid (ds/dt > 0). The isentropic expansion of the dry fluid certainly ends in the superheated phase (Chen, Goswami, & Stefanakos 2010). If the fluid was too dry then it results in additional cooling load on condenser/cooler due to presence of substantial superheat in the turbine by fluid. (e.g., Benzene, n-butane, isopentane).

Isentropic working fluid: The entropy of isentropic fluid remains constant irrespective of any change in saturation temperature. The slope on critical point is infinite for an isentropic fluid (ds/dt=0). The fluid stays in saturated vapor state after an isentropic expansion (e.g., Toulene, R11, R12 3, R245fa, and Fluorinal 85).$dS/dT < 0$, the slope of the vapor saturation curve is negative. Entropy of wet fluid increases with decreased saturation temperature. Isentropic expansion of wet working fluid always ends in the two-phase region. Superheating of wet fluid is essential to prevent blade damage and reduction in isentropic efficiency of the turbine caused by saturated liquid. Therefore, a minimum quality factor of 85% must be required at the turbine outlet (Bao & Zhao 2013) (e.g., H_2O, ammonia, R707, R502).

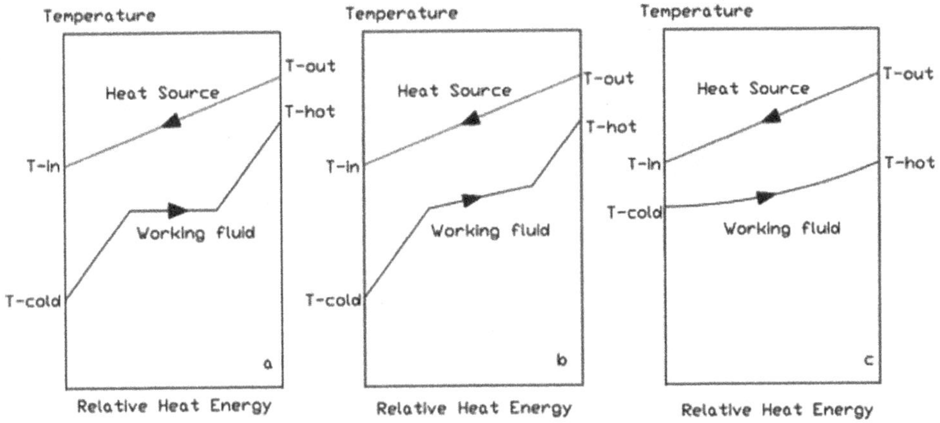

Figure 4.3 Pinching by different fluids in heat exchanger (a) pure fluid, (b) zeotropic fluid, and (c) supercritical fluid.

The relation between interactions of working fluid in heat exchanger (hot and cold streams) is represented in Figure 4.3. Supercritical fluids have a tendency to make a better thermal match than pure and zeotropic fluids. Pinch point technology is a methodology which may ensure optimal design of heat exchanger networks, coolers and heaters along with the minimal operational cost and capital (Rokni 2016; Baronci et al. 2015).

4.2 HEAT SOURCES SUITABLE WITH S-CO_2 CYCLE

4.2.1 Concentrating solar power (CSP) sources

The detailed terms refer to the concentration of solar energy (solar radiations) by using reflectors (mirrors) to power a thermodynamic cycle for electricity generation. CSP comprises technologies such as: central receiver (power towers), parabolic trough collector, concentrated linear Fresnel reflector, and solar dish/engine systems. Heat transfer fluids (HTF) such as oil, salt, or steam are selected for the transfer of solar energy to the power cycle (power block). Each fluid employed in this power system has its own operating limits, i.e., molten salt has an upper limit of ~600°C. Supercritical CO_2 is able to be utilized for both purposes, heat transfer fluid as well as working fluid of cycle (Turchi et al. 2013). The s-CO_2 cycle is an advanced power cycle which provides a high-efficiency power conversion system to be coupled with solar thermal power plants operating at maximum cycle's temperature range of 600°C or above (Turchi, Stekli, & Bueno 2017). The s-CO_2 cycle is capable of putting up dry cooling that is desirable for CSP application in arid regions.

4.2.2 Nuclear reactors

Supercritical CO_2 power cycles were initially envisioned for nuclear power plants. As a heat source, nuclear reactors supplies heat at high temperature over a limited temperature range. Nuclear reactors such as gas-cooled, sodium (Na)-cooled and lead (Pb)-cooled fast reactor systems are appropriate heat sources for s-CO_2 power cycle (Liu, Wang, & Huang 2019). An advanced nuclear reactor (sodium-cooled fast reactor) integrated with s-CO_2 recompression cycle power cycle delivers a high efficiency near 43% as s-CO_2 power systems are highly suitable to operate at 510–525°C outlet temperatures from sodium-cooled fast reactors. The utilization of an s-CO_2 power cycle with advanced nuclear reactors can also lessen the cost of electricity (Sienicki & Moisseytsev 2017). s-CO_2 was used in British reactors operated at 650°C as core exit temperatures, reaching a thermal efficiency of ~50%. The supercritical CO_2 cycle showed great potential for significant lessening in the capital investment of reactor assisted power plants, which is currently the hindrance toward their deployment (Dostal, Driscoll, & Hejzlar 2004).

4.2.3 Waste heat recovery (WHR)

4.2.3.1 Industrial waste heat

Waste heat sources are break into three general categories as high grade heat (more than 650°C), low grade heat (232°C or lower), and medium/intermediate grade heat (232–650°C) (Liu, Wang, & Huang 2019; Sajwan, Sharma, & Shukla 2020). Waste heat comes from various industries like petroleum refining, chemical industries, furnaces related to mining and metallurgical industries. Waste heat source may be a stream of hot water or any other fluid-like steam, hot flue gases, cooling unit of compressed export gas. Metallurgical and ceramics industries including glass and brick manufacturers can produce waste heat in the temperature range of 300–400°C (Sarkar 2015).

4.2.3.2 Internal combustion engine (ICE)

The temperature of engine exhaust is ~150°C; the coolant of the cooling system of the engine can also be utilized as heat source (Sarkar 2015). Modern marine propulsion ICEs are generally equipped with waste heat recovery equipment.

4.2.3.3 Fuel cells

Fuel cells such as molten carbonate and solid oxide fuel cells operate at a range of exhaust temperatures which are 620–660°C and 800–1,000°C, respectively. Exhaust is often indicated as waste heat, which subsequently can be recuperated by utilizing a power cycle.

4.2.4 Geothermal energy

Heat energy can be extracted from hot springs and geysers. This is referred to as geothermal energy, generally available in a temperature range of 50–350°C. Heat sources above 220°C are suitable for electricity generation. Low temperature often ends in lower efficiency. Heat energy is flowing from the center to the crust at an energy flow rate of 44.2 TW (Sarkar 2015).

4.2.5 Coal

Coal-fired thermal power plants are the prime source of electricity generation. CO_2 emissions in flue gases of power plants reach almost two-fifths of total carbon emissions (Liao et al.

2019). Coal-based power plants are greatly significant in reducing energy losses and preventing global warming. An s-CO_2 cycle coupled to a coal-based power plant showed 41.3% as net efficiency when worked at 620°C (Le Moullec 2013).

4.2.6 Biomass

Biomass includes woods, energy crops, waste or by-products of crops, plants and algae, municipal solid waste, wastes from the processing and food industry. Supercritical CO_2 cycle achieves maximum efficiency of 36% (in conversion from biomass to electricity), which accounts almost 10% more than that for the currently existing range of biomass-based power generation plants (Manente & Lazzaretto 2014).

4.2.7 Cryogenic fuel

Liquified natural gas was oxy-combusted to power a novel semi-closed supercritical CO_2 cycle of its own kind known as an Allam cycle. Moreover, this cycle also produces CO_2 by oxy-combustion which is later used by CO_2 turbines. The cycle achieves high efficiency (>50%) operated near 1,000°C as turbine inlet temperature.

4.3 SUPERCRITICAL CO_2 AS WORKING FLUID

The working fluid possesses specific properties in the supercritical phase which exhibit an intermediate behavior between that of liquid state and gaseous state. Particularly fluids in supercritical phase display density like liquid, viscosity like gas, and diffusivity which is intermediate to that of a liquid and a gas. CO_2 is abundant, low priced, unreactive, reliable, non-flammable, non-toxic, non-explosive, and ease of accessibility makes it desirable for use as working fluid (Musgrove, Ridens, & Brun 2017).

Carbon dioxide reaches its critical point by attaining specific conditions as represented by Figure 4.4. Moreover, ambient temperature of environment and critical temperature of CO_2 are almost similar, making it an adequate working fluid to be utilized in Brayton power cycles as it can be incorporated into manifold ranges of power cycles.

Utilizing s-CO_2 as working fluid can contribute to a reduction in CO_2 emissions, as it has the potential to contribute towards green energy. Carbon dioxide exhibits properties such as critical point at ambient conditions and high density in supercritical state comparison with working fluids other than s-CO_2 for supercritical cycles. Properties of CO_2 show vast significant variations near the critical point. Moreover, supercritical CO_2 is also a good solvent due to properties like high solubility and diffusivity. As per the thermophysical properties of CO_2, heat transfer characteristics of supercritical phase are better than CO_2 in the state of liquid or gas. CO_2 is also employed in certain applications other than power generation such as sterilization of various substances, enhanced oil recovery, and various manufacturing activities (Musgrove, Ridens, & Brun 2017).

There are some major problems related to supercritical CO_2. The first problem is the uneven changes in thermophysical properties adjoining the critical point, causing the local flow field overlapping of the fluid. So, accurate computation of physical properties of s-CO_2 near the critical point became crucial. Another problem is the occurrence of a two-phase region at the inlet of the compressor which can cause damage to compressor blades. Two-phase regions may occur due to reduction in terms of pressure and temperature lower than critical point triggered by acceleration in the flow field at the blade's leading edge.

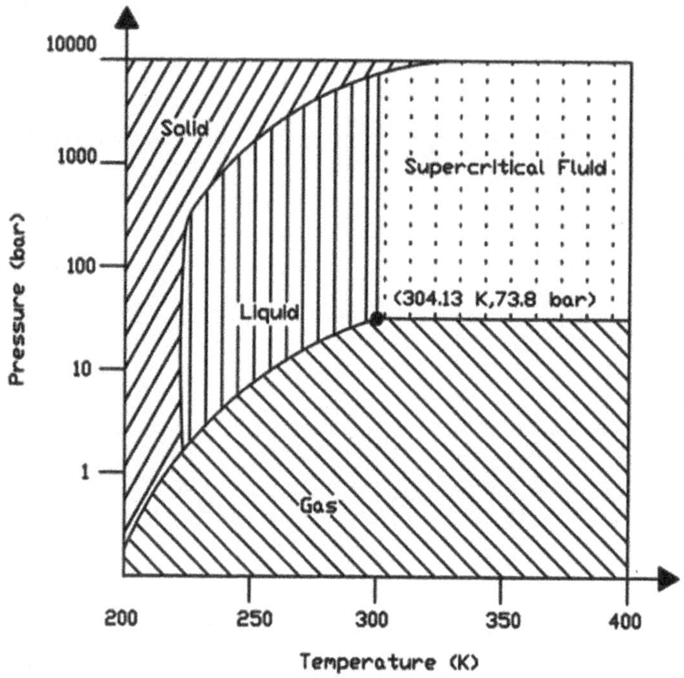

Figure 4.4 Critical point of carbon dioxide.

From the thermodynamic perspective, higher cycle efficiency is attained with temperatures as lower as possible at which the heat is rejected. Unsuitability of several other working fluids also gives s-CO_2 an extra advantage over them; SO_2 is highly toxic and highly corrosive for its utilization as a supercritical working fluid. Hydrocarbons and chlorofluorocarbons with critical temperatures near 30–40°C are generally employed in refrigeration and air-conditioning and geothermal plants, prime reasons against their usage in required applications are high radiation instability, flammability, and ozone layer damage potential (Dostal, Driscoll, & Hejzlar 2004).

4.4 MERITS OF SUPERCRITICAL CO_2 AS WORKING FLUID

CO_2 possesses comparative inertness for the concerned range of temperature and appropriate value of its critical pressure and temperature. It does not require recycling like other fluids.

It is economically affordable and abundant without any toxicity, flammability, and corrosiveness. It is less hazardous to the environment than other working fluids since it has no potential for ozone depletion or century-long global warming.

Supercritical CO_2 possesses high fluid density which enables extremely compact design of turbomachinery and heat exchanger technology. Exhaust heat exchanger can be positioned directly with high temperature heat sources, which is not so possible with organic Rankine cycle (ORC) (Kacludis et al. 2021).

It can contribute to environmental improvement from greenhouse gas reduction. It reduces water consumption as it is suitable for dry cooling in arid environments.

4.5 THERMOPHYSICAL PROPERTIES OF SUPERCRITICAL CO_2

Relation between density and particular temperature of CO_2 near critical region is represented in Figure 4.5. Different isobars are showing the variations in density due to change in temperature. The effect of temperature on specific heat is shown by isobars in Figure 4.6. These sharp fluctuations of thermophysical properties extremely influence the performance of the cycle which is discussed in the results section.

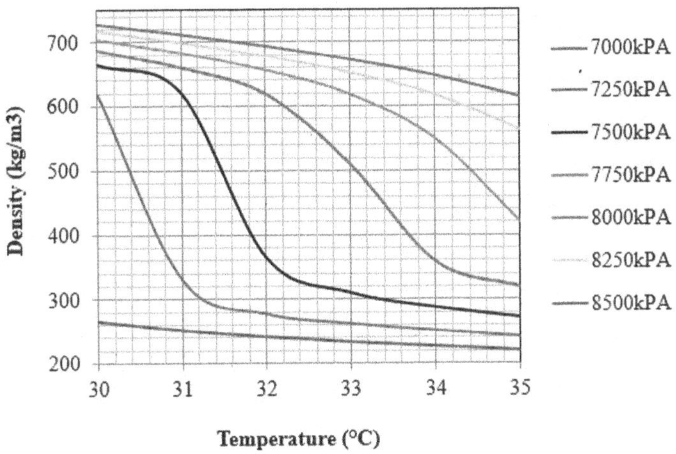

Figure 4.5 Variation in density near critical point of CO_2.

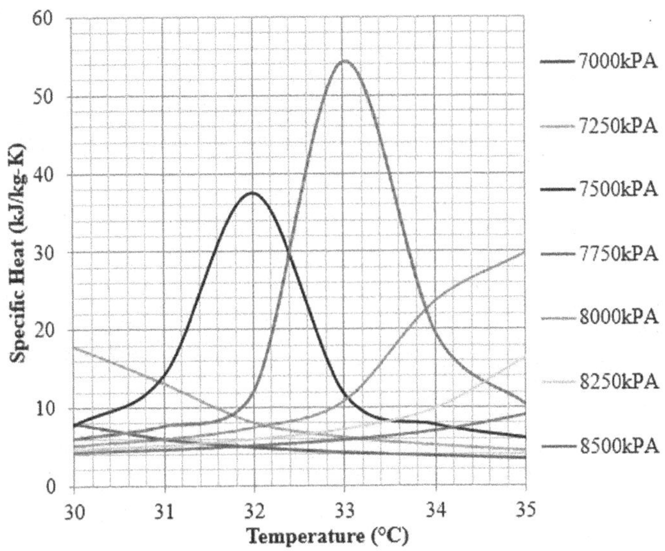

Figure 4.6 Variation in specific heat near critical point of CO_2.

4.6 LAYOUTS OF DIFFERENT SUPERCRITICAL CO_2 CYCLE CONFIGURATIONS

The supercritical cycle (Figure 4.7) shows the following characteristics: improved thermal efficiency, low value of volume/power ratio, almost nil erosion of turbomachinery's blade and single-phase fluid throughout the process of cooling which is desirable for practical application. Electrical generation even portable, space power generation (electrical), shaft powers for marine propulsion are some applications of a supercritical power cycle (Feher 1968).

Simple recuperation (Figure 4.8) improves the net thermal efficiency of the thermodynamic cycle by utilizing the heat that is going to be rejected. Compression is split into two stages in intercooling cycle arrangement (Figure 4.9) with working fluid cooling between the stages, to reduce the work required for compression. Another common approach to increase work output is reheating between the turbine stages (Figure 4.10) (Moisseytsev & Sienicki 2008).

The supercritical CO_2 recompression cycle reaches high efficiency; it is significantly more compact and less. Additionally, it achieves efficiency equivalent to helium Brayton cycle

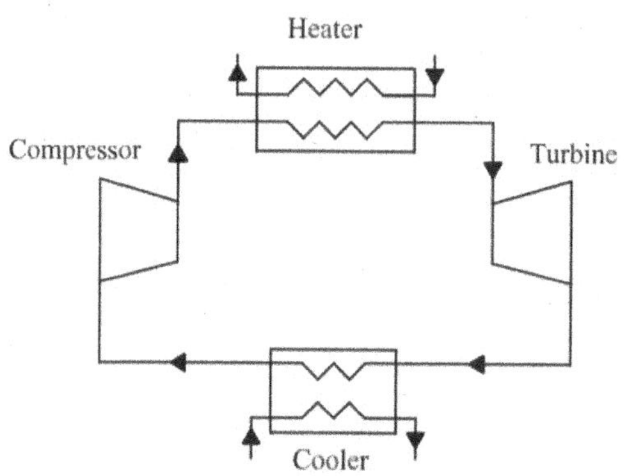

Figure 4.7 Simple supercritical Brayton cycle.

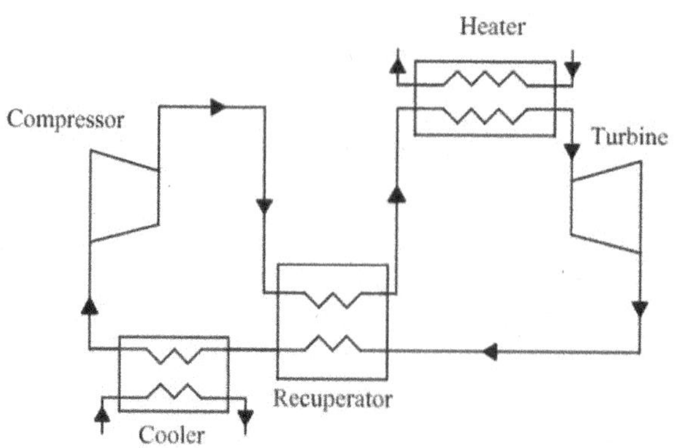

Figure 4.8 Simple recuperated supercritical Brayton cycle.

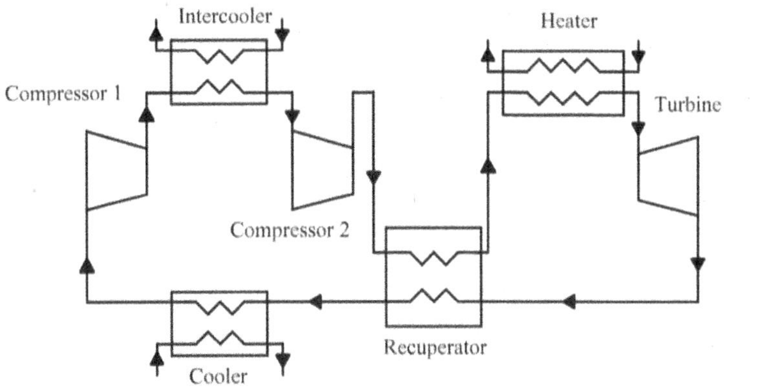

Figure 4.9 Supercritical CO_2 Brayton with intercooling.

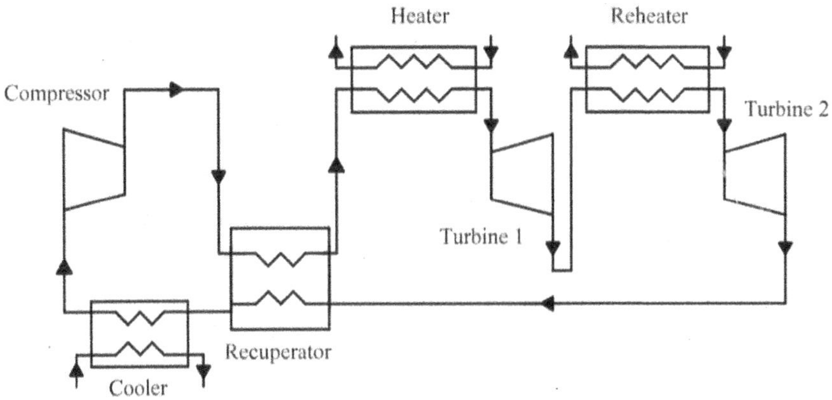

Figure 4.10 Supercritical CO_2 Brayton with reheating.

Table 4.1 Effect of basic modifications on cycle performance

Modification	*Work output*	*Thermal efficiency*
Regeneration/recuperation	No change	Increment
Intercooling	Increment	Decrement
Reheating	Increment	Decrement
Recuperated intercooling	Increment	Increment
Recuperated reheating	Increment	Increment
Reheating plus intercooling	Increment	Decrement
Recuperated reheating plus intercooling	Increment	Increment

operated at much greater temperatures. The supercritical CO_2 cycle attains an efficiency of 46% at 550°C, which is equivalent to the helium Brayton cycle operating at 800°C even after accounting all of the losses (Dostal, Driscoll, & Hejzlar 2004). The thermodynamic analysis performed by Angelino compared the performance of different cycle variations and configurations to show that those with recompression and reheat achieve the highest efficiency among the rest in the highest cycle temperature range of 500–800°C (Angelino 1968; Musgrove & Wright 2017).

4.7 REVIEW OF LITERATURE

Recent studies are concluded in this literature review for the supercritical CO_2 cycle and its allied technologies. Nevertheless, the concept of the CO_2 Brayton cycle was initially studied by Angelio (1968) and Feher (1968). Supercritical CO_2 turbomachinery and components were compact in design because of the high density of working fluid. The supercritical CO_2 cycle has a tendency to be configured in various layouts achieving a particular efficiency (first and second law) and work output. Many layouts of the s-CO_2 cycle (Table 4.1) were analyzed on varying certain operation conditions like temperature and pressure. Dostal, Driscoll, and Hejzlar (2004) proposed and extensively analyzed the thermodynamics of all possible modifications in the s-CO_2 cycle for nuclear energy as heat input. Analysis of various layouts such as Brayton cycle with intercooling, reheating, recompression, regeneration, split expansion, and several possible combinations of these modifications have been evaluated by various researchers. Nuclear reactors, CSP, and WHR are the most commonly used heat sources along with heat rejected by fuel cells and a few with geothermal and biomass-based heat sources for this cycle. The latest studies show that the rejected heat of the supercritical CO_2 cycle can be utilized for desalination of water, absorption cycle, domestic or industrial heating, and as a heat source of bottoming cycle. The supercritical CO_2 cycle was also analyzed in various studies for optimal design of turbomachinery, components, recuperator – especially printed circuit heat exchanger, and heat exchangers.

Le Moullec (2013) performed thermodynamic analysis and compared various configurations. Analysis was performed on constant temperatures at inlet of compressor (32°C) and turbine (550°C) by varying outlet pressure of compressor and pressure ratio of turbine. Pre-compression cycle at compressor outlet pressure 10 MPa achieves thermal efficiency ~44%. Recompression cycle achieves highest thermal efficiency of ~45% at optimal pressure ratio, minimum cycle pressure of 7.7 MPa corresponds to turbine inlet pressure of 20 MPa. Change in optimal pressure ratio can decrease efficiency significantly. Split-expansion achieves lower thermal efficiency than a recompression cycle due to reduction of pressure in the reactor. A partial cooling cycle shows thermal efficiency like a recompression cycle, which can be improved by regeneration but suffers from pinch point in regenerators. Results show that for compressor outlet pressure of 20MPa, the most effective refinement to a simple Brayton cycle was the addition of pre-compression, as higher pressure gives better results with flow dividing the recompressor. Authors investigated several configurations of the supercritical CO_2 cycle, which consist of pre-compression, partial cooling, and recompression. The recompression cycle was most efficient and was further investigated by utilizing mathematical modelling. The simulation results show that with 550°C and 20 MPa as turbine inlet conditions, the cycle can achieve efficiency of 44%.

Chai and Tassou (2019) carried out a numerical modeling to evaluate the thermo hydraulic performance of the s-CO_2 cycle for printed circuit heat exchanger (PCHE). PCHE was a recuperative heat exchanger recommended to be employed in the supercritical CO_2 cycle because it could withstand high temperature, high pressure, showed improved heat transfer, and was more compact. Results show an improved heat transfer rate (PCHE) of 393.5 W in comparison with the conventional heat exchanger (96.2 W). Fiaschi et al. (2014) evaluated and compared four different systems for heat recovery from furnaces which comprises several configurations of s-CO_2 cycle, ORC, and regenerative air Brayton. Results show that the ORC was adequate for heat recovery from smaller furnaces and the air Brayton-Joule cycle was reflected suitable for bigger furnaces. In contrast, the s-CO_2 cycle delivers the highest output of power for the entire range of furnace sizes.

Recently, researchers have been exploring solar energy for powering the s-CO_2 cycle. Khan et al. (2019) conducted a study on s-CO_2 recompression with reheating Brayton power cycle coupled with solar parabolic dish collector as heat source. The performance of thermal oil-based nanofluids is analyzed for working in a solar cavity receiver integrated with the s-CO_2 cycle. The study also investigates the variation in maximum cycle pressure along with pressure ratio for optimization of thermal and exergetic efficiency of thermodynamic cycle. Results reveal that the proposed cycle reaches a thermal efficiency of 47.65 %. Also, the thermal oil-based nanofluid Al_2O_3 has first and second law efficiency of 33.73% and 36.27%, which is higher than TiO_2- and CuO-based nanofluids. Linares et al. (2020) investigated different cycle layouts with intercooling and reheating through both dry and wet cooling for a supercritical CO_2 Brayton power cycle. Working with printed circuit heat exchangers in a dry cooling environment was generally suggested for its appropriateness to operate at high-pressure conditions. A cavity-type receiver was selected to minimize the radiation heat losses on high temperatures. $MgCl_2$/KCl/NaCl molten salts were employed due to their higher constancy up to 800°C. Investigation results recommend the wet cooling for reheating with 54.6% efficiency. The intercooling with reheating is suitable for dry cooling. Wang et al. (2018) studied a solar-thermal energy system by using a supercritical CO_2-based recompression cycle as a power block and a thermal energy storage system coupled with power block. Heat transfer fluid (HTF) and storage medium was a molten halide salt to be operated at a relatively high temperature. Authors analyzed thermodynamic performance with comparison between halide and nitrate salts. Power system with halide salt achieves 19.16–22.03% (solar-to-electric efficiency) in four demonstrative days, higher than existing tower central receiver-based concentrated power plants.

Khatoon and Kim (2020) contribute a completely developed approach for heliostat, solar receiver, power block, and thermal energy storage (TES); theoretically analyzed performance of CO_2-based topping and bottoming power conversion cycle integrated with TES on basis of work done, first and second law of thermodynamics. A CO_2 recompression Brayton cycle was employed as topping cycle and a transcritical CO_2 Brayton cycle as bottoming cycle for recovery of heat rejected by topping cycle, to improve efficiency by waste heat recovery. Sharan, Neises, and Turchi (2019) studied advanced thermodynamics of various layouts of closed-loop s-CO_2 cycles as a power block for concentrating solar thermal sources. In comparison with superheated or supercritical steam Rankine cycles, s-CO_2 offers much higher cycle efficiency at equivalent temperatures for solar thermal applications. Due to higher density of s-CO_2, supercritical CO_2 power blocks have less complexity, less weight and volume compared to Rankine cycle.

Garg, Kumar, and Srinivasan (2013) studied concentrated solar power-assisted supercritical CO_2 cycles and compared for certain operating conditions. Results conclude that the thermal efficiency linearly increases with low side pressure in the subcritical and transcritical cycles. The s-CO_2 cycle is capable of achieving thermal efficiency of approximately 30% even at a lower source temperature (500°C). Sharan, Neises, and Turchi (2019) examined a thermal desalination system driven by rejected heat (>70°C) of supercritical carbon dioxide cycle. To generate power together with fresh water production, authors proposed a multi-effect distillation (MED) coupled with an s-CO_2 cycle. Parametric analysis was conducted to obtain optimal design parameters. Supercritical CO_2 cycle efficiency was unaffected by MED, a comparison was also done with an MED-integrated conventional Rankine cycle. Optimization results show that an s-CO_2 power plant reaches 49.2% efficiency with integration of MED and daily production of 2,813 m^3/day of distilled water.

A system optimization consisting of solar thermal energy-based supercritical CO_2 cycles was performed by Padilla et al. (2016). Optimum operational conditions were obtained by multi-objective thermodynamic optimizations performed on four configurations by considering the effect of reheating and input parameters. Overall, outcomes show that at inlet temperature of 700–750°C, all configurations of solarized system achieved optimum exergy efficiency and cycle pressure depending on reheat condition exists in the range of 24.2– 25.9 MPa. Yang, Yang, and Duan (2020) proposed and analyzed the part-load performance of a power system comprised of an s-CO_2 cycle as power block, a solar power tower as heat source, and molten salt thermal storage. Comparison between several configurations was analyzed under part-load conditions. Outcomes of analysis shows that cycle configured with reheating achieves higher efficiency than regeneration cycle, while under part-load conditions, higher performance was attained by recompression and intercooling cycles than reheating cycle. The intercooling cycle performed more efficiently than the recompression cycle only when the load exceeds 60%. The recompression cycle performed better when proportion of actual generation to maximum generation was lower than 62.5%.

Wang et al. (2020) evaluated a method of combining direct air cooling with for a parabolic trough solar-based s-CO_2 recompression cycle. A TES system along with heat exchangers and turbomachinery were also examined on different parameters. The results show that the pinch point of recuperators was significantly affected by recompression ratio and its optimum value falls with an increment in pressure at entrance of compressor and shaft speed of turbomachinery.

4.8 METHODOLOGY

4.8.1 Cycle description and input parameters

The temperature of s-CO_2 stream is initially heated up in a heater and later in a reheater (refer Figure 4.11). The expansion of working fluid takes place in high pressure and low pressure (HP/LP) turbines which convert thermal energy into useful work. During first expansion, the maximum cycle pressure decreases to intermediate pressure and stream attains lowest pressure after expansion in an LP turbine (Figures 4.12 and 4.13). High-temperature

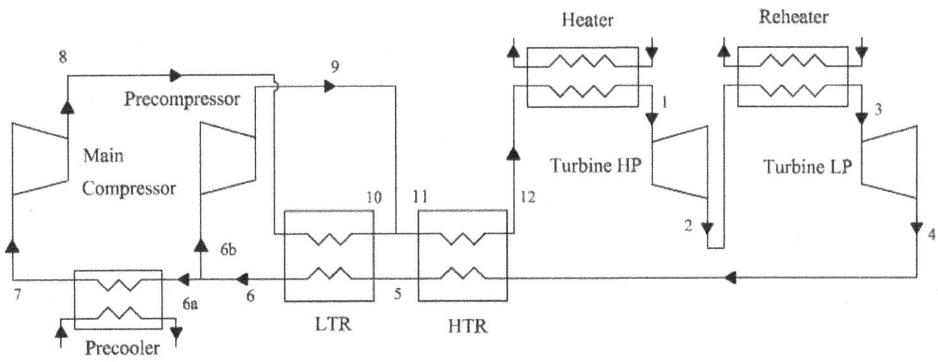

Figure 4.11 Configuration of supercritical carbon dioxide recompression with reheating power conversion cycle.

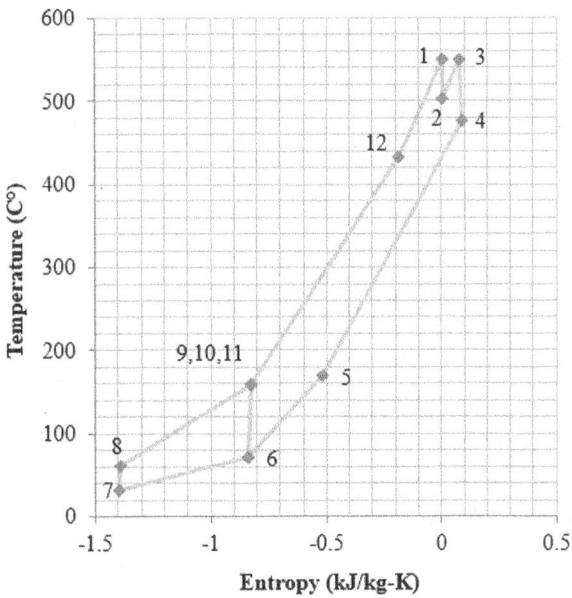

Figure 4.12 Temperature vs. entropy plot of supercritical CO_2 recompression with reheat cycle.

regenerator and low-temperature regenerators (HTR/LTR) are used to attain higher efficiencies by internal recuperation. Compression work is separated between principal compressor and recompressor/precompressor. There is a split up between precooler (6a) and recompressor (6b) after a cold stream of LTR. Further, the temperature of the stream reduces to minimum cycle temperature (compressor inlet temperature, CIT) by precooler before compression in the main compressor (MC). Whereas, the compressed stream by the recompressor (RC) rejoins the other compressed stream between HTR and LTR.

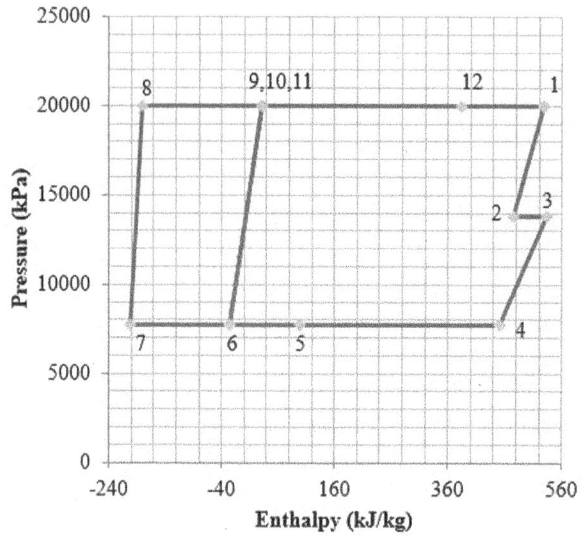

Figure 4.13 Pressure vs. enthalpy plot of supercritical CO_2 recompression with reheat cycle.

Table 4.2 Standard cycle design considerations and input parameters

Input parameter	*Symbol*	*Value*
Turbine inlet temperature	*TIT* (T_I)	550 °C
Main compressor inlet temperature	*CIT* (T_7)	32 °C
Compressor efficiency	η_C	0.90
Turbine efficiency	η_T	0.93
LTR effectiveness	ε_{HTR}	0.97
HTR effectiveness	ε_{LTR}	0.97
Maximum cycle pressure	TIP,P_max,PI	20,000 KPa
Pressure ratio	PR	2.6

4.8.1.1 *Assumptions*

1 All processes are based on steady state conditions.
2 Expansion and compression in turbomachinery is purely isentropic in nature.
3 No loss of pressure occurred in pipes and heat exchanger.
4 Identical turbine inlet temperatures for both of the turbines.
5 Intermediate pressure is the average of maximum and minimum cycle pressure.
6 All cycle components are insulated including pipes.
7 Difference between temperature at the receiver and turbine inlet temperature is 250 K.

4.8.2 Mathematical model

A mathematical model is prepared to simulate recompression with reheating s-CO_2 cycle using thermophysical properties of CO_2 based on Span and Wagner correlations (Span & Wagner 2009) . The values of particular state points for cycle components are calculated on the basis of mass-energy and exergy balance equations by utilizing the input parameters given in Table 4.2.

Relation between PR, $P_{\max}$ and $P_{\min}$,

$$PR = \frac{P_{\max}}{P_{\min}} \tag{4.1}$$

The turbomachinery performance is estimated with an isentropic model using equations (4.2) to (4.5).

4.8.2.1 Turbine

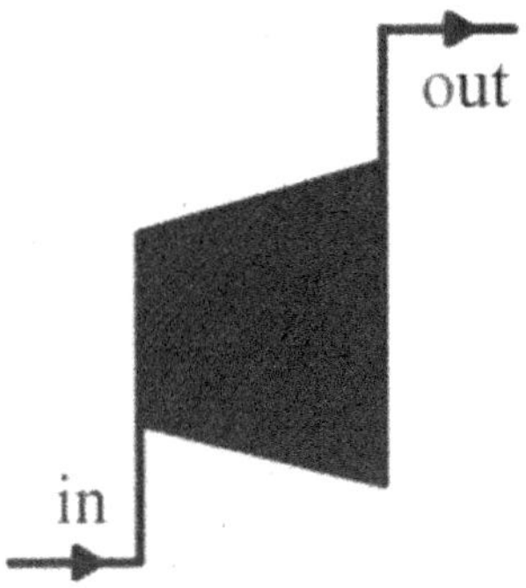

$$W_T = \dot{m}(h_{in} - h_{out}) \tag{4.2}$$

$$\eta_T = \frac{(h_{in} - h_{out})}{(h_{in} - h_{out,is})} \tag{4.3}$$

4.8.2.2 Compressor

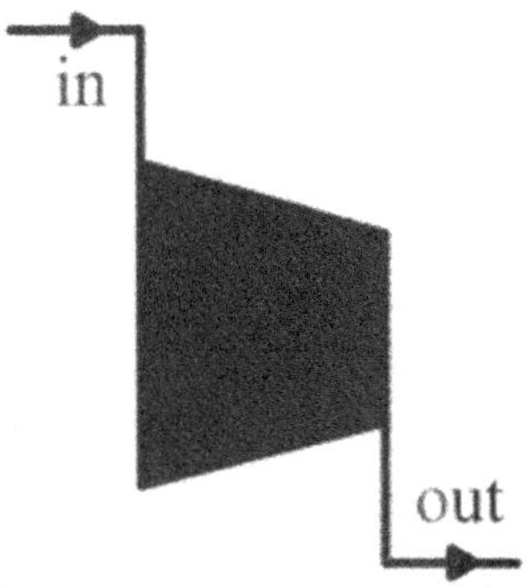

$$W_C = \dot{m}(h_{out} - h_{in}) \tag{4.4}$$

$$\eta_C = \frac{(h_{out,ise} - h_{in})}{(h_{out} - h_{in})} \tag{4.5}$$

4.8.3 Exergy model

The exergetic model of s-CO_2 cycle is derived from the second law of thermodynamics. Modeling has been done by considering assumptions and input parameters, several fundamental exergy balance equations are used to determine the state points. Furthermore, the exergetic destruction rate is calculated to obtain the exergetic efficiency of the system. A detailed set of equations for each component is presented (Mohammadi, Fallah, & Mahmoudi 2019; Liu, Wang, & Huang 2019).

Specific exergy of a state point is determined by

$$E_i = \dot{m}\left((h_i - h_0) - T_0(s_i - s_0)\right) \tag{4.6}$$

4.8.3.1 Exergy balance equations for components of cycle

Exergy destruction by high-pressure turbine

$$E_{HP} = E_1 - E_2 - W_{HP} \tag{4.7}$$

Exergy destruction by low-pressure turbine

$$E_{LP} = E_3 - E_4 - W_{LP} \tag{4.8}$$

Exergy destruction by main compressor

$$E_{MC} = E_7 - E_8 + W_{MC} \tag{4.9}$$

Exergy destruction by recompressor

$$E_{RC} = E_{6b} - E_9 + W_{RC} \tag{4.10}$$

Exergy destruction by precooler

$$E_{PC} = E_{6a} - E_7 \tag{4.11}$$

Exergy destruction by HTR

$$E_{HTR} = E_{12} - E_{11} + E_4 - E_5 \tag{4.12}$$

Exergy destruction by LTR

$$E_{LTR} = E_5 - E_6 + E_8 - E_{10} \tag{4.13}$$

Exergy destruction by heater and reheater

$$E_{HE} = E_{12} - E_1 + Q_{HE}\left(1 - \frac{T_0}{T_H}\right) \tag{4.14}$$

$$E_{RE} = E_3 - E_2 + Q_{RE}\left(1 - \frac{T_0}{T_H}\right) \tag{4.15}$$

Total cycle exergy destruction

$$\dot{I} = E_{HP} + E_{LP} + E_{HTR} + E_{LTR} + E_{PC} + E_{MC} + E_{RC} + E_{RE} + E_{RE} \tag{4.16}$$

Exergy input of cycle

$$E_{in} = Q_s\left(1 - \frac{T_0}{T_m}\right) \tag{4.17}$$

Heat supplied

$$Q_{HE} = \dot{m}(h_1 - h_{12}) \tag{4.18}$$

$$Q_{RE} = \dot{m}(h_3 - h_2) \tag{4.19}$$

$$Q_{in} = Q_{HE} + Q_{RE} \tag{4.20}$$

Net work done

$$W_{net} = W_{HP} + W_{LP} - W_{MC} - W_{RC} \tag{4.21}$$

Second law efficiency

$$\eta_{II,cycle} = \frac{W_{net}}{\dot{I} + W_{net}} \tag{4.22}$$

4.9 RESULTS AND DISCUSSION

The parametric evaluation is conducted for examining the trends in exergetic efficiency and exergetic destruction rate with variation in certain variables. These variables are input parameters which include: compressor inlet temperature, turbine inlet temperature, pressure ratio, maximum cycle pressure (turbine inlet pressure, TIP) and intermediate cycle pressure (IP). Each parameter influences performance of certain components or the overall cycle's performance in its own particular manner which is discussed in this chapter. The parametric analysis is conducted within the following range of variables.

$$550\,^{\circ}C \le TIT \le 950\,^{\circ}C \tag{4.23}$$

$$32\,^{\circ}C \le CIT \le 50\,^{\circ}C \tag{4.24}$$

$$2.2 \le PR \le 3 \tag{4.25}$$

$$20000\,kPa \le TIP \le 28000\,kPa \quad (4.26)$$

$$8000\,kPa \le IP \le 18000\,kPa \quad (4.27)$$

4.9.1 Effect of input parameters on exergetic destruction rate of individual component

4.9.1.1 *Effect of compressor inlet temperature*

Exergetic destruction rate of various components with respect to compressor inlet Temperature is represented by Figure 4.14. Irreversibility of both turbines remains constant as CIT does not affect the conditions at inlet and outlet of both turbines. Irreversibility of main compressor and recompressor increases and slightly decreases respectively with increase in CIT. Highest destruction rate is shown by LTR at lowest CIT and PC exhibits highest irreversibility at 47°C. Irreversibility of LTR and HTR decrease while irreversibility of PC increases with raising CIT for the reason that Precooler inlet temperature (T_6) increases rapidly even after a little increase in CIT from 32°C which increase the net difference between T_6 and CIT.

4.9.1.2 *Effect of turbine inlet temperature*

The effect of turbine inlet temperature on the exergetic destruction rate of various components is shown in Figure 4.15. HTR has the highest destruction rate except at 550°C which increases with turbine inlet temperature because the temperature at HTR inlet, T_4 also rises with turbine

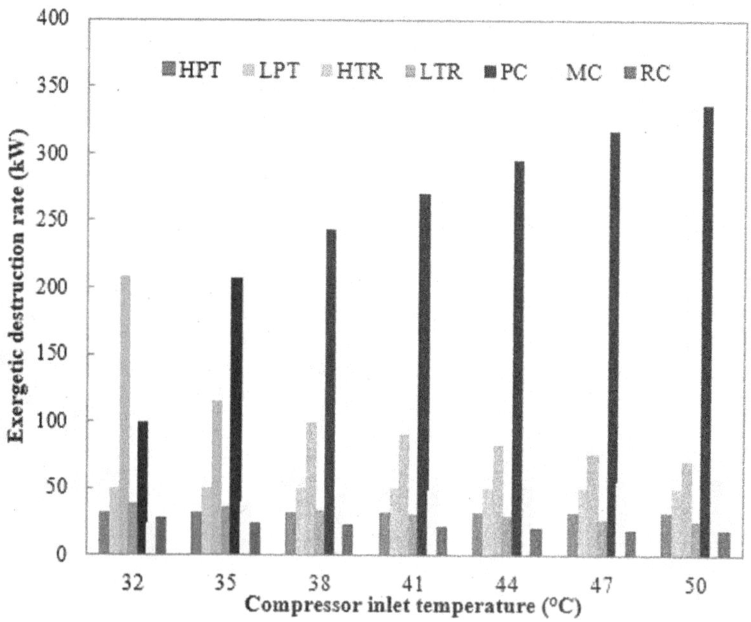

Figure 4.14 Effect of compressor inlet temperature on exergetic destruction rate of cycle components.

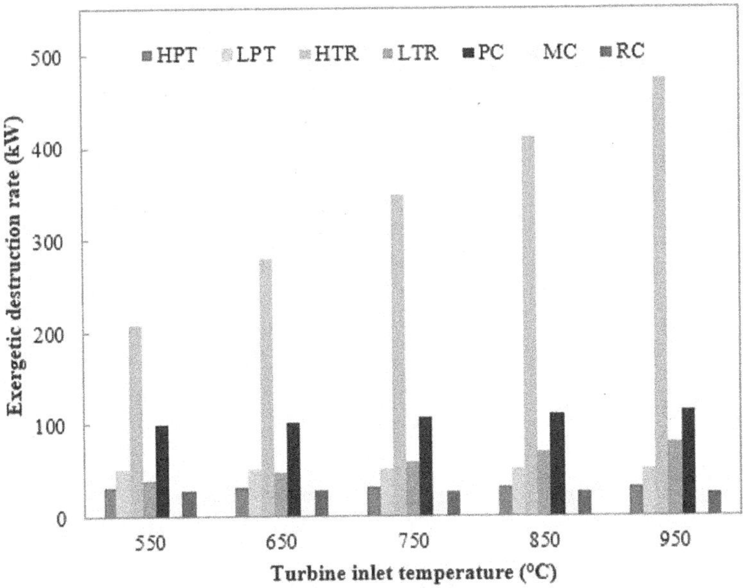

Figure 4.15 Effect of turbine inlet temperature on exergetic destruction rate of cycle components.

inlet temperature. The trends reveal that turbine inlet temperature (TIT) has a massive impact on both recuperators. However, the effect on LTR and PC is very minor which increases with TIT. Exergetic destruction rate of turbomachinery remains almost constant with limited effect because difference between exergetic flow rates at inlet and outlet remains almost same throughout the given range of TIT.

4.9.1.3 Effect of pressure ratio

The influence of pressure ratio on the exergetic destruction rate of various components is represented by Figure 4.16. Result reveals that the effect of PR is very noticeable as it affects each component. The graph shows that the minimum optimum destruction rate occurs approximately at 2.6. The exergetic destruction rate of turbomachinery increases with rise in PR, although net work done also increases. PR affects a lot in heat exchangers in the exergetic destruction rate in HTR and LTR rises and falls respectively as increase in PR. The value of exergetic destruction rate initially decreases then reaches minimum and rises again for PC. An optimum minimum value of the exergetic destruction rate of PC exists near 2.6 because value of minimum cycle pressure reaches below critical pressure and cycle becomes transcritical at higher PR.

4.9.2 Effect of various input parameters on exergetic efficiency

4.9.2.1 Effect of TIT

The effect of turbine inlet temperature on both efficiencies and split ratio are shown in Figure 4.17. The first and second law efficiency rises as turbine inlet temperature increases.

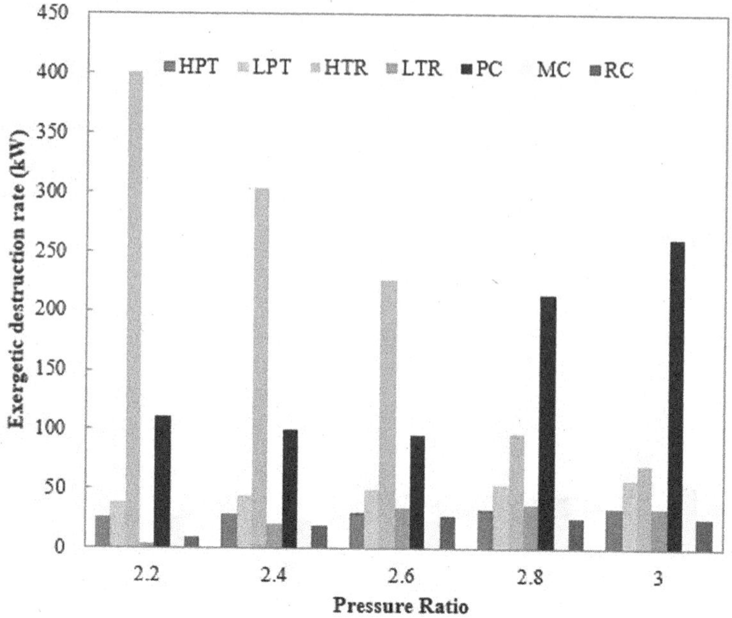

Figure 4.16 Effect of pressure ratio on exergetic destruction rate of cycle components.

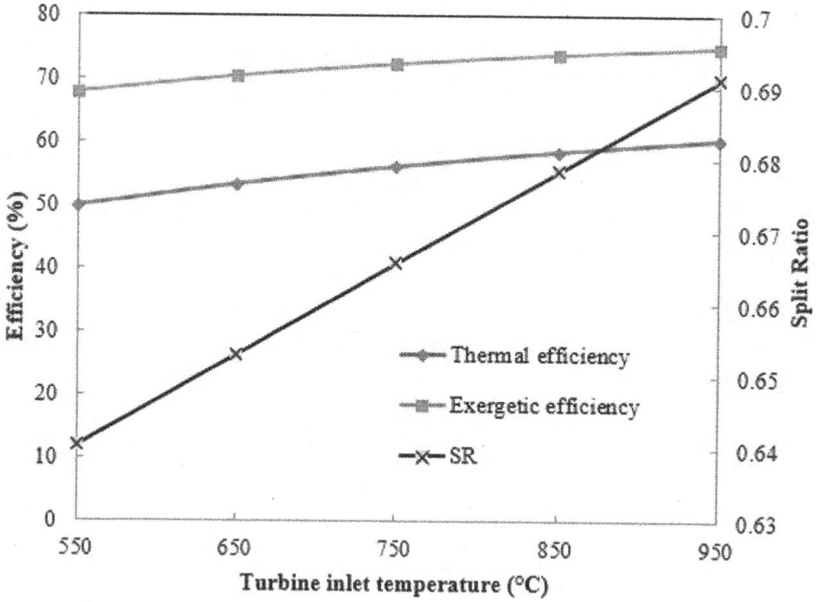

Figure 4.17 Effect of TIT on thermal efficiency, exergetic efficiency and split ratio.

SR also increases with turbine inlet temperature due to higher temperature at RC inlet (T_6), which influences the compressor in increased work done by them. So, the mass flow rate through RC is decreased and that of PC stream is increased which results in a rise in split ratio.

4.9.2.2 Effect of compressor inlet temperature at different turbine inlet temperature

The influence of compressor inlet temperature at various turbine inlet temperatures on exergetic efficiency of cycle is shown in Figure 4.18. The exergetic destruction rate is proportional to the turbine inlet temperature, however it decreases with a rise in CIT. Lower CIT is preferred for the operating cycle because it reduces the requirement of work done by compressor which improves net work done and overall efficiencies. Moreover, maximum efficiency is achieved at CIT=32°C as it is immediately above the critical point and thermophysical properties of CO_2 at 32°C are considered to be most suitable for the compression process. The turbine produces higher work done output as the turbine inlet temperature increases, which later improves the cycle efficiencies. Although the compressor's work done increases with CIT increment, which subsequently reduces the cycle efficiencies.

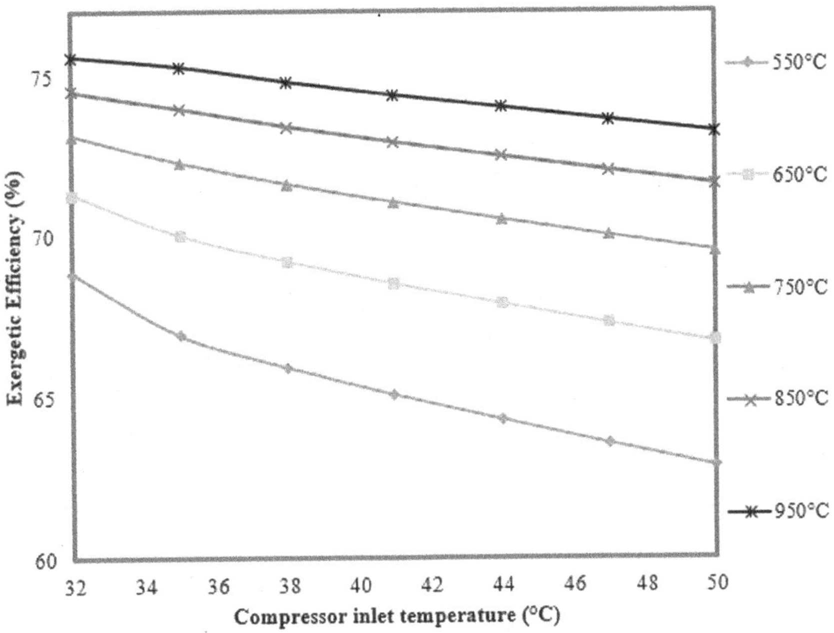

Figure 4.18 Effect of compressor inlet temperature at different turbine inlet temperature on exergetic efficiency of cycle.

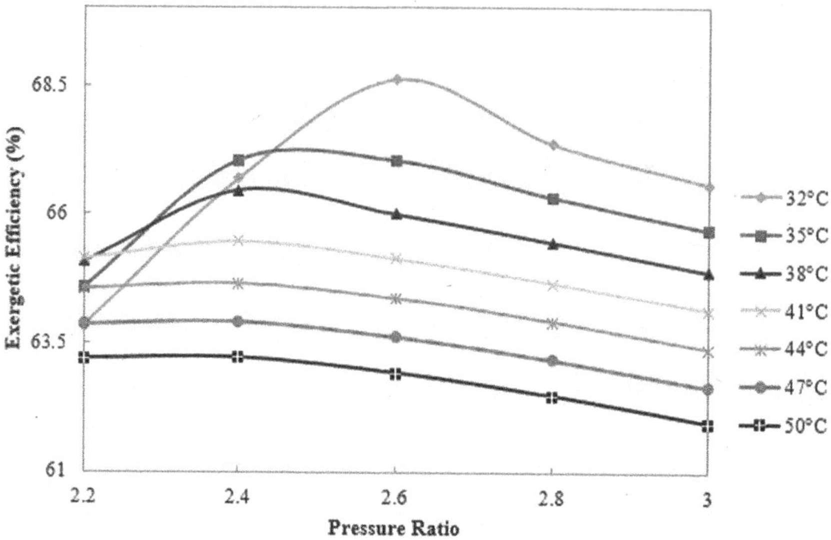

Figure 4.19 Effect of pressure ratio at various compressor inlet temperature on exergetic efficiency of cycle.

4.9.2.3 *Effect of pressure ratio at various compressor inlet temperature*

The variation in exergetic efficiency due to influence of pressure ratio at different compressor inlet temperature is represented by Figure 4.19. At higher CIT above 39°C, the exergetic efficiency decreases with increase in CIT. Although an optimum value of PR exists for CIT up to 35°C. However it decreases at higher PR because the value of minimum cycle pressure reaches below critical pressure and the cycle becomes transcritical. This is the reason for fall in efficiency after a particular optimum value for lines showing 32°C and 35°C.

4.9.2.4 *Effect of pressure ratio at different turbine inlet temperature*

The variation in exergetic efficiency due to the influence of pressure ratio at different turbine inlet temperatures is represented by Figure 4.20. At lower turbine inlet temperature up to 750°C an optimum value of PR exists near 2.6 and decreases rapidly thereafter because the cycle became transcritical in nature. At a higher turbine inlet temperature, efficiency increases with the rise in PR. Higher pressure ratios result in higher output by turbines but also increase the work done by compressors Therefore an optimum value of net work done exists and is similar for efficiency at lower turbine inlet temperature. However, efficiency increases at higher PR for higher values of TIT because TIT and PR both are proportional to work done by turbines.

4.9.3 Effect of various input parameters on performance of turbomachinery

4.9.3.1 *Effect of turbine inlet pressure or maximum cycle pressure*

The relation between maximum cycle pressure (turbine inlet pressure) and performance of turbomachinery is represented in Figure 4.21. Work done by both turbines remains constant

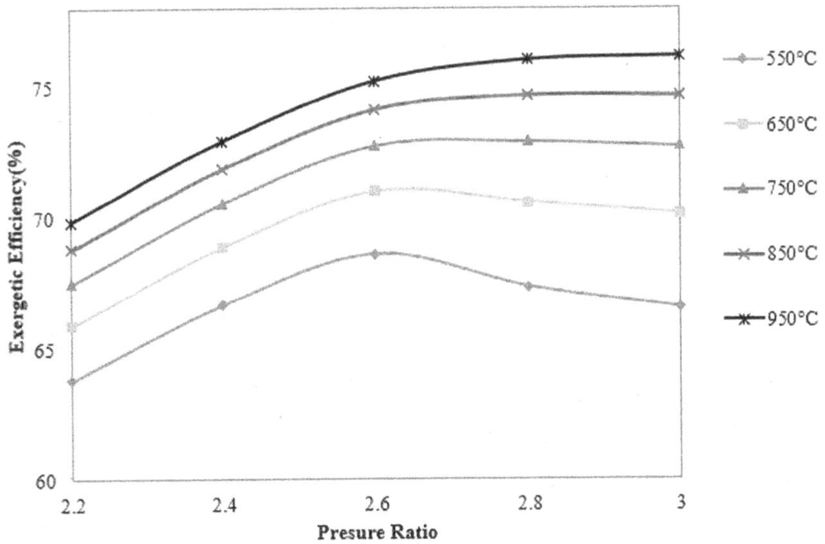

Figure 4.20 Effect of pressure ratio at different turbine inlet temperature on exergetic efficiency of cycle.

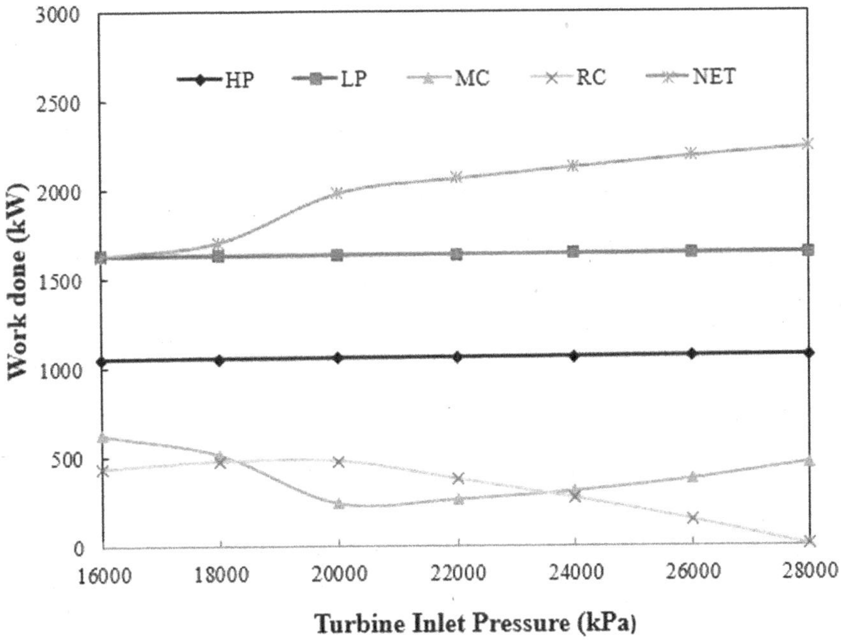

Figure 4.21 Effect of turbine inlet pressure on the work done by turbomachinery in a recompression (with reheat) supercritical CO_2 Brayton cycle.

as TIP increases because the difference between enthalpy of entry and exit points of turbines remains same for a given range of TIP. For work done by the main compressor the curve first decreases, reaches minimum, and then increases, and the curve of work done for pre-compressor first increases, reaches maximum, and then decreases.

4.9.3.2 *Effect of intermediate pressure*

Figure 4.22 depicts the relationship between intermediate pressure and turbomachinery performance. Work done by HP turbines is inversely proportional to intermediate pressure and work done by LP turbines is directly proportional to intermediate pressure. Work done by mains and pre-compressors remains constant at all given values of intermediate pressure.Net work done initially increases marginally, reaches maximum to 1,983 kW at intermediate pressure of 12,000 kPa and then decreases slightly. It is observed that at low intermediate pressure of 8,000 kPa, work done by LP turbines is almost nothing and the optimum value of intermediate pressure exists near the intersection of the curve of work done by both turbines.

4.9.3.4 *Effect of pressure ratio*

The relation between pressure ratio and performance of turbomachinery is represented in Figure 4.23. Work done by both turbines is directly proportional to the pressure ratio. However net work done initially increases, reaching a maximum of 1,980 kW at PR=2.6. It is observed that PR=2.6 exists as an optimum pressure ratio. Work done by the main compressor first

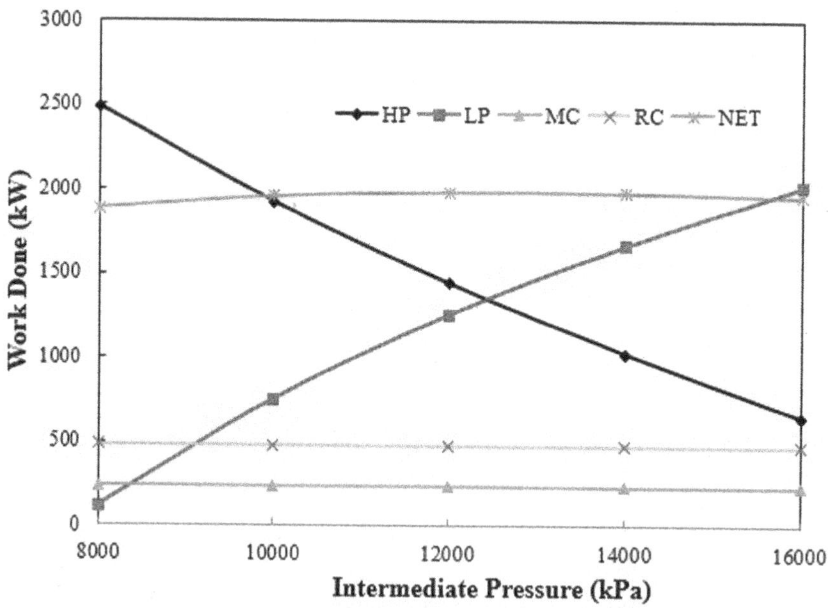

Figure 4.22 Effect of intermediate pressure on the work done by turbomachinery in cycle.

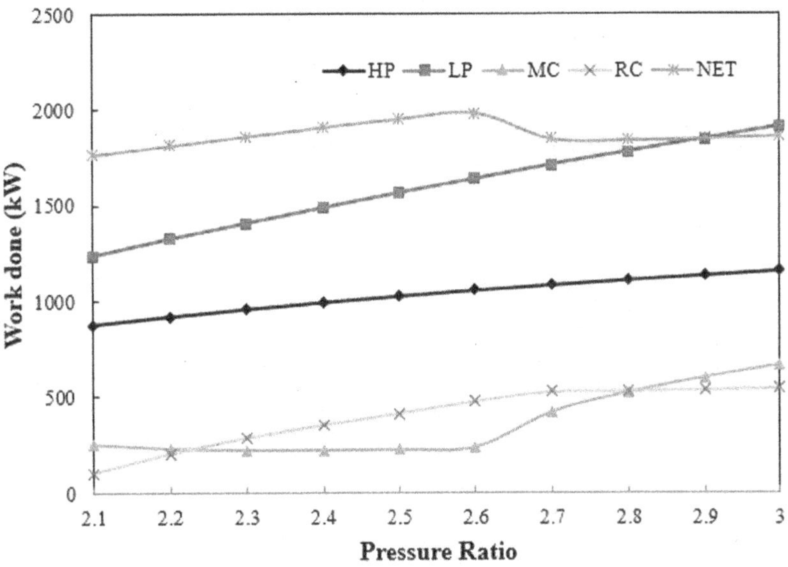

Figure 4.23 Variation of pressure ratio on the work done by turbomachinery in cycle.

decreases slightly, reaches minimum, and then starts inclining. Whereas, the curve of the pre-compressor initially rises, reaches maximum at PR=2.8 and then decreases minutely as it tends to be constant. Work done by recompressor is more than main compressor in pressure ratio within the range of 2.2 to 2.8.

4.10 SUMMARY

In this chapter, according to the results of evaluation, the following conclusions are made:

- For the supercritical CO_2 Brayton cycle, the compressor inlet temperature should be as low as possible and near to critical point. If compressor inlet temperature declines below critical point then two-phase region may occur inside compressor, which will surely cause damage to the blade. Overall, the thermal efficiency of the cycle is directly proportional to the turbine inlet temperature.
- SR should remain optimum or near to optimum value otherwise either the main compressor or the recompressor will stall if single shaft turbomachinery is used.
- An optimum PR exists for the cycle at which it reaches maximum exergetic efficiency for particular operating parameters. Turbine work done is linearly directly proportional to PR. While main compressor work done first decreases, reaches minimum, and then increases, and work done for recompressor first increases, reaches maximum, and then almost remains constant.
- PCIT rises with an increment in turbine inlet temperature. High PCIT suitable for WHR exists at higher compressor inlet temperature. Moreover, PCIT is also directly proportional to PR and PCIT rises rapidly for PR at which P_{min} exists near critical point.

The WHR system may be integrated with a cycle having higher PCIT which improves overall cycle efficiency.

- In contrast to irreversibility, heat exchangers PC, HTR, and LTR exhibit a higher destruction rate than turbomachinery. Turbomachinery is slightly affected by turbine inlet temperature and compressor inlet temperature. But PR, TIP, and intermediate pressure have a wide impact on turbomachinery.
- Net work done increases with TIP but thermal efficiency decreases. Work done by the turbine remains constant with change in TIP. While main compressor work done first decreases, reaches minimum, and then increases, and work done for recompressor first increases, reaches maximum and then decreases.
- An optimum value of intermediate pressure exists at which the cycle shows the highest thermal efficiency and net work done. LP turbine work done is directly proportional to IP, and HP turbine work done is inversely proportional to IP. Work done by both compressors remains constant and unaffected by IP. Intermediate pressure can affect the irreversibility of turbomachinery, especially the turbines.
- In general, the current assessment has validated the potential of supercritical CO_2 recompression with reheat cycle in combination with solar thermal energy resources in terms of the second law of thermodynamics.

REFERENCES

Angelino, Gianfranco 1968. "Carbon dioxide condensation cycles for power production." *Journal of Engineering for Power* 90(3): 287–295. doi:10.1115/1.3609190.

Bao, Junjiang, & Li Zhao. 2013. "A review of working fluid and expander selections for organic Rankine cycle." *Renewable and Sustainable Energy Reviews* 24: 325–342. doi:10.1016/j.rser.2013.03.040.

Baronci, Andrea, Giuseppe Messina, Stephen J. McPhail, & Angelo Moreno. 2015. "Numerical investigation of a MCFC (molten carbonate fuel cell) system hybridized with a supercritical CO_2 Brayton cycle and compared with a bottoming organic Rankine cycle." *Energy* 93: 1063–1073. doi:10.1016/J.ENERGY.2015.07.082.

Cengel, Yunus A., Michael A. Boles, & Mehmet Kanoglu. 2011. *Thermodynamics: An Engineering Approach*. New York: McGraw-Hill Education.

Chai, Lei, & Savvas A. Tassou. 2019. "Numerical study of the thermohydraulic performance of printed circuit heat exchangers for Supercritical CO_2 Brayton cycle applications." *Energy Procedia* 161: 480–488. doi:10.1016/J.EGYPRO.2019.02.066.

Chen, Huijuan, D. Yogi Goswami, & Elias K. Stefanakos. 2010. "A review of thermodynamic cycles and working fluids for the conversion of low-grade heat." *Renewable and Sustainable Energy Reviews* 14(9): 3059–3067. doi:10.1016/J.RSER.2010.07.006.

Dostal, Vaclav, Michael J. Driscoll, & Pavel Hejzlar. 2004. "A supercritical carbon dioxide cycle for next generation nuclear reactors." Advanced Nuclear Power Technology Program. https://web.mit.edu/22.33/www/dostal.pdf.

Feher, Ernest G. 1968. "The supercritical thermodynamic power cycle." *Energy Conversion* 14(9): 3059–3067. https://doi.org/10.1016/0013-7480(68)90105-8.

Fiaschi, Daniele, Adi Lifshitz, Giampaolo Manfrida, & Duccio Tempesti. 2014. "An innovative ORC power plant layout for heat and power generation from medium- to low-temperature geothermal resources." *Energy Conversion and Management* 88: 883–893. doi:10.1016/J.ENCONMAN.2014.08.058.

Garg, Pardeep, Pramod Kumar, & Kandadai Srinivasan. 2013. "Supercritical carbon dioxide Brayton cycle for concentrated solar power." *The Journal of Supercritical Fluids* 76: 54–60. doi:10.1016/J.SUPFLU.2013.01.010.

Kacludis, A., S. Lyons, D. Nadav, & E. Zdankiewicz. 2021. "Waste heat to power (WH2P) applications using a Supercritical CO_2-based power cycle." *Aidea.org*.

Khan, Muhammad Sajid, Muhammad Abid, Hafiz Muhammad Ali, Khuram Pervez Amber, Muhammad Anser Bashir, & Samina Javed. 2019. "Comparative performance assessment of solar dish assisted s-CO_2 Brayton cycle using nanofluids." *Applied Thermal Engineering* 148: 295–306. doi:10.1016/J.APPLTHERMALENG.2018.11.021.

Khatoon, Saboora, & Man Hoe Kim. 2020. "Performance analysis of carbon dioxide based combined power cycle for concentrating solar power." *Energy Conversion and Management* 205: 112416. doi:10.1016/J.ENCONMAN.2019.112416.

Le Moullec, Yann. 2013. "Conception of a pulverized coal fired power plant with carbon capture around a supercritical carbon dioxide Brayton cycle." *Energy Procedia* 37: 1180–1186. https://doi.org/10.1016/j.egypro.2013.05.215.

Liao, Gaoliang, Lijun Liu, Jiaqiang E, Feng Zhang, Jingwei Chen, Yuanwang Deng, & Hao Zhu. 2019. "Effects of technical progress on performance and application of supercritical carbon dioxide power cycle: A review." *Energy Conversion and Management* 199: 111986. doi:10.1016/J.ENCONMAN.2019.111986.

Linares, José I., María J. Montes, Alexis Cantizano, & Consuelo Sánchez. 2020. "A novel supercritical CO_2 recompression Brayton power cycle for power tower concentrating solar plants." *Applied Energy* 263: 114644. doi:10.1016/J.APENERGY.2020.114644.

Liu, Yaping, Ying Wang, & Diangui Huang. 2019. "Supercritical CO_2 Brayton cycle: A state-of-the-art review." *Energy* 189: 115900. doi:10.1016/J.ENERGY.2019.115900.

Manente, Giovanni, & Andrea Lazzaretto. 2014. "Innovative biomass to power conversion systems based on cascaded Supercritical CO^2 Brayton cycles." *Biomass and Bioenergy* 69: 155–168. doi:10.1016/J.BIOMBIOE.2014.07.016.

Mohammadi, Zahra Beig, Mohsen Fallah, & S. M. Seyed Mahmoudi. 2019. "advanced exergy analysis of recompression Supercritical CO_2 cycle." *Energy* 178: 631–643. doi:10.1016/J.ENERGY.2019.04.134.

Moisseytsev, Anton, & James J. Sienicki. 2008. "Performance improvement options for the supercritical carbon dioxide Brayton cycle." www.osti.gov/biblio/935094.

Musgrove, Grant, Brandon Ridens, & Klaus Brun. 2017. "Physical properties." In Klaus Brun, Peter Friedman, & Richard Dennis (eds.), *Fundamentals and Applications of Supercritical Carbon Dioxide (sCO_2) Based Power Cycles*, pp. 23–40. Duxford: Woodhead Publishing. doi:10.1016/B978-0-08-100804-1.00002-5.

Musgrove, Grant, & Steven Wright. 2017. "Introduction and background." In Klaus Brun, Peter Friedman, & Richard Dennis (eds.), *Fundamentals and Applications of Supercritical Carbon Dioxide (sCO_2) Based Power Cycles*, pp. 1–22. Duxford: Woodhead Publishing. doi:10.1016/B978-0-08-100804-1.00001-3.

Padilla, Ricardo Vasquez, Yen Chean Soo Too, Regano Benito, Robbie McNaughton, & Wes Stein. 2016. "Multi-objective thermodynamic optimisation of supercritical CO_2 Brayton cycles integrated with solar central receivers." *International Journal of Sustainable Energy* 37(1): 1–20. doi:10.1080/14786451.2016.1166109.

Rokni, Masoud. 2016. "Introduction to pinch technology." https://orbit.dtu.dk/en/publications/introduction-to-pinch-technology.

Sajwan, Kshitiz, Meeta Sharma, & Anoop Kumar Shukla. 2020. "Performance evaluation of two medium-grade power generation systems with CO_2 based transcritical Rankine cycle (CTRC)." *Distributed Generation & Alternative Energy Journal* 35(2): 111–138. doi:10.13052/DGAEJ2156-3306.3522.

Sarkar, Jahar. 2015. "Review and future trends of Supercritical CO_2 Rankine cycle for low-grade heat conversion." *Renewable and Sustainable Energy Reviews* 48: 434–451. doi:10.1016/J.RSER.2015.04.039.

Sharan, Prashant, Ty Neises, & Craig Turchi. 2019. "Thermal desalination via supercritical CO_2 Brayton cycle: Optimal system design and techno-economic analysis without reduction in cycle efficiency." *Applied Thermal Engineering* 152: 499–514. doi:10.1016/J.APPLTHERMALENG.2019.02.039.

Sienicki, James J., & Anton Moisseytsev. 2017. "Nuclear power." In Klaus Brun, Peter Friedman, & Richard Dennis (eds.), *Fundamentals and Applications of Supercritical Carbon Dioxide (sCO$_2$) Based Power Cycles*, pp. 339–391. Duxford: Woodhead Publishing. doi:10.1016/B978-0-08-100804-1.00013-X.

Span, Roland, & Wolfgang Wagner. 2009. "A new equation of state for carbon dioxide covering the fluid region from the triple-point temperature to 1100 K at pressures up to 800 MPa." *Journal of Physical and Chemical Reference Data* 25(6): 1509. doi:10.1063/1.555991.

Turchi, Craig S., Joseph Stekli, & Pablo C. Bueno. 2017. "Concentrating solar power." In Klaus Brun, Peter Friedman, & Richard Dennis (eds.), *Fundamentals and Applications of Supercritical Carbon Dioxide (sCO$_2$) Based Power Cycles*, pp. 269–292. Duxford: Woodhead Publishing. doi:10.1016/B978-0-08-100804-1.00011-6.

Turchi, Craig S., Zhiwen Ma, Ty W. Neises, & Michael J. Wagner. 2013. "Thermodynamic study of advanced supercritical carbon dioxide power cycles for concentrating solar power systems." *Journal of Solar Energy Engineering, Transactions of the ASME* 135(4). doi:10.1115/1.4024030.

Wang, Xiaohe, Qibin Liu, Jing Lei, Wei Han, & Hongguang Jin. 2018. "Investigation of thermodynamic performances for two-stage recompression supercritical CO2 Brayton cycle with high temperature thermal energy storage system." *Energy Conversion and Management* 165: 477–487. doi:10.1016/J.ENCONMAN.2018.03.068.

Wang, Xurong, Xiaoxiao Li, Qibin Li, Lang Liu, & Chao Liu. 2020. "Performance of a solar thermal power plant with direct air-cooled supercritical carbon dioxide Brayton cycle under off-design conditions." *Applied Energy* 261: 114359. doi:10.1016/J.APENERGY.2019.114359.

Yang, Jingze, Zhen Yang, & Yuanyuan Duan. 2020. "Part-load performance analysis and comparison of supercritical CO$_2$ Brayton cycles." *Energy Conversion and Management* 214: 112832. doi:10.1016/J.ENCONMAN.2020.112832.

Chapter 5

Integrated fuel cell hybrid technology

Abhinav Anand Sinha, Tushar Choudhary, and Mohd. Zahid Ansari

PDPM Indian Institute of Information Technology, Design and Manufacturing, Jabalpur, India

5.1 INTRODUCTION

Rapid urbanization and industrial revolution have resulted in an unprecedented rise in energy demand, which is currently met by fossil fuel power plants (Kwan et al. 2020) Nowadays, coal, oil, and natural gas are the world's main source of electricity, accounting for more than 75% of overall production, and they will remain so in the upcoming years (Yu et al. 2008). The combustion of fossil fuels results in greenhouse gas emissions which ultimately cause global warming (Weldu & Assefa 2016). Economic development heavily depends on consumption of fossil fuels (Awodumi & Adewuyi 2020). This exploitation of fossil fuels in the form of combustion, generating greenhouse gases, creates a global challenge in the field of energy security and climate change (Choudhary 2017a). Clean or green energy technology helps to save the environment from this overexploitation of fossil fuels (Shukla et al. 2018). A long-term solution to this problem is to make use of non-conventional energy sources instead of conventional sources such as fossil fuels. Fuel cells can be considered as an alternative or non-renewable source for power generation after combustion. Out of those available – proton exchange membrane fuel cells (PEMFCs), alkaline fuel cells (AFCs), phosphoric acid fuel cells (PAFCs), molten carbonate fuel cell (MCFCs), and solid oxide fuel cells (SOFCs) – only SOFCs have the capability to operate at high temperatures and allow the existing setup that are uses for the fuel cell heating application, which further move towards lower emissions and improved efficiency (Blum et al. 2011).

SOFCs are solid-state fuel cells with no mobile parts whose combined thermal and electric efficiencies can be reached greater than 90% and power obtained 99.99% of the time as they are not limited by a Carnot efficiency unlike fossil-fueled technologies (Stambouli & Traversa, 2002). Reforming reactions in SOFCs can be classified into two categories. First is internal reforming cells (IRCs) and second is external reforming cells (ERCs). Furthermore ERCs can be subcategorized into two more parts, direct and indirect reformer cells (Bavarsad 2007). Internal reforming eliminates the need for a separate reforming unit by joining the endothermic reaction of steam reforming and the reaction of exothermic oxidation in a single unit. Internal reforming is of two types: one is direct and the second is indirect internal reforming (IIR). It effectively distinguishes the reforming and process of electrochemical processes by using the heat released by the cell stack, either by transfer of heat through radiation (Larminie, Dicks, & McDonald 2003) or by immediate physical interaction between the hardware of cell and the reforming unit. The hydrocarbon steam-fuel mixture is introduced directly into the anode compartment in direct internal reforming (DIR), and the reformed fuel on the porous, nickel-based anode sheet. The benefits of direct internal reforming over other approaches include

DOI: 10.1201/9781003213741-5

lower capital costs for additional equipment and improved heat transfer efficiency inside the cell stack (Pirkandi et al., 2012). This chapter includes a brief history and current status of fuel cells also different fuel cell integrating technologies like gasification-SOFC, pressurized SOFC-GT, non-pressurized SOFC-GT, SOFC-CHP, SOFC-trigeneration, SOFC-desalination, SOFC-GT-absorption chillers, and SOFC-PV. Fuel cells seem to have great potential in terms of zero-emissions generation of electricity and improved energy security. In order to enhance the output of SOFCs, research is being carried out regarding integration between two energy converting devices. This chapter explains the concept of fuel cell integrating technologies and its scope.

5.2 RESEARCH METHODOLOGY

This chapter gives a brief and systematic review of the hybridization of fuel cells with different existing configurations. Among various fuel cells, the SOFC is one of the most suitable for power generation because of its waste at high temperature. In general, SOFCs are unable to function at maximum fuel utilization (0.7). As a result, the unutilized fuel at the outlet of the SOFC is further combusted in the combustion chamber and then cooled by expanding it into a gas turbine. So in this way, the waste of SOFCs or unutilized fuel in SOFCs is further available for power generation and generates more power (Saebea 2013).

Al-Khori, Bicer, and Koç (2020) discussed the three primary stages of a literature review: identify, screen, and analyze the studies/research added along with the respective topic. These three are further classified into several sub-steps as mentioned in Figure 5.1.

The first stage is identification, which includes the keywords of SOFCs and SOFC-hybrid technologies. In the beginning the single term fuel cell, SOFC was searched from Google Scholar, Science Direct, Springer. However, this search gave too many results, including papers not related to the topic which we consider in this study. Therefore, we searched for combinations of topics like gasification-SOFC, SOFC-GT, SOFC-PV, SOFC-trigeneration, SOFC desalination, etc.

The second stage is related to screening; here we read abstract, results, and conclusions, and screened out the various methodologies used by the researchers. This style of studying

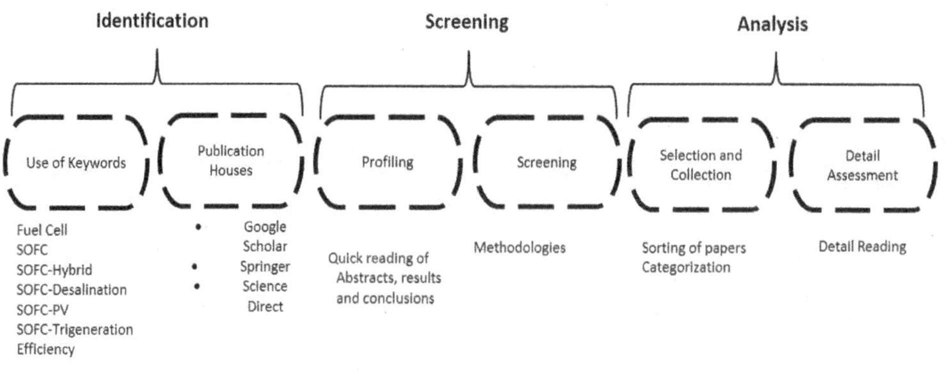

Figure 5.1 Steps to carry out a literature review.

papers helps to focus towards more relevant reviews. The third stage consists of selection and collection of data and finally analysis. After reading a number of papers, sorting is done and content analyzed. The selected articles were analyzed in-depth and the important facts highlighted.

5.3 DESCRIPTION OF FUEL CELL

A fuel cell is working on an electrochemical reaction and such system that can transform electrical energy from the fuel's chemical energy, without any combustion, resulting water as a by-product in a clean and efficient manner (Vigier et al. 2006). Although hydrogen is the primary fuel for low to moderate temperature fuel cells, H_2, CH_4, CO, and their mixtures have been found to be the best suitable fuel cells for a high temperature range (Mohammed et al. 2019). SOFCs are operating at high temperature to generate superior quality of heat as a by-product which can be further utilized for cogeneration or application of combined cycle (Ellamla et al. 2015)

SOFC analysis is conducted for two types of electrolytes: SOFC-H^+ conducts hydrogen ions and SOFC-O_2 conducts oxygen ions. The electrochemical reaction process for each form of SOFC is depicted in Figure 5.2. The hydrogen ion (H^+) and oxygen ion (O_2) will pass through the electrolyte of a SOFC. From the cathode side the oxygen gas molecules pass through the electrolyte and at the anode side it reacts with the hydrogen gas particles, in the oxygen ion-conducting electrolyte (SOFC-O^{2-}). Steam is formed at the anode side when an oxygen ion reacts with a proton, as shown in Figure 5.2(a). From the anode side, the hydrogen gas molecules move through the H^+-conducting electrolyte (SOFC-H^+), and at the cathode they react with the oxygen gas molecules. As a result, as shown in Figure 5.2(b), steam is generated and exits from the cathode side. The electrons travel through the external circuit in order to create voltage (Mojaver et al. 2020).

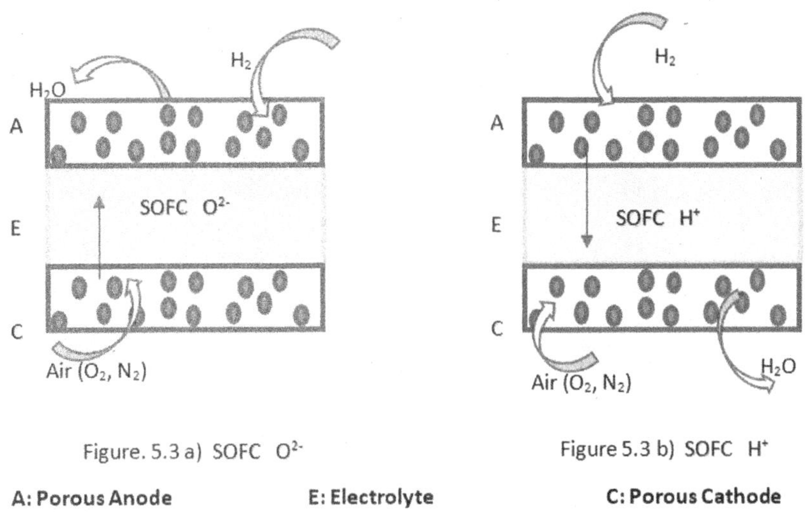

Figure 5.2 Diagram of electrochemical reaction.

Table 5.1 Comparison of SOFC-H_2, SOFC syngas, and SOEC

	SOFC		
Parameters	*H2 – based*	*Syngas-based*	*SOEC*
Temperature at inlet	800°C	800°C	800°C
Temperature at outlet	900°C	900°C	700°C
J	8,203.58 A/m2	8,203.58 A/m2	8,203.58 A/m2
W	12 MW	10.9 MW	18.2735 MW
V	0.75 V	0.68 V	1.14 V
Energy efficiency ratio	40%	37%	76%
Exergy efficiency ratio	41%	36%	86%
Total flow rate (H_2O+ H_2 +CH_4 +CO + CO_2)	0.12 kmol/s	0.141 kmol/s	0.12 kmol/s
Exergy destruction	5,686 kW	5,295 kW	2,847

Source: Botta et al. (2018).

Botta et al. (2018) performed energy and exergy analysis for the three base-cases SOFC H_2-based, SOFC syngas-based, and SOEC configurations. At pressure 1.2 bar, area is 1950.38 m^2 and the utilization factor for fuel cell and electrolyzer cell is 0.75 mol/mol for all the three cases.

5.4 INTEGRATED TECHNOLOGIES

5.4.1 Gasification-SOFC

Concerns about global warming and global climate change prompt attempts to minimize CO_2 and greenhouse gas (GHG) emissions by increasing the utilization of non-conventional energy sources and improving energy efficiency (Kulak, Graves, & Chatterton 2013). When solar, wind, and biomass energy are used in a renewable energy mix, biomass or organic material can be used as a flexible, regulated energy source that can be supplied in greater amounts when wind and solar energy supply is limited (Heidenreich & Foscolo 2015). Biological or thermochemical processes are used to turn biomass into consumer goods (Lin & Tanaka 2006). Until entering the anode (Figure 5.3), the feeding syngas is combined with "recirculated anode exhaust gas and steam in a SOFC (point of mixing 1)". While supplying syngas, the only remaining CO that hasn't been absorbed by the electrochemical reaction (2) needs to be converted by the chemical reaction like water gas shift (WGS) into H_2. Syngas and water at their feeding temperatures T_{syn} and T_{H20} are heated in the heat exchanger, HE with recirculated anode exhaust gas to blending temperature, T_1, while preheated air to its inlet temperature of cathode, $T_{air,in,ca,}$ from its feeding temperature, T_{air}, at the cost of the burner B, flue gases. At temperatures $T_{swraeg,in,a}$ and $T_{air,in,ca}$, a combination of syngas, steam, recirculated anode exhaust gas, and air join the SOFC's anode and cathode, respectively. At particular average operating temperature, satisfactory work of electrolytes is observed and leaves the respective side of electrode. Fuel gas (CO or H_2) in the burner is sufficiently oxidized, exhaust gas utilized at cathode as oxidizing agent, which consists of O_2. $T_{gas,out,B}$,

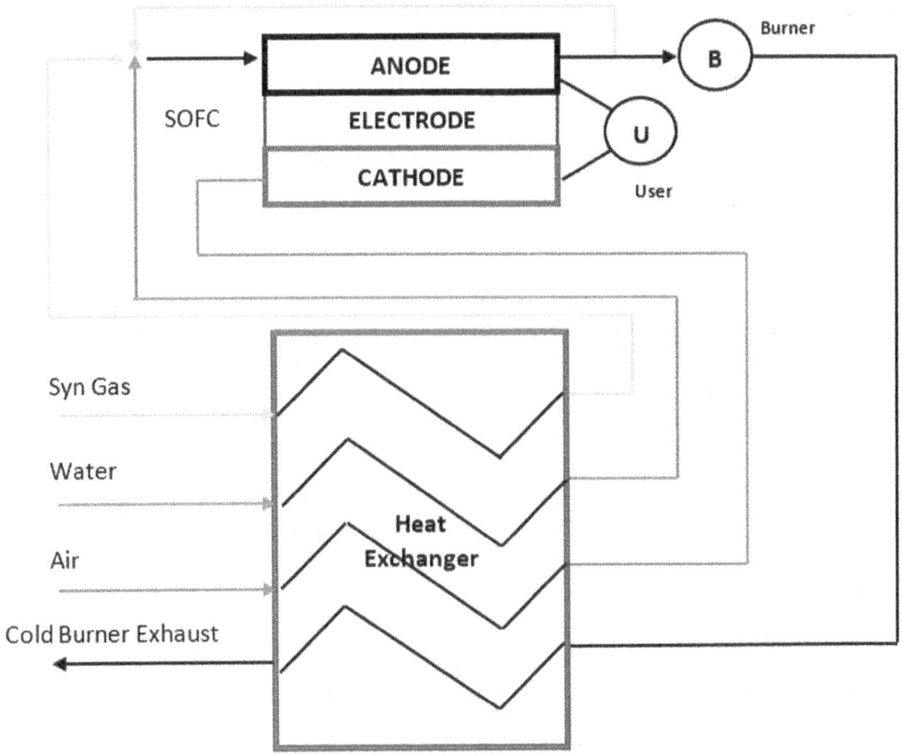

Figure 5.3 Schematic of SOFC-gasification. (Source: Puig-Arnavat, Bruno, & Coronas 2010.)

is the oxidation temperature, burner releases the gas which is combusted, which escapes the HE at temperature $T_{gas,out,HE}$.

Syngas is a combination of various gases such as CH_4, H_2, CO, CO_2, N_2, char, ash, tar, oil, and small amount of water vapor (Kelly-Yong et al. 2007). Temperatures of gasification are usually high enough that no hydrocarbons other than methane can exist in any substantial amount, both thermodynamically and functionally. The various stages (Molino, Chianese, & Musmarra 2016) of the biomass gasification process are shown in Figures 5.4, 5.5 and 5.6.

Gasification produces syngas, contains high levels of sulfur, tar, and other pollutants, resulting in SOFC degradation. As a result, according to the SOFC impurity standards, cleaning of syngas is required. This cleaned syngas is introduced into the SOFC and processed further for electricity generation. The anode recirculation ratio is modified to avoid carbon deposition. The drained syngas and air streams reach the afterburner, where the leftover fuel is burned and raises the temperature of the streams.

The heat exchanger receives this burned gas mixture and heats the entered air in the SOFC. In the case of gasification of steam, the heat exchanger leaves a mixture of gas and provides heat to the gasifier and steam generator, as well as the dryer. The gas is then released into the atmosphere (Colpan et al. 2010). The current modeling employs a model of SOFC based on

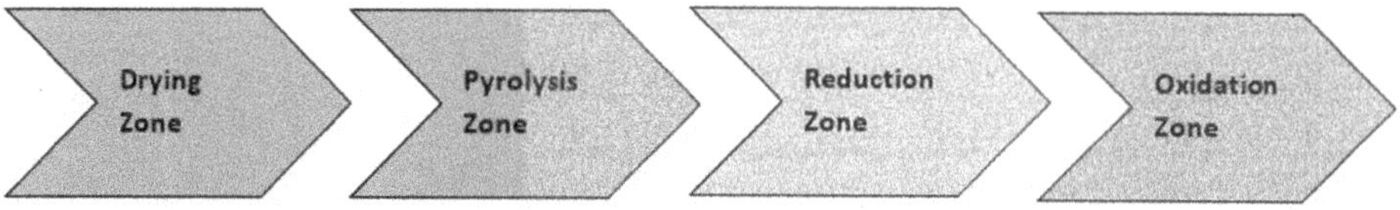

Figure 5.4 Up draft gasification process.

(1) Oxidation (exothermic stage), (2) Drying (endothermic stage), (3) Pyrolysis (endothermic stage), (4) Reduction (endothermic stage)

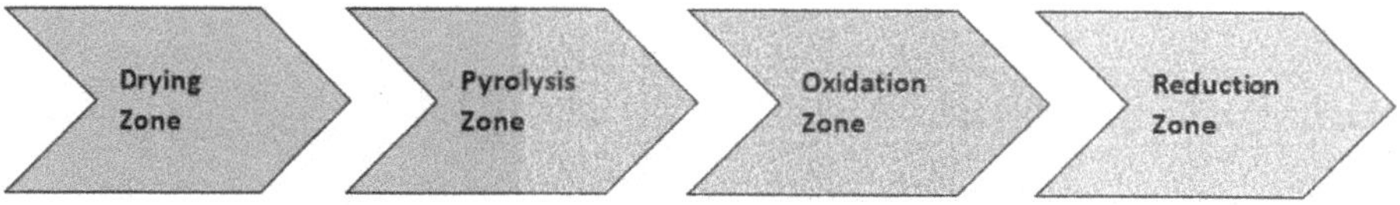

Figure 5.5 Down draft gasification process.

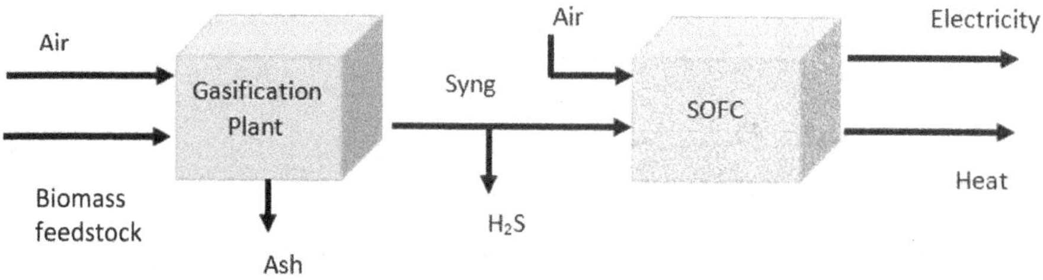

Figure 5.6 SOFC-gasification process.

heat transfer and for the remaining part a thermodynamic model is created, additionally an exergy analysis was performed for each component. According to the results, the best performance of electricity is achieved by using steam as a gasification agent (41.8%), the lowest fuel consumption efficiency (50.8%), exergetic efficiency (39.1%), and power-to-heat ratio (4.649) (Colpan et al. 2010).

5.4.2 SOFC-GT

A typical direct thermal coupling scheme is the combined SOFC and Brayton cycle. Integrating SOFCs with gas turbines can improve efficiency, particularly in big stationary applications (Al-Khori, Bicer, & Koç 2020). McLarty, Brouwer, and Samuelsen (2014) proposed a novel MCFC and SOFC GT hybrid system approach, utilizing steady state models to analyze energy and/or exergy as well. At 1.2 MW, at present MCFC and μ-turbine technologies reach 74.4% efficiency (LHV). On synthesis gas, SOFC–GT technology achieves >75% efficiency (LHV) (McLarty, Brouwer, & Samuelsen 2014).

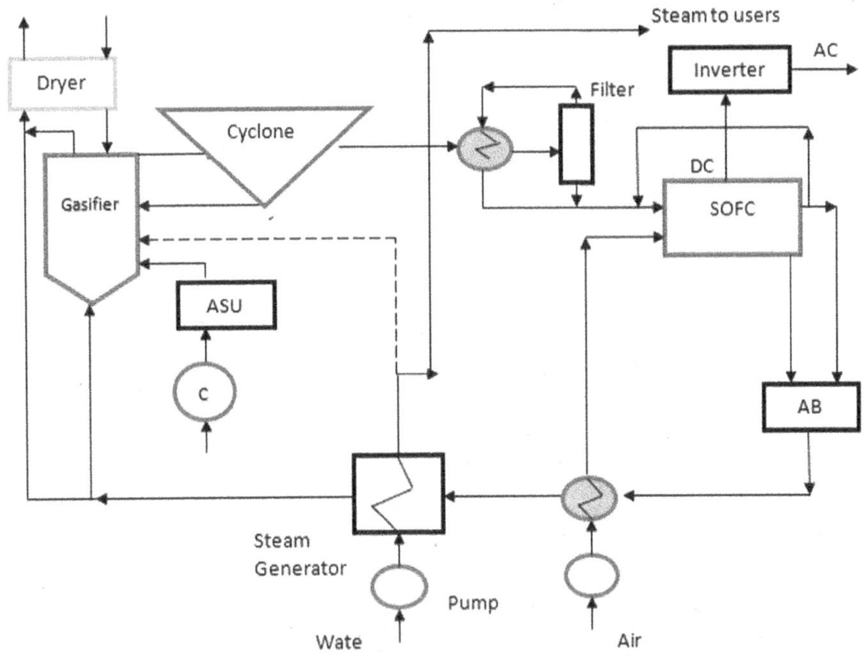

Figure 5.7 Schematic of integrated BG-SOFC.

A computational study was carried out in order to optimize the design of an integrated SOFC-GT power generator, with a focus on the effect of fuel consumption inside the system performance and installed costs. Over a large range of operations, this hybrid model gained impressive electricity generation efficiencies (>70%) (60% to 90% utilization of fuel), and decreased the size of fuel cell stack fuel cell proportion to the utilization of fuel. With SOFC supplying only 65% of the fuel, peak efficiency was discovered. Hybrid designs (Figure 5.7) with SOFC output capacity of 40–60% had analogous economics (Oryshchyn et al. 2018).

5.4.3 Pressurized SOFC-GT

Type 1, SOFC-GT performed at pressurized condition, shown in Figure 5.8, and Type 2 SOFC-GT performed at unpressurized conditions, shown in Figure 5.9. In a pressurized process, high pressure air is redirected to the oxygen reduction at SOFC cathode. Pirkandi et al. (2012) have found from thermo-economic modeling studies that when purchasing, installing, and starting a hybrid system with one pressurized SOFC is a priority, such hybrid systems are preferable to those with two stack fuel cells or one atmospheric fuel cell.

At the cathodic end, ambient air is drawn into the compressor and compressed. Air passes into a recuperator after compression, where its temperature is increased by using the high temperature exhaust gas from the turbine. It is then added to the SOFC cathode side, whereas on anode side it electrochemically interacts with the hydrogen, transforming into electricity from these chemical energy while also generating a significant amount of fuel. The heat which

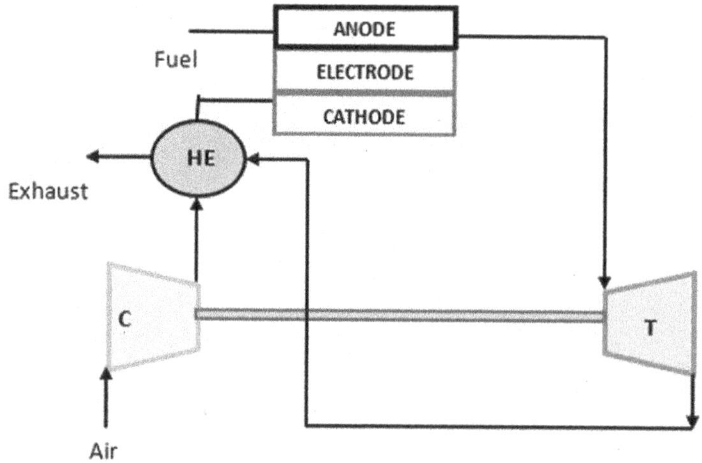

Figure 5.8 Pressurized SOFC-GT.

is remaining in SOFCs can be utilized to heat the compressed air and fuel, as well as the processing reaction, and sustain the endothermic fuel (Ferrari et al. 2017; Uechi, Kimijima, & Kasagi 2004).

The SOFC's exhausts are redirected to an additional system of combustion, where the remaining fuel is burned to significantly stream temperature rises in the turbine. Combustible gas enters into the turbine for further expansion to give mechanical work. The compressors are driven by a portion of the mechanical work, while the remainder is converted to electric power through a generator. The waste heat in the flow is recovered from the exhaust gas by transmitting energy in thermal form to the flow of air on the lower temperature side of the recuperator; complete performance of the system also increases.

5.4.4 Non-pressurized SOFC-GT

Figure 5.9 depicts a non-pressurized SOFC integrated with the gas turbine. Air at the inlet of the SOFC is sourced from the exhaust of the gas turbine in this process. The SOFC stack is designed to work at ambient pressure.

The minimum temperature at the inlet of the SOFC cannot be reached due to expansion of gas through the turbine because the working temperature of the fuel cell is higher. Furthermore, since the exhaust temperature gas is typically larger than that of the pressurized SOFC system, waste heat recovery is less reliable. Williams, Siddle, and Pointon (2001) use an operating temperature of 650°C for the purpose of their study, and achieve around 55% efficiency for non-pressurized systems and 60.9% for SOFC pressurized by recuperated gas turbine. This hybrid scheme's optimum performance is around 5% lower than a pressurized SOFC device with zero extra fuel burned in the combustor. Because the temperature of the air at the fuel cell inlet is lower than that of the gas turbine exhaust and additional pressure losses are introduced by extra heat exchangers, a booster is required in the air system to overcome the pressure losses in the stack and recuperator.

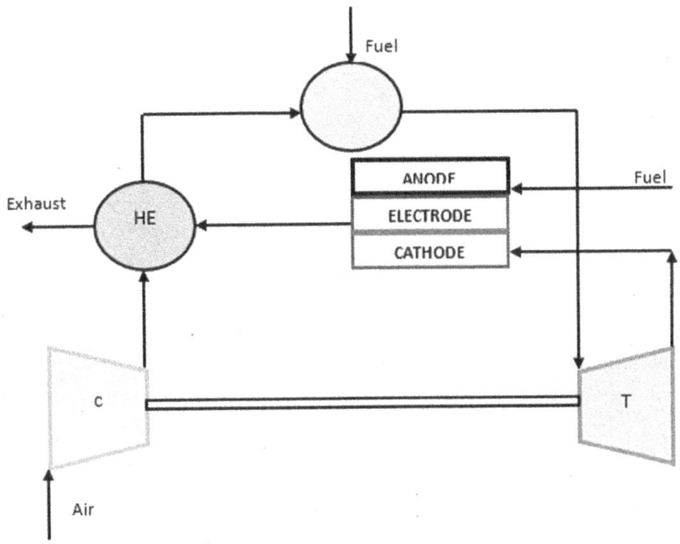

Figure 5.9 Non-pressurized SOFC-GT.

5.4.5 SOFC-CHP

Among fuel cell technologies, SOFC-CHP is preferred for energy production (Hosseinpour et al. 2020). The SOFC operates at higher temperatures (typically 700–1,000°C) (Wang et al. 2021) and among the different varieties of fuel cells is the most convincing for combined heat and power (CHP) generation. Using boilers, most of the heat of a SOFC is generally recovered for water heating and/or cooling in an absorption chiller in a SOFC-based CHP device (Wang et al. 2021). Alternatives to the vacuum thermionic generator (VTG) and the cascading thermoelectric generator (CTG) and cooler have recently been considered for waste heat recovery (Zhang et al. 2019). Figure 5.10 shows the layout of the tubular SOFC-CHP system. The proposed system is tubular-based SOFC stack with direct internal reforming (DIR), a fuel and air compressor, an afterburner chamber, a water pump, and three recuperators.

The heating, cooling, and electrical loads in a building all need electricity, which is expected to be provided by the proposed SOFC-CHP system. The heat received from the third heat exchanger will also be used to heat a building and provide disinfecting hot water in the cold season. This produced heat energy is utilized in an absorption chiller during the summer, to meet a building's cooling capacity as well as provide clean water.

5.4.6 SOFC-trigeneration

Trigeneration means it is a combined effect of three different processes such as combined cooling, heating, and power (CCHP), is a method of maximizing the chemical energy of a fuel to generate power while also utilizing the heat from the exhaust (Al-Sulaiman, Hamdullahpur, & Dincer 2011). At the same time, by using absorption or desiccant cooling

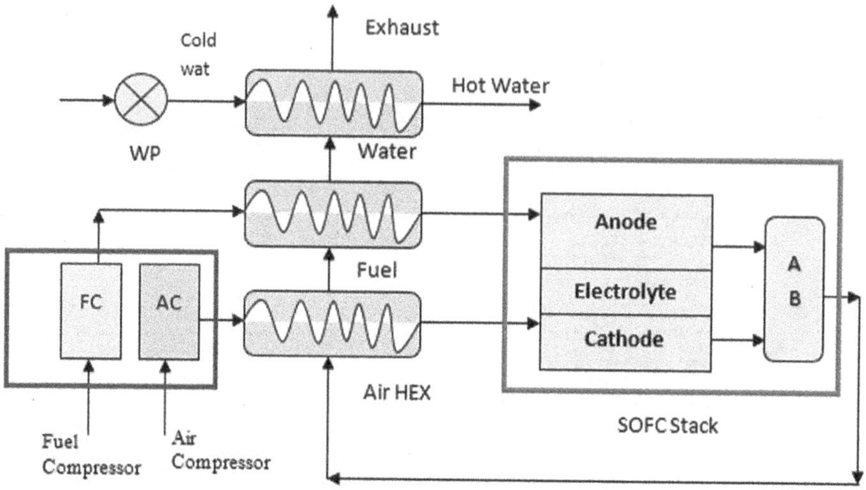

Figure 5.10 Schematic of SOFC-CHP. (Source: Zhang et al. 2019.)

can be achieved, reducing the amount of electricity consumed by a conventional air conditioner. As a result, it is possible to increase overall productivity to nearly 90%. Al-Sulaiman have performed comparisons of the three trigeneration systems. SOFC, biomass, and solar-trigeneration systems are the three systems. Out of these three system, the electrical performance is maximum for SOFC-trigeneration hybrid systems. On comparing among different types of hybrid tri-generation system, biomass-trigeneration system and solar-trigeneration system, have higher trigeneration efficiencies than the SOFC-trigeneration system (Al-Sulaiman, Hamdullahpur, & Dincer 2011).

The following is a definition of the flow in the ORC, as shown in Figure 5.11. As a saturated liquid, the fluid exits the desorber. The pump then raises the saturated liquid's pressure. The working fluid is then introduced as a liquid to the evaporator and exits as vapor. In a turbine the expansion of organic fluids takes place and then generates mechanical energy and further into electrical energy.

The electricity-generating generator, attached with the shaft of the turbine, is rotated using mechanical energy. The turbine leaves the working fluid at high temperature which is further used in HE for the heating operation.

The heat exchanger in the heating phase rejects heat in order to supply heat to the heating load. Through the desorber organic fluid then passes as saturated vapor. Providing the single-effect absorption chiller's cooling load, the heat is absorbed by the desorber. As a saturated liquid, organic fluid then leaves the desorber once again. The flow streams move water or a mixture of LiBr and H_2O between both the components of this chilling cycle. As a result of the thermal energy at the input into the desorber, water is evaporated from the solution of LiBr and H_2O and then enters into the condenser, and rejects heat from it. As a result, the temperature of water decreases and emerges as a saturated liquid from the condenser. The water is transferred to the evaporator at low temperature. The evaporator supplies the cooling load. Water then leaves the evaporator and flows into the absorber. The LiBr and water mixture

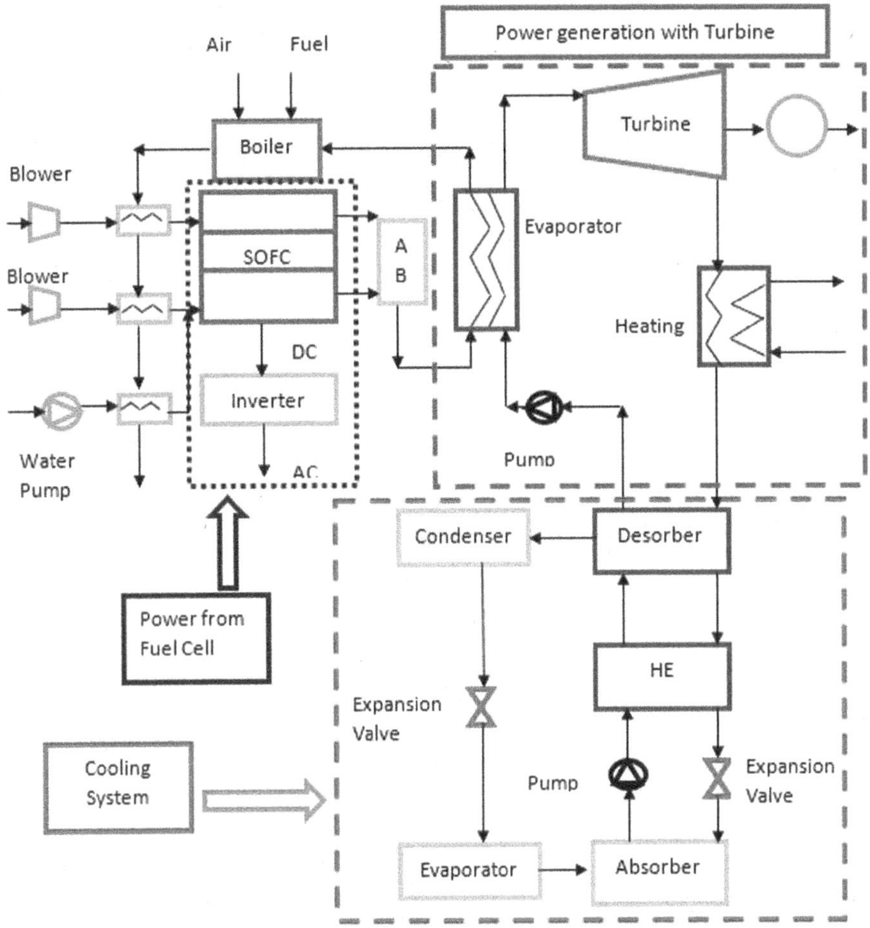

Figure 5.11 Schematic of the SOFC-trigenration. (Source: Al-Sulaiman, Hamdullahpur, & Dincer 2011.)

is blended with the water. The mixture is supplied to the HE after leaving the absorber. The mixture then passes through the HE and into the desorber. In the desorber, the mixture is heated and some of the water evaporates. The mixture then escapes the desorber with a high value of Li-Br content and enters the HE to obtain heat as a result of the evaporation of water. It then leaves the HE and is forced down to the absorber (Al-Sulaiman, Hamdullahpur, & Dincer 2011).

5.4.7 SOFC-GT-absorption chillers

A system consists of an absorption chiller that cools by absorbing heat rather than using electricity to run a mechanical compressor. Figure 5.12 depicts a schematic diagram of a typical single-stage chiller device. The chiller is powered by waste heat from the SOFC-GT system

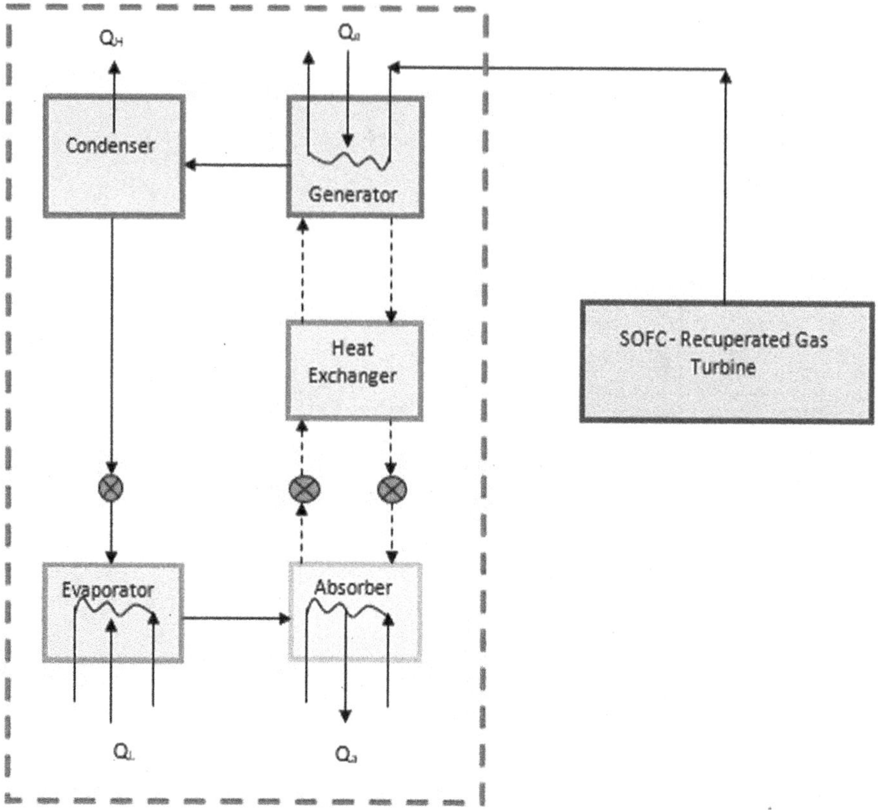

Figure 5.12 Schematic diagram of hybrid absorption chiller – SOFC-GT system.

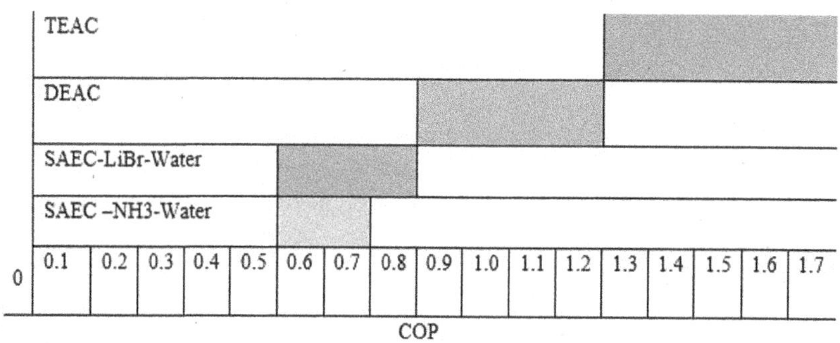

Figure 5.13 COP range for SAEC, DAEC, and TAEC. (Source: Tse et al. 2011.)

Notes: SAEC – single effect absorption chiller; DAEC – double effect absorption chiller; TEAC – triple effect absorption chiller.

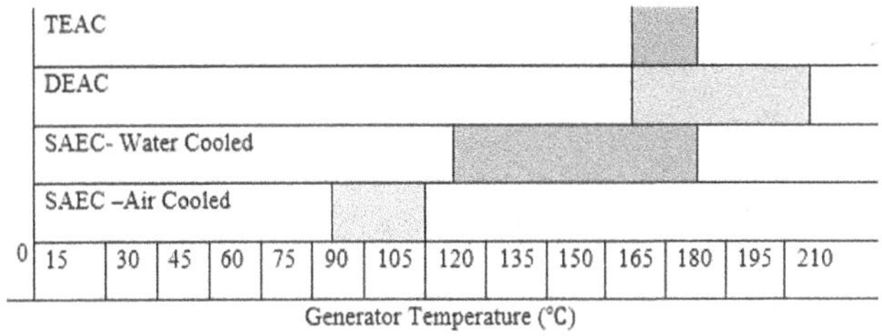

Figure 5.14 Generator temperature range for SAEC, DAEC, and TAEC. (Source: Tse et al. 2011.)

exhaust stream; and generate chilled water by transferring the heat to the generator (Q_{gen}) to cool air at the HVAC's CHP coil (Q_l) (Tse et al. 2011).

Higher coefficient of performance (COP) is observed with double and triple effect absorption chillers (Figure 5.13). The generator temperature is high for higher absorption chillers (Figure 5.14). More expensive materials are required to work at higher temperatures. Also the complexity and cost of the system increases with increasing the number of condensers and absorbers.

5.4.8 SOFC-PV

SOFC-PV is a new combination of clean or green energy and conventional energy systems. Its working is a bridge to the future. Hybridization of PV power systems (Figure 5.15) is one of the most promising integrations of SOFC (Al-Khori, Bicer, & Koç 2020). The generation of clean and green electricity, as opposed to traditional power generation, is one of the key benefits of this development (İnci & Türksoy 2019). In this setup, the SOFC is paired with PV that runs for both, in light and in dark to deliver the necessary amount of electricity.

Since the initialization and shutdown processes are lengthy, the SOFC unit must be kept working at all times. This reduces the fuel cell's expected lifespan. Throughout the day, the essential energy is provided by PV while the SOFC will function at a low fuel rate to maintain the hot standby mode. The SOFC unit would increase the fuel (natural gas) during the night and when PV is not operational to produce the necessary demand (de Anda & Eduardo 2016).

To satisfy the SOFC stack demand, the air and fuel must be pressurized and heated. To do so, a HE is used to pass heat from the SOFC heat output to the fuel and air. To pressurize the air and fuel, small compressors are used and some heat is to be required to satisfy the SOFC stack. For achieving this, the heat transfer from SOFC heat output is used to increase the fuel and air temperatures by introducing a heat exchanger. And small compressors are introduced to pressurize the air and fuel. Because of the system's high working temperatures, only CO_2 and virtually no other harmful gases (such as carbon monoxide (CO), SOx, NOx, and some volatile organic compounds, and hydrocarbons which are generally releases during combustion process) are released during SOFC operation and if emissions do occur they are in insignificantly small amounts.

The source of flue gas is the only one, which is predominantly CO_2, which accounts for the PV-gross SOFC's emissions of greenhouse gases (GHG) (Al-Khori, Bicer, & Koç 2021). The levelized price of power in the SOFC-PV case is 0.11 USD/kWh, and it can also be as low as 0.04 USD/kWh in the SOFC-PV case (projected scenario) (Al-Khori 2020).

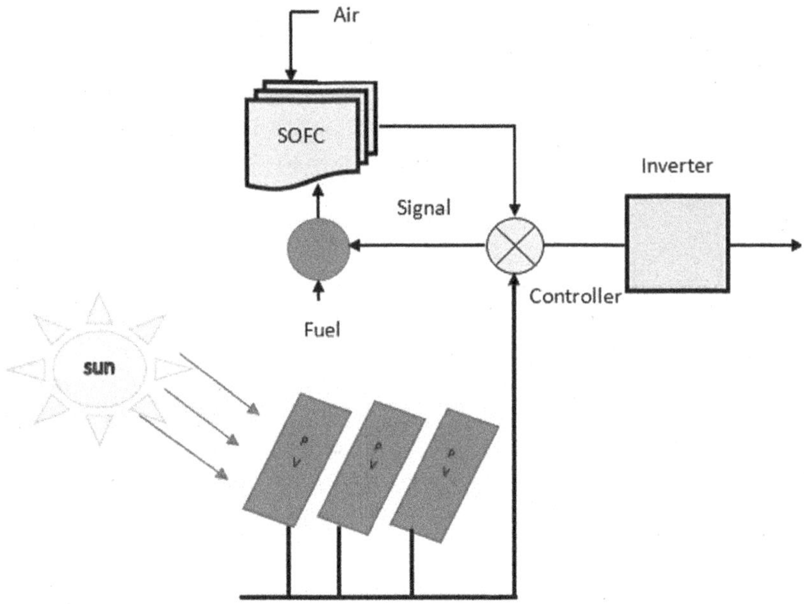

Figure 5.15 Schematic of PV-SOFC system. (Source: Al-Khori, Bicer, & Koç 2016.)

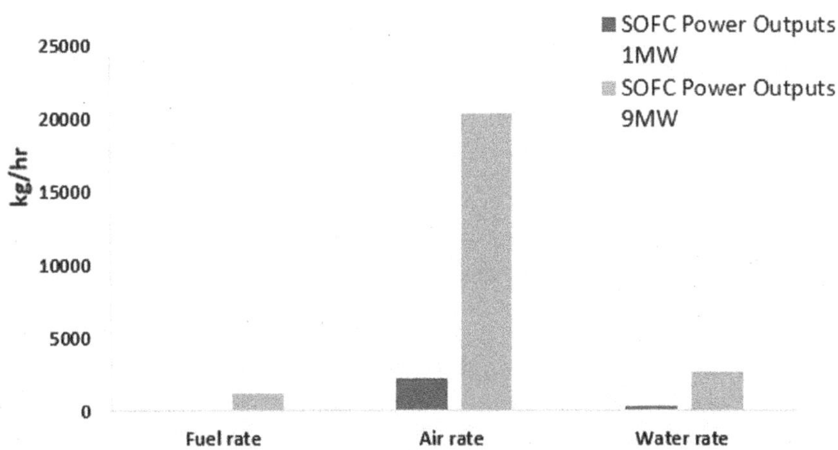

Figure 5.16 Fuel, air, and water rates for the PV-SOFC operation. (Source: Al-Khori 2020.)

Figure 5.16 shows the important set of constraints for 11MW. The natural gas, fresh air, and water needed for a 1 MW and 9 MW SOFC production, as well as the min. and max. power outputs based on the environment and the time of year.

5.4.9 Future scope and challenges

- The major challenges in the field of full commercialization of fuel cells are fabrication cost, robustness, and reliability (Bauen, Hart, & Chase 2003).
- The environmental benefits of using SOFCs are that there is zero harmful emissions of any gases. Poor startup and lengthy cooling time are its major disadvantages (Mekhilef, Saidur, & Safari 2012).
- For stationary power, the lifetime reliability and cost of commercialization are the two major challenges. For stationary power 40,000 working hours of SOFCa is available in the market with 10% degradation (Lee et al. 2015).
- The usage of SOFCs is based on lowering operating temperatures to 400–600°C, which usually entails use of such strong electrolytes, and these findings need further attention from scholars.
- India is also contributing in a reduction of carbon emission of 25–30% until 2025 (Sinha, Shukla, & Prasad 2016). For supporting this, India must move towards an emission-free power generation system. And for this hydrogen is one of the best choices for power generation. But it is not easy to handle it so systems are required to handle hydrogen during transportation and fuel cells are one of the best options.
- Hybrid power plants based on microturbine SOFCs are also being developed to enhance total energy productivity and financial viability.
- The use of nanotechnology in materials is becoming more prevalent in fabrication, architecture, and scale-up of SOFCs. For example, electrodes of nanoscale and nanostructured structures.
- Beyond the existing techniques of pressing, tape casting, freeze casting, dip coating, screen printing, and so on, to produce reusable and budget processing strategies for various SOFC components, such as 2D/3D printing with enhanced reproducibility and easy, fast or efficient configurability (Al-Khori, Bicer, & Koç 2020).

However, in order for SOFC systems to be successfully integrated with oil and gas activities, the following technical and research issues must be resolved by further analysis:

1 SOFC convergence techno-economic-environmental analysis in multiple scenarios.
2 SOFC coupled with oil and gas operations is also being examined for long-term viability.
3 Plants for renewable energies are being coupled with oil and gas activities.
4 Waste gas is being used to produce electricity and other valuable resources.

There are issues related to meeting the needs of the economy, society, and environment that need additional effort and studies to ensure energy supply for sustainable growth. When it comes to energy resources, people expect a high degree of efficiency. These energy supplies, on the other hand, must be provided in a more dependable, safe, and environmentally sustainable way, which would necessitate a significant shift in technology, infrastructure, and people's views have all changed. This transition will only be feasible with government support and regulation, as well as the actions of corporate and social leaders to ensure that these policies are enforced.

5.5 CONCLUSIONS AND FUTURE CHALLENGING PROSPECTS

Fuel cells seem to have great potential in terms of zero-emissions generation of electricity and improved energy security. In order to improve the performance of SOFCs, research is being carried out by integration between two energy converting devices. This hybridization is a highly attractive method to enhance the performance of pre-existing systems. This chapter includes brief discussion on different fuel cell integrating technologies such as gasification-SOFC, pressurized SOFC-GT, non-pressurized SOFC-GT, SOFC-CHP, SOFC-trigeneration, SOFC-desalination, SOFC-GT-absorption chillers, and SOFC-PV. In general, fuel use for SOFC is H_2, which is explosive, and a unique system is required for storing hydrogen-based energy. High temperature-based fuel cells are made up of some ceramic type materials. In high temperature fuel cells all three (anode, electrolyte, and cathode) are in solid state. To increase the performance of SOFC integrated systems, two SOFC may be connected in series. The unused fuel of the first SOFC is directly utilized in the second SOFC. And the second SOFC must be smaller in size as compared to the first one. Future work should focus on investigating other experimental parameters of the system, such as types of fuel, pressure ratios, and working temperature. The capital cost is high for SOFC-based integrated systems, which hinders commercialization of such systems. In future, cost-based analysis and life cycle assessment are important parameters that require more research. This can be improved by recycling and increasing the reliability of fuel cells. Many hybrid systems such as SOFC-heat pumps and SOFC-GT-desalination, still don't have any experimental investigation – the scarcity of such hybrid systems needs some real investigation. Such implementation based on the above discussions of hybrid systems can open the door for researchers in many aspects, like fuel, materials, configure uration, pressure ration, temperature, and applied voltage.

ABBREVIATIONS

AB	Afterburner	J	Current density
AC	Air compressor	LHV	Lower heating value
AFC	Alkaline fuel cell	MCFC	Molten carbonate fuel cell
APU	Auxiliary power unit	MED	Multiple effect distillation
BG	Biomass gasification	MSF	Multistage flash
BoP	Balance of plants	PAFC	Phosphoric acid fuel cell
C	Compressor	PEMFC	Proton exchange membrane fuel cell
CHP	Combined heat and power	PV	Photovoltaic
COP	Coefficient of performance	SCR	Steam to carbon ratio
CFG	Cascading thermoelectric generator	SEC	Specific energy consumption
DIR	Direct internal reforming	SED	Single effect distillation
ERC	External reforming cell	SOEC	Solid oxide electrolysis cell
FC	Fuel cell	SOFC	Solid oxide fuel cell
GT	Gas turbine	SWRO	Sea water reverse osmosis
GHG	Greenhouse gas	T	Turbine
HE	Heat exchanger	V	Voltage
HVAC	Heating ventilation and air conditioning	W	Work flow rate
IIR	Indirect internal reforming	WP	Water pump
IRC	Internal reforming cell	VTG	Vacuum thermionic generator

REFERENCES

Al-Khori, K. M. 2020. *Sustainability Assessment for the Integration of Solid Oxide Fuel Cell (SOFC) Technologies in Natural Gas Processing Plant*. Doctoral dissertation, Hamad Bin Khalifa University, Qatar.

Al-Khori, K., Y. Bicer, & M. Koç. 2020. "Integration of solid oxide fuel cells into oil and gas operations: needs, opportunities, and challenges." *Journal of Cleaner Production* 245, 118924

Al-Khori, K., Y. Bicer, & M. Koç. 2021. "Comparative techno-economic assessment of integrated PV-SOFC and PV-battery hybrid system for natural gas processing plants." *Energy* 222: 119923.

Al-Sulaiman, F. A., F. Hamdullahpur, & I. Dincer. 2011. "Performance comparison of three trigeneration systems using organic Rankine cycles." *Energy* 36(9): 5741–5754.

Awodumi, O. B., & A. O. Adewuyi. 2020. "The role of non-renewable energy consumption in economic growth and carbon emission: Evidence from oil producing economies in Africa." *Energy Strategy Reviews* 27: 100434.

Bauen, A., D. Hart, & A. Chase. 2003. "Fuel cells for distributed generation in developing countries: An analysis." *International Journal of Hydrogen Energy* 28(7): 695–701.

Bavarsad, P. G. 2007. "Energy and exergy analysis of internal reforming solid oxide fuel cell–gas turbine hybrid system." *International Journal of Hydrogen Energy* 32(17): 4591–4599.

Blum, L., R. Deja, R. Peters, & D. Stolten. 2011. "Comparison of efficiencies of low, mean and high temperature fuel cell systems." *International Journal of Hydrogen Energy* 36(17): 11056–11067.

Botta, G., R. Mor, H. Patel, & P. V. Aravind. 2018. "Thermodynamic evaluation of bi-directional solid oxide cell systems including year-round cumulative exergy analysis." *Applied Energy* 226: 1100–1118.

Choudhary, T. 2017a. "Novel and optimal integration of SOFC-ICGT hybrid cycle: Energy analysis and entropy generation minimization." *International Journal of Hydrogen Energy* 42(23): 15597–15612.

Choudhary, T. 2017b. "Thermodynamic assessment of advanced SOFC-blade cooled gas turbine hybrid cycle." *International Journal of Hydrogen Energy* 42(15): 10248–10263.

Colpan, C. O., F. Hamdullahpur, I. Dincer, & Y. Yoo. 2010. "Effect of gasification agent on the performance of solid oxide fuel cell and biomass gasification systems." *International Journal of Hydrogen Energy* 35(10): 5001–5009.

de Anda, L., & J. Eduardo. 2016. *Conceptual Study on the Energy Independence of Fuel Cell Cogeneration Systems Using Solar Energy*. Doctoral dissertation.

De Lorenzo, G., & P. Fragiacomo. 2015. "Energy analysis of an SOFC system fed by syngas." *Energy Conversion and Management* 93: 175–186.

Ellamla, H. R., I. Staffell, P. Bujlo, B. G. Pollet, & S. Pasupathi. 2015. "Current status of fuel cell based combined heat and power systems for residential sector." *Journal of Power Sources* 293: 312–328.

Ferrari, M. L., U. M. Damo, A. Turan, & D. Sánchez. 2017. *Hybrid Systems Based on Solid Oxide Fuel Cells*. New York: John Wiley & Sons.

Heidenreich, S., & P. U. Foscolo. 2015. "New concepts in biomass gasification." *Progress in Energy and Combustion Science* 46: 72–95.

Hosseinpour, J., A. Chitsaz, L. Liu, & Y. Gao. 2020. "Simulation of eco-friendly and affordable energy production via solid oxide fuel cell integrated with biomass gasification plant using various gasification agents." *Renewable Energy* 145: 757–771.

İnci, M., & Ö. Türksoy. 2019. "Review of fuel cells to grid interface: Configurations, technical challenges and trends." *Journal of Cleaner Production* 213: 1353–1370.

Kelly-Yong, T. L., K. T. Lee, A. R. Mohamed, & S. Bhatia. 2007. "Potential of hydrogen from oil palm biomass as a source of renewable energy worldwide." *Energy Policy* 35(11): 5692–5701.

Kulak, M., A. Graves, & J. Chatterton. 2013. "Reducing greenhouse gas emissions with urban agriculture: A life cycle assessment perspective." *Landscape and Urban Planning* 111: 68–78.

Kwan, T. H., Katsushi, F., Shen, Y., Yin, S., Zhang, Y., Kase, K., & Yao, Q. (2020). "Comprehensive review of integrating fuel cells to other energy systems for enhanced performance and enabling polygeneration." *Renewable and Sustainable Energy Reviews* 128: 109897.

Larminie, J., A. Dicks, & M. S. McDonald. 2003. *Fuel Cell Systems Explained.* Chichester, UK: John Wiley.

Lee, Y. D., K. Y. Ahn, T. Morosuk, & G. Tsatsaronis. 2015. "Environmental impact assessment of a solid-oxide fuel-cell-based combined-heat-and-power-generation system." *Energy* 79: 455–466.

Lin, Y., & S. Tanaka. 2006. "Ethanol fermentation from biomass resources: current state and prospects." *Applied Microbiology and Biotechnology* 69(6): 627–642.

McLarty, D., J. Brouwer, & S. Samuelsen. 2014. "Fuel cell–gas turbine hybrid system design part I: Steady state performance." *Journal of Power Sources* 257: 412–420.

Mekhilef, S., R. Saidur, & A. Safari. 2012. "Comparative study of different fuel cell technologies." *Renewable and Sustainable Energy Reviews* 16(1): 981–989.

Mohammed, H., A. Al-Othman, P. Nancarrow, M. Tawalbeh, & M. E. H. Assad. 2019. "Direct hydrocarbon fuel cells: A promising technology for improving energy efficiency." *Energy* 172: 207–219.

Mojaver, P., A. Chitsaz, M. Sadeghi, & S. Khalilarya. 2020. "Comprehensive comparison of SOFCs with proton-conducting electrolyte and oxygen ion-conducting electrolyte: Thermoeconomic analysis and multi-objective optimization." *Energy Conversion and Management* 205: 112455.

Molino, A., S. Chianese, & D. Musmarra. 2016. "Biomass gasification technology: The state of the art overview." *Journal of Energy Chemistry* 25(1): 10–25.

Oryshchyn, D., N. F. Harun, D. Tucker, K. M. Bryden, & L. Shadle. 2018. " Fuel utilization effects on system efficiency in solid oxide fuel cell gas turbine hybrid systems." *Applied Energy* 228: 1953–1965.

Pirkandi, J., M. Ghassemi, M. H. Hamedi, & R. Mohammadi. 2012. "Electrochemical and thermodynamic modeling of a CHP system using tubular solid oxide fuel cell (SOFC-CHP)." *Journal of Cleaner Production* 29: 151–162.

Puig-Arnavat, M., J. C. Bruno, & A. Coronas. 2010. "Review and analysis of biomass gasification models." *Renewable and Sustainable Energy Reviews* 14(9): 2841–2851.

Saebea, D., Y. Patcharavorachot, S. Assabumrungrat, & A. Arpornwichanop. 2013. "Analysis of a pressurized solid oxide fuel cell–gas turbine hybrid power system with cathode gas recirculation." *International Journal of Hydrogen Energy* 38(11): 4748–4759.

Shukla, Anoop Kumar, Sandeep Gupta, Shivam Pratap Singh, Meeta Sharma, & Gopal Nandan. 2018. "Thermodynamic performance evaluation of SOFC based simple gas turbine cycle." *International Journal of Applied Engineering Research* 13(10): 7772–7778. www.ripublication.com/ijaer18/ijaerv13n10_69.pdf.

Sinha, A. A., A. Shukla, & R. B. Prasad. 2016. "A review on CSP technologies with heat transfer fluids used in Indian power plants." In *2016 21st Century Energy Needs-Materials, Systems and Applications (ICTFCEN)*, pp. 1–6. New Jersey: IEEE.

Stambouli, A. B., & E. Traversa. 2002. "Solid oxide fuel cells (SOFCs): A review of an environmentally clean and efficient source of energy." *Renewable and sustainable energy reviews* 6(5): 433–455.

Tse, L. K. C., S. Wilkins, N. McGlashan, B. Urban, & R. Martinez-Botas. 2011. "Solid oxide fuel cell/gas turbine trigeneration system for marine applications." *Journal of Power Sources* 196(6): 3149–3162.

Uechi, H., S. Kimijima, & N. Kasagi. 2004. "Cycle analysis of gas turbine–fuel cell cycle hybrid micro generation system." *Journal of Engineering for Gas Turbines and Power* 126(4): 755–762.

Vigier, F., S. Rousseau, C. Coutanceau, J. M. Leger, & C. Lamy. 2006. "Electrocatalysis for the direct alcohol fuel cell." *Topics in Catalysis* 40(1–4): 111–121.

Wang, Y., L. Wehrle, A. Banerjee, Y. Shi, & O. Deutschmann. 2021. "Analysis of a biogas-fed SOFC CHP system based on multi-scale hierarchical modeling." *Renewable Energy* 163: 78–87.

Weldu, Y. W., & G. Assefa. 2016. "Evaluating the environmental sustainability of biomass-based energy strategy: Using an impact matrix framework." *Environmental Impact Assessment Review* 60: 75–82.

Williams, G. J., A. Siddle, & K. Pointon. 2001. "Design optimisation of a hybrid solid oxide fuel cell and gas turbine power generation system." www.osti.gov/etdeweb/biblio/20249899.

Yu, K. M. K., I. Curcic, J. Gabriel, & S. C. E. Tsang. 2008. "Recent advances in CO_2 capture and utilization." *ChemSusChem: Chemistry & Sustainability Energy & Materials* 1(11): 893–899.

Zhang, H., W. Kong, F. Dong, H. Xu, B. Chen, & M. Ni. 2017. "Application of cascading thermoelectric generator and cooler for waste heat recovery from solid oxide fuel cells." *Energy Conversion and Management* 148: 1382–1390.

Zhang, H., J. Wang, F. Wang, J. Zhao, H. Miao, & J. Yuan. 2019. "Performance assessment of an advanced triple-cycle system based upon solid oxide fuel cells, vacuum thermionic generators and absorption refrigerators." *Energy Conversion and Management* 193: 64–73.

Chapter 6

CHP coupled with a SOFC plant

Nima Norouzi

Department of Energy Engineering and Physics, Amirkabir University of Technology (Tehran Polytechnic), Tehran, Iran

Yashar Peydayesh

School of Industrial Engineering, Tabriz University, Tabriz, Iran

Saeed Talebi and Maryam Fani

Department of Energy Engineering and Physics, Amirkabir University of Technology (Tehran Polytechnic), Tehran, Iran

6.1 INTRODUCTION

Fuel cells provide better opportunities for power generation because of their greater efficiency and lower pollution. Solid oxide fuel cells (SOFCs) are the most often used fuel cells, either alone or as part of a cycle. The coupling of a SOFC and a gas turbine is the most prevalent and widely employed in high-efficiency, low-pollution power generation cycles (Park & Kim 2006; Zhang et al. 2011). The first step in determining the performance of SOFC-GT cycles is to model their thermodynamics. Fuel cells employ a variety of geometries and fuels (Zhang et al. 2011). The nature of fuel has significant effects on the environmental behavior and thermodynamics of fuel cells. It should be mentioned that using methane or natural gas to improve the process is required. There are several fuel-improvement approaches (Mortazaei & Rahimi 2016).

It is required to investigate the chemical and electrochemical reactions, as well as the relationships governing all of the cycle components, in order to simulate the hybrid cycles of solid oxide fuel cell and gas turbine (Singh et al. 2005). Concepts described in Anastasiadis et al. (2017) have been used to get these relationships. The potential loss is in Mortazaei and Rahimi (2018), and its details are in Singh et al. (2005). We live in a world where oil and gas and coal reserves are depleting and high fuel consumption has accelerated global warming. Unfortunately, fossil fuels cannot be reduced overnight, as about 80–90% of the world's energy needs are met this way. But there is a question: What can be done now that renewable energy such as solar or wind energy cannot play a major role in the energy supply? One solution is to move some of our plants to different systems called combined heat and power (CHP) plants. Simultaneous electricity and heat generation enables better fossil fuel efficiency for energy supply with about 15–40% energy storage. These power plants perform well both in terms of cost reduction and reducing environmental impact (Najafi et al. 2014).

Conventional power plants usually generate electricity through an inefficient process. Fossil fuels such as coal, gasoline, or natural gas release heat energy in a bulky furnace during the combustion process. This heat is used to boil and evaporate water to turn the steam from a turbine, move the generator's generator, and generate electricity by rotating the generator.

DOI: 10.1201/9781003213741-6

The problem with this cycle is that a lot of energy is wasted at each stage. Boiled water, which is supposed to start the turbine, must be cooled in very large cooling towers in the open air, which consumes a lot of energy (Chatrattanawet et al. 2018).

Now think about transferring the generated heat to the residential, commercial, and industrial areas as hot water instead of allowing the generated heat to leave the cooling towers without any use. This is the basic idea behind cogeneration (CHP): recovering the heat generated in the power generation process and using it for local buildings. While conventional power plants waste the extra heat generated, a CHP power plant simultaneously provides electricity and hot water to the consumer (Owebor et al. 2019).

The actual efficiency of a CHP plant depends on how it supplies the heat generated. The CHP power plant's efficiency will be maximized if the power plant is close to consumer buildings. In other words, the CHP plant will perform best as a type of decentralized power supply with a large number of consumer sources close by (Roy et al. 2019). Also, as electricity flows through the wires to reach the consumer, energy loss is reduced due to reduced resistance as the distance between the power plant and the consumer decreases. In this way, offices, schools, hotels, hospitals, and even residential buildings can build a small-scale CHP or micro-CHP power plant to supply the hot water and electricity they need and even inject additional electricity into the grid (see Figure 6.1).

From a theoretical point of view, it is easy to build a CHP power plant by sending the heat lost from a power plant to local consumer buildings, but in practice, CHP power plants produce energy in a completely different way using different heating machines. Smaller CHP plants typically use internal combustion engines (such as gasoline engines in cars and diesel engines in trucks) to turn on power generators and heat exchangers to use the heat generated to heat water. But in larger power plants, high-efficiency gas turbines and steam turbines are used. Future CHP plants are likely to use fuel cells that run on hydrogen gas (Valencia, Benavides, & Cárdenas 2019; Lisbona, Uche, & Serra 2005).

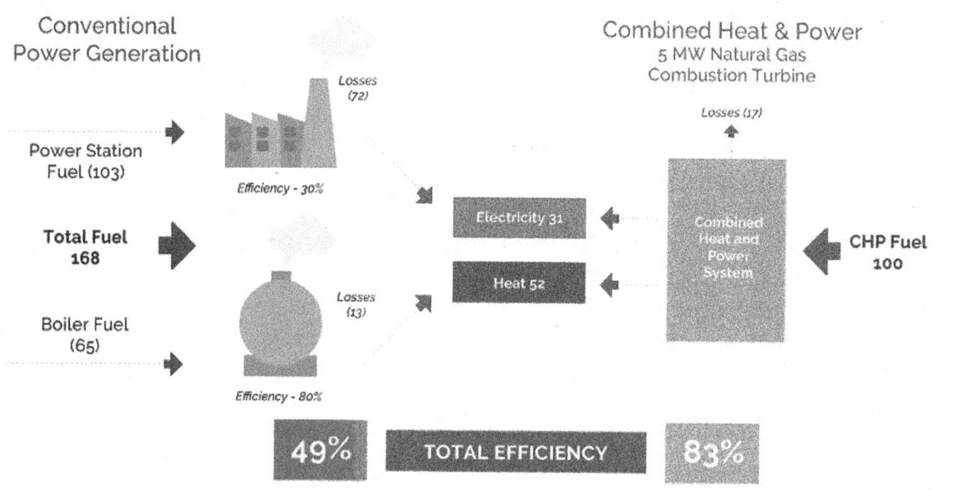

Figure 6.1 CHP cogeneration efficiency.

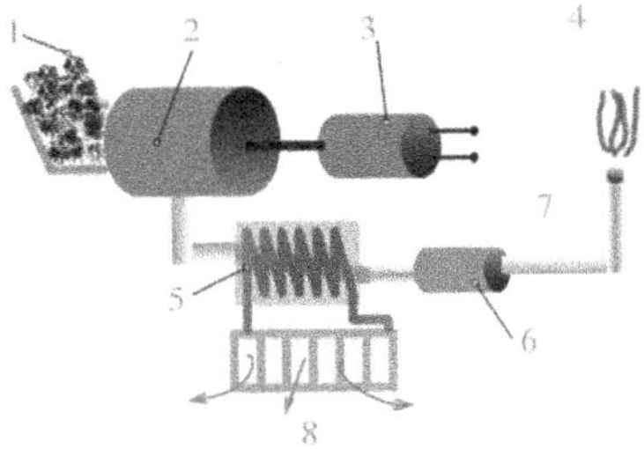

Figure 6.2 A CHP plant diagram. (Source: Reyhani et al. 2016.)

But all this does not mean that CHPs are new and untested ideas. The world's first known power plant, built in 1882 by Thomas Edison in New York, was originally designed as a CHP. The plant provided heat and power to Manhattan buildings. For many years, the idea of cogeneration plants has been neglected due in the rush to build low-cost coal-fired power plants. Now, when environmental issues are becoming increasingly challenging, we need to rediscover and expand these ideas (Eveloy, Rodgers, & Qiu 2016).

Figure 6.2 shows a simplified diagram of the main components of a typical micro-PC cogeneration unit. In practice, there are several heat exchangers, sound and vibration reducers that have been abandoned here to help make the process simple to understand (Ghirardo et al. 2011).

The steps specified in this figure are as follows:

- Fuel (coal, natural gas, oil, or biomass) enters the system.
- The engine (about the size of the engine of a four-cylinder car) burns this fuel under a normal combustion process.
- A power generator is connected to the motor shaft and rotated by it.
- Several kilowatts of electricity are generated depending on the system's capacity, which can be used as a conventional energy source or emergency storage.
- Exhaust gases from the electric motor flow in one or more heat exchangers, through which most of their heat energy is taken.
- A catalytic converter (like the one in a car) takes some of the gas pollution.
- Relatively clean exhaust gases come out of a long chimney or pipe.
- The cold water that enters the heat exchanger is heated by these gases' heat and reaches a higher temperature. If this water is hot enough, it enters directly into the radiator and is used for heating.

The efficiency benefits of these power plants, as mentioned earlier, remain, but the environmental benefits must also be taken into account. If we prevent the combustion of any amount of fossil fuels, we have prevented the production of large amounts of carbon dioxide and its entry into the atmosphere, which slightly reduces the intensity of global warming. Reducing fossil fuels' combustion helps reduce air pollution and related problems and reduces water pollution and acid rain. Replacing older power plants with much smaller cogeneration plants reduces our dependence on centralized power grids and

Figure 6.3 A micro CHP plant.

our vulnerability to global power outages (see Figure 6.3). Like conventional power plants, these power plants run on any fuel such as oil, gas, or biomass (Ahmadi, Pourfatemi, & Ghaffari 2017).

Of course, CHP plants also have disadvantages, including the fact that their construction technology is still complex and expensive, and therefore building a cogeneration plant requires more initial capital. Of course, the cost of energy savings offsets this investment, but the start-up cost of building such plants is still higher than conventional power plants (Bao et al. 2018). Also, the costs of maintaining CHP power plants are usually higher. Another problem with these power plants is that small-scale power plants produce more expensive electricity than large-scale power plants. Most importantly, these power plants strengthen our need and dependence on fossil fuels. Some critics argue that CHP is less efficient than new alternative technologies such as geothermal heat pumps (Beyrami et al. 2019). These technologies, which have also been developed on a larger scale, are a good way to meet climate change challenges. The heat generated in conventional power plants is not wasted as we imagine; waste heat is an essential part of the high-efficiency power generation cycle in power plants, and optimizing this cycle to reduce heat loss will reduce power generation (Jehandideh, Hassanzade, & Shakib 2020). The problem is that electricity is a more useful and practical form of energy for us. Investors see the problem differently: the cost of generating electricity from your small CHP unit is less than the cost of purchasing that amount of electricity from the grid, and the other advantage is a backup source to supply power (Hosseini et al. 2013).

The solid oxide fuel cell (SOFC) uses a ceramic electrolyte (a combination of zircon (Zr) and yttrium (Y) with the formula $Zr_{0.92}Y_{0.08}O_{1.96}$ instead of a liquid electrolyte. The anode electrode is an amalgam of cyanide nickel, and the cathode electrode is made of lanthanum manganese. The operating temperature of the cell is 2–4°C (Bejan, Tsatsaronis, & Moran 1996). The solid electrolyte is coated with electrodes made of special porous materials (Pirkandi, Mahmoodi, & Amanloo 2015). At the high operating temperatures of the SOFC fuel cell, oxygen ions (negatively charged) pass through the crystal lattice. When a hydrogen-containing gas fuel passes through the anode, a negatively charged stream of oxygen ions passes through

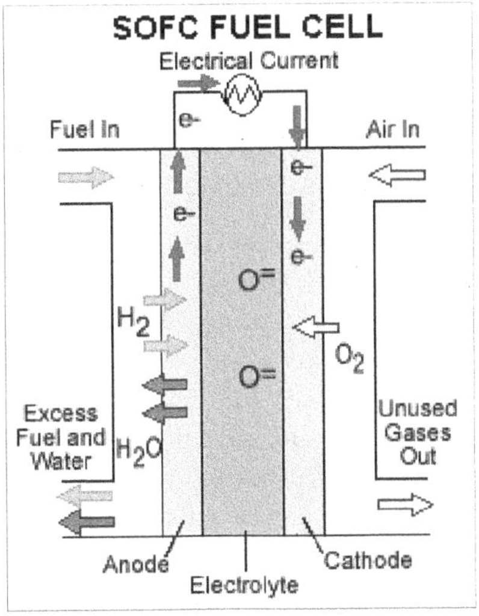

Figure 6.4 Schematics of a solid oxide fuel cell.

the electrolyte to oxidize the fuel. The oxygen stored at the cathode is usually taken from the air. The electrons generated at the anode pass through an external charge (external circuit) and go to the cathode (Ni 2012; Cheddie 2011).

This completes an electrical circuit and generates electricity. Solid oxide fuel cells do not require external reformers. The overall performance of the SOFC can be seen in Figure 6.4 (Choudhary & Sahu 2017, 2019; Choudhary et al. 2018; Shukla et al. 2018).

This chapter aims to introduce a cogeneration system based on a tubular SOFC. To accurately study the performance of the fuel cell, first, an electrochemical model for the fuel cell is presented, and after calculating the voltage and output power of the cell, separate thermal and thermodynamic analyses are performed on it. Simultaneous performance of three electrochemical, thermal, and thermodynamic analyses can obtain accurate and complete results of fuel cell performance in the proposed cogeneration system. In the final part, the proposed system is compared from an economic point of view with simultaneous production of gas microturbines, and the related results are presented.

6.2 THERMODYNAMIC MODELING

There are three retrievals. The fuel of the system is natural gas, which contains 97% methane, 1.5% carbon dioxide, and 1.5% nitrogen, and the composition of air is 21% oxygen and 79% nitrogen (Alanne et al. 2006). Part of the electrical power generated by the fuel cell is used to drive two compressors and a pump, and the rest is the net electrical power output from the system. The system is designed to recover the heat of dense fuel and air before entering the fuel tank. Natural gas is converted to pure hydrogen after passing through the recovery and after going to the fuel cell in the anode section by the reforming process (Al-Sulaiman, Dincer, & Hamdullahpur 2010a). Hydrogen obtained from natural gas reacts with oxygen in the air that passes through another recycler and enters the fuel cell. The exhaust gases from the cell then enter the combustion chamber and react with excess air. The hot output products enter the three introduced recoveries along the way. The third extractor produces hot water in the temperature range of 25°C–90°C (Al-Sulaiman, Dincer, & Hamdullahpur 2010b).

The energy consumption required in the building includes all the thermal, refrigeration, and electrical energy of the building – which the system must be able to supply (El-Emam, Dincer, & Naterer 2012). The heat obtained from fuel cells and recyclers can provide the heating load and sanitary spa of a building in winter, and in summer, it can be used to create cold and provide cooling load in a chiller and provide sanitary hot water for the building (Khani et al. 2016). The electrical energy produced in the fuel cell can also be used to supply the building with an electrical charge such as a lighting system, electrical peripherals (such as compressors, pumps, compression chillers, etc.). To study and analyze the results obtained from this cycle, for a sample building of 100 m in an area with a temperate climate, calculations of heating and cooling load, sanitary spa load, and electrical power calculations required for the building were performed (Ranjbar et al. 2014). According to the calculations, the amount of heating load required for the mentioned building was estimated at 17 kW, and its electric load was estimated at 8 kW. The hypotheses are also described below (Ivers-Tiffée, Weber, & Herbstritt 2001; Dusastre & Kilner 1999):

In modeling and analyzing the proposed system (shown in Figure 6.5), the following hypotheses have been considered.

- Gas leakage inside the system is excluded.
- Fluid flow is considered in all components of the stable cycle.
- The behavior of all gases in the cycle is assumed to be ideal gas.
- The temperature distribution, pressure, and chemical components inside the cell are ignored.
- The fuel inside the fuel cell is converted to hydrogen by internal reforming.

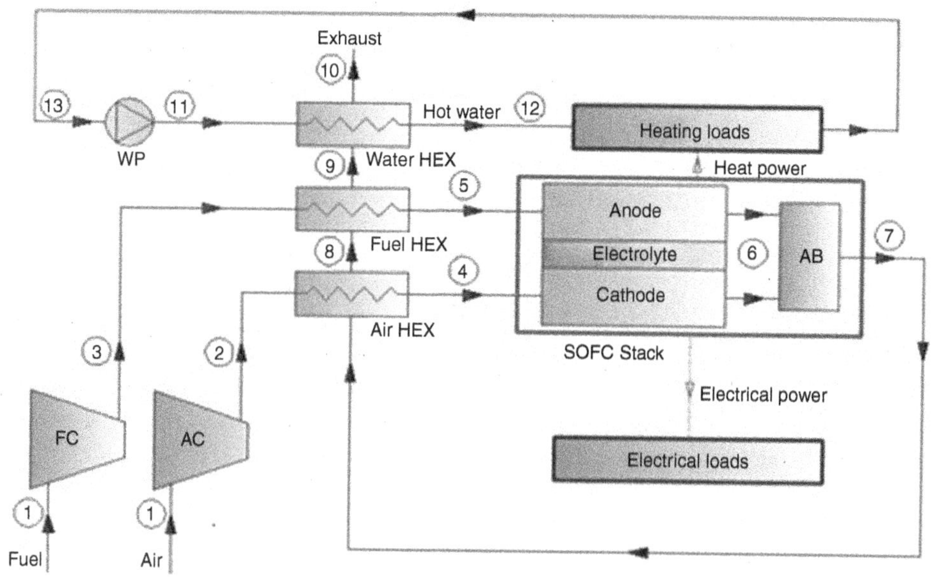

Figure 6.5 Schematic of the cycle.

6.3 METHODS AND MATERIALS

First, using the existing relations, the performance of all components used in the proposed system is introduced, and then separately and under stable conditions, the above system is analyzed and studied. To perform the written computer program, the thermodynamic and electrochemical performance of the system has been investigated by changing several effective parameters such as battery temperature and operating pressure, battery flow density, and air to fuel ratio entering the system (Padulles, Ault, & McDonald 2000; Matsuzaki & Yasuda 2000).

The SOFC is modeled using the tubular fuel cell interactions and information from Tsatsaronis (2007). The following are the relations used for improving and upgrading the fuel:

$$CH_4 + H_2O \leftrightarrow H_2 + CO \tag{6.1}$$

$$CO + H_2O \leftrightarrow H_2 + CO_2 \tag{6.2}$$

$$H_2 + \frac{1}{2}O_2 \leftrightarrow H_2O \tag{6.3}$$

All chemical processes are presumed to be in equilibrium, and the constants of equilibrium are represented using the formulas below (Aljundi 2009; Hammond & Stapleton 2001):

$$K_{reforming} = \frac{x_{H_2}^2 x_{CO}}{x_{CH4} x_{H2O}} P^2 \tag{6.4}$$

$$K_{shifting} = \frac{x_{H2} x_{CO2}}{x_{CO} x_{H2O}} \tag{6.5}$$

$$lnK = AT^4 + BT^3 + CT^2 + DT + E \tag{6.6}$$

A, B, C, and D are constant values that appear in both the modification and fuel upgrade tables (Table 6.1).

The Nernst equation used to calculate the optimum voltage of a fuel cell.

$$V_N = V_0 - \frac{RT}{2F} lnln\left(\frac{x_{H_2O}}{x_{H_2}\sqrt{x_{O_2}}}\right) \tag{6.7}$$

Table 6.1 Coefficients used in Equation (6.6)

Constant	*Reforming*	*Shifting*
A	−2.6312e-11	5.47e-12
B	1.2406e-07	−2.5748e-08
C	−0.00022523	0.000046374
D	0.19503	−0.03915
E	−66.1395	13.2097

Sources: Aljundi (2009); Hammond and Stapleton (2001).

However, because of the irreversibility, the actual voltage of the fuel cell is lower than the Nernst voltage. Significant losses, activation, and concentration are the three basic categories of irreversibilities. As a result, the actual voltage of the cell can be calculated as follows (Talebi & Norouzi 2020; Norouzi, Talebi, & Najafi 2021):

$$V = V_N - V_{ohm} - V_{act} - V_{con} \tag{6.8}$$

The following equation can be used to calculate significant potential loss (Norouzi 2020; Fani, Norouzi, & Ramezani 2020):

$$V_{ohm} = \left(R_{contact} + \Sigma \rho_k L_k\right)_i \tag{6.9}$$

Wherein:

$$\rho_a = 8.114e - 6expexp\left(\frac{600}{T}\right)$$

$$\rho_e = 2.94e - 6expexp\left(\frac{10350}{T}\right)$$

$$\rho_c = 2.94e - 6expexp\left(\frac{-1392}{T}\right)$$

$$\rho_i = 125.6e - 6expexp\left(\frac{4690}{T}\right)$$

The activation losses is estimated by the equation given below (Norouzi 2021a, 2021b):

$$V_{act} = V_{act,a} + V_{act,c} = \frac{RT}{F}\left(\frac{i}{2i_{o,a}}\right) + \frac{RT}{F}\left(\frac{i}{2i_{o,c}}\right) \tag{6.10}$$

Concentration potential drop is estimated as follows:

$$V_{con} = V_{con,a} + V_{con,c} = -\frac{RT}{2F}\left[lnln\left(1 - \frac{i}{i_{as}}\right) - lnln\left(1 + \frac{x_{H2}i}{x_{H2O}i_{as}}\right)\right] + \frac{RT}{4F}lnln\left(1 - \frac{i}{i_{cs}}\right) \tag{6.11}$$

Current restrictions at the cathode and anode are also specified in reference (Norouzi 2021b).

The power output of the fuel cell is shown by:

$$W_{FC} = IV \tag{6.12}$$

The thermodynamic modeling of the fuel cell is as follows (Norouzi & Talebi 2020):

$$\left(\sum_{anode} m_i h_i + \sum_{cathod} m_i h_i\right)_{in} = W_{FC} + \left(\sum_{anode} m_i h_i + \sum_{cathod} m_i h_i\right)_{out} + Q_{FC} \tag{6.13}$$

The following general relation is applied to calculate exergy losses for each component of the cycle:

$$E_D = Q\left(1-\frac{T_0}{T}\right) - W + \sum_{in} E - \sum_{out} E \tag{6.14}$$

The total exergy rate is given as:

$$E = E_{ch} + E_{ph} \tag{6.15}$$

The thermodynamic modeling of air compressor is based on isentropic efficiency and compressor pressure ratio (Norouzi, Kalantari, & Talebi 2020):

$$T_{out} = T_{in}\left(1+\frac{1}{\eta_{AC}}\left[r_c^{\frac{\gamma_a-1}{\gamma_a}} - 1\right]\right) \tag{6.16}$$

The exit temperature of gas turbine is calculated as (Norouzi et al. 2021a):

$$T_{out} = T_{in}\left(1+\eta_{GT}\left[1-\left(\frac{P_{in}}{P_{out}}\right)^{\frac{1-\gamma_g}{\gamma_g}}\right]\right) \tag{6.17}$$

The fuel cell and the natural gas exhaust gases react in the combustion chamber. The combustion products' molar flow rate and their temperature can be obtained by applying the energy balance (Norouzi et al. 2021b).

The thermodynamic modeling of preheaters is given as (Khajehpour, Norouzi, & Fani 2021):

$$\eta_{AP} m_g \left(h_{in} - h_{out}\right)_g = m_a \left(h_{in} - h_{out}\right)_a \tag{6.18}$$

At a fixed pressure, Boyer heat recovery forms saturated vapor (P_{main}). In HRSG performance, the pinch point is crucial (Norouzi et al. 2021c).

HRSG was modeled using the following equations:

$$T_{out} = T_{sat} + PP \tag{6.19}$$

$$m_{steam}\left(h_{out} - h_{in}\right)_{steam} = m_g C_{Pg}\left(T_{in} - T_{out}\right)_g \tag{6.20}$$

The net output work and exergy efficiency of the entire cycle can be calculated using the calculations below (Norouzi & Kalantari 2021):

$$W_{net} = W_{FC} + W_{GT} - W_{AC} - W_{Pump} \tag{6.21}$$

$$\eta_{total} = \frac{W_{net} + E_{steam}}{1.06 m_{fuel} LHV} \tag{6.22}$$

Considering that the economic discussion of energy production systems is an important issue in their selection, in this section, the relations related to the pricing of the proposed system are given. Pre-design investment costs are usually estimated using outdated pricing information (Norouzi 2021c). This PEC_{ref} reference information is updated using an appropriate cost index to show current PEC_{new} prices.

$$PEC_{new} = PEC_{ref} \frac{I_{new}}{I_{ref}} \tag{6.23}$$

In the above relation, I_{new} and I_{ref} represent the new cost index related to price estimation and the cost index in the reference year, respectively. Figure 6.6 shows the graph of the cost index in recent years (Ramezani et al. 2020).

The equipment price of the proposed system will be obtained using equations (6.25) to (6.35), which are established for the year 1994. As mentioned above, the diagram in Figure 6.6 is used to update these prices. To update the prices according to Equation (6.24), you must first divide the amount for the year in question by the amount for the year 1994 and then multiply the result by the price of the device. After calculating the price of each piece of equipment, the total price of this system can be obtained by summing them together in different working conditions (Tashakori-Miyanroudi et al. 2020; Vahed et al. 2020).

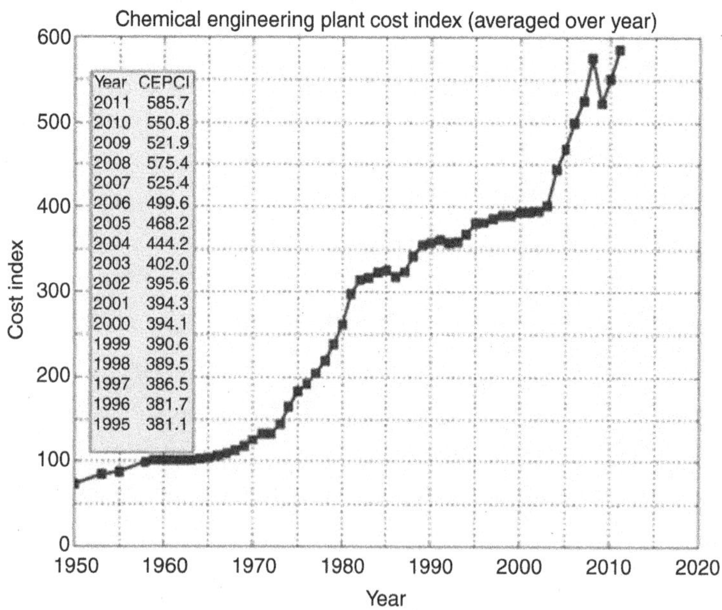

Figure 6.6 Graph of the cost index.

Determining the price of compressors

Price of air compressor (Zhang et al. 2011; Rahimi 2018)

$$PEC_{Comp} = \left(\frac{71.1m_a}{0.9-\eta_{sc}}\right)\left(\frac{P_2}{P_1}\right)ln\left(\frac{P_2}{P_1}\right) \tag{6.24}$$

Price of fuel compressor (Singh et al. 2005)

$$PEC_{Comp} = \left(\frac{71.1m_f}{0.9-\eta_{sc}}\right)\left(\frac{P_3}{P_1}\right)ln\left(\frac{P_3}{P_1}\right) \tag{6.25}$$

Determining the price of fuel (Anastasiadis et al. 2017)

The price of fuel cell mass

$$PEC_{SOFC} = A_{SOFC}\left[2.96T_{cell} - 1970\right] \tag{6.26}$$

Inverter price

$$PEC_{inverter} = 1e+5\left[\frac{W_{cell}}{500}\right]^{0.7} \tag{6.27}$$

Price of peripheral battery equipment

$$PEC_{SOFC,aux} = 0.1PEC_{SOFC} \tag{6.28}$$

Determining the price of a combustion chamber

$$PEC_{ab} = \left(\frac{46.08m_{ab}}{0.995-\frac{P_7}{P_6}}\right)\left[1+exp\,exp\left(0.018T_7 - 26.4\right)\right] \tag{6.29}$$

Pricing of retrievers (Anastasiadis et al. 2017)

$$PEC_{ah} = 4122\left[\frac{m_g\left(h_7 - h_8\right)}{18\Delta T_{lm,ah}}\right]^{0.6} \tag{6.30}$$

$$PEC_{fh} = 4122\left[\frac{m_g\left(h_8 - h_9\right)}{18\Delta T_{lm,fh}}\right]^{0.6} \tag{6.31}$$

$$PEC_{wh} = 4122\left[\frac{m_g\left(h_9 - h_{10}\right)}{18\Delta T_{lm,wh}}\right]^{0.6} \tag{6.32}$$

Pump pricing (Anastasiadis et al. 2017)

$$PEC_{pump} = 442\left[\dot{W}_p\right]^{0.71} 1.41 f_\eta \tag{6.33}$$

$$f_\eta = 1 + \left(\frac{0.2}{1-\eta_p}\right) \tag{6.34}$$

In this section, the total revenue requirement (TRR) method is used to analyze the system, which originates from the proposed process in the Electricity Generation Research Center. This method calculates all project costs, including the amount of return on investment. In this method, based on economic assumptions and calculating the purchase price of equipment and fuel, the final required revenue is calculated from year to year. Finally, all costs, including maintenance costs and fuel costs, are leveled annually during the system operation period (Anastasiadis et al. 2017).

6.4 RESULTS AND DISCUSSION

According to the equations mentioned in the previous sections, a computer program has been written in EES software to analyze the problem. System input information includes air flow rate and fuel input to the system and battery operating pressure. In the first part of this code, the non-linear reforming and electrochemical equations and the thermal equations of the cell are solved simultaneously, and the desired results, including combinations of output components, temperature, voltage, and power in the fuel cell, are obtained. In the second part, the whole hybrid system is analyzed from a thermodynamic and economic point of view, and the efficiency, production capacity, entropy production rate, and price of the proposed system are obtained.

To validate the prepared code, a certain sample must compare the results of this prepared code with existing laboratory tests. Due to the lack of experimental results in the field of solid oxide fuel cells fed with methane, in this research, using the experimental results performed by Bao et al. (2018) the code prepared in this research has been validated (see Figure 6.7).

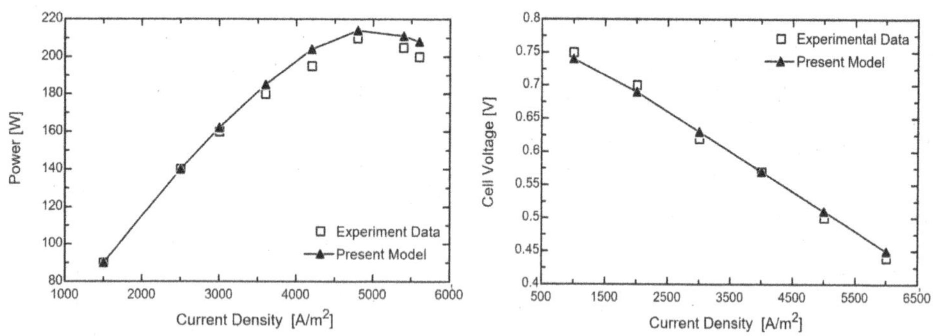

Figure 6.7 Validation of the system comparing the simulation results and experimental reference results: (a) power; (b) cell voltage.

Laboratory results were obtained based on fuel, consisting of 89% hydrogen and 11% water (Anastasiadis et al. 2017). As shown in the polarization curve of Figure 6.7b and the power of Figure 6.7a, the consistency of the experimental results and the numerical results show the accuracy of the prepared code. In this experiment, the temperature and working pressure of the cell is considered to be about 1,000°C and 1 bar, respectively. The operating temperature and pressure of the cell and the current density in it, along with the ratio of airflow to the fuel entering the system, are considered several parameters affecting the performance of the proposed cogeneration system (Najafi et al. 2014). The aim is to investigate the effect of these parameters on the electrical and thermal power generated in this system. To analyze the problem accurately, first, the complete electrochemical calculations of the fuel cell are performed, and then the whole proposed system will be analyzed from a thermodynamic point of view. As mentioned before, the fuel cell used in this research is a tubular solid oxide type, the specifications of which along with other specifications of the proposed cogeneration system are presented in Table 6.2 (Radmanesh & Hadi 2019).

In Figure 6.8, the changes in the output power relative to its current density in two different cases are investigated. In Figure 6.8(a), by keeping the cell pressure constant, its performance at different operating temperatures is investigated, and in Figure 6.8(b), by keeping the temperature constant, its function at different operating pressures is obtained. As shown in

Table 6.2 Assumptions of system performance parameters

Parameter	*Value*	*Parameter*	*value*
Area of each cell	1,036 cm^2	Total system efficiency	73.6%
The length of each cell	150 cm	Electrical efficiency of the system	49%
The diameter of each cell	2.2 cm	Thermal efficiency of the system	24.6%
Number of cells	5,749	Pressure chamber pressure drop	3%
Current density	3,000 A/m^2	Recuperator pressure drop	2%
Dubai Air	100 kmol/h	Efficiency of air and fuel retrievals	70–85%
Dubai fuel	63.63 kg/h	Water recovery efficiency	75%
Fuel consumption ratio	0.85	Isentropic efficiency of the fuel compressor	90%
Cell pressure drop	5%	Isentropic efficiency of air compressor	70%

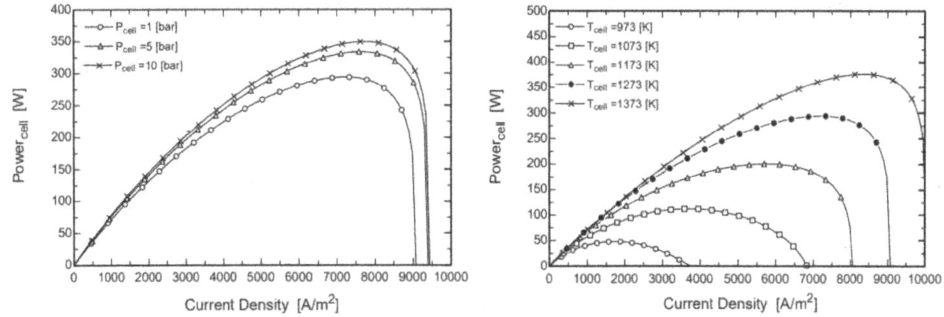

Figure 6.8 (a) Changes in cell power relative to current density at operating pressure 1 bar. (b) Changes in cell production capacity relative to density at operating temperature of 1,200 K.

Figure 6.8(a), increasing the cell's temperature increases its voltage and thus increases its efficiency, and at high current densities reaches its maximum output power. Although it seems that the optimal design value occurs at the maximum point of the above diagrams, its proximity to the area of concentration drop due to the cell is extremely dangerous (Chatrattanawet et al. 2018). As shown in Figure 6.8, with the increase of the working pressure of the cell, its performance improves, and the decrease of the concentration voltage at higher current densities shows its effect. A very important point that should be considered in the design of the cell is that the current density of the cell at the design point should be in the area of drops due to resistance and away from the concentration area so that the system can generate its electrical power stably.

To analyze the performance of the proposed cogeneration system, in Figures 6.9(a) and 6.9(b), the changes in electrical and thermal power generated along with the electrical, thermal, and system efficiencies of the system at different pressures and operating temperatures of the cell are obtained. As stated in the analysis of relationships related to the electrochemical process of the cell, the operating temperature of the cell is one of the effective factors in changing its performance. As shown in Figure 6.9(a), with increasing cell operating temperature, the electrical power generated in this system increases to a certain extent. Increasing the electrical power will reduce the heat generated in the cell, which will reduce the heat generated in the third recovery. The main cause of decreasing electrical power and increasing thermal power at high temperatures is the drop in concentration voltage. Figure 6.9(b) shows that according to the electrical and thermal needs, the temperature range of the cell can be determined. For example, if a building unit needs a high electrical load and a low thermal load, a high cell operating temperature will be considered, and if the opposite is the case, the use of solid oxide fuel cells with a low operating temperature will be developed. Figure 6.9 shows the electrical, electrical, and thermal efficiencies of the proposed system in the range of different cell temperatures. As the operating temperature of the cell increases, the electrical and overall efficiency of the system will increase, and its thermal efficiency will decrease. This increase and decrease of efficiencies are to a certain temperature limit, and then they will go in the opposite direction with increasing temperature, which can be considered a result of the negative effect

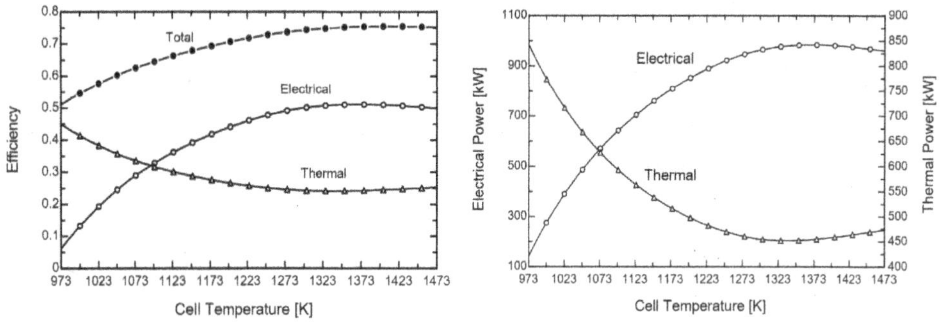

Figure 6.9 (a) Changes in electrical and thermal power of the system. (b) Changes in the proposed system's electrical, thermal, and total efficiency concerning the battery's operating temperature at 25°C.

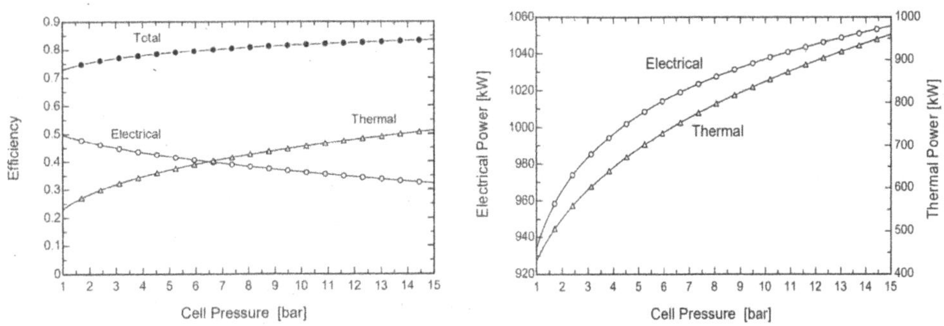

Figure 6.10 (a) Changes in electrical and thermal power of the system. (b) Changes in the electrical, thermal, and total efficiency of the proposed system relative to the operating pressure of the cell at the proposed operating temperature relative to the operating pressure of the cell at a working temperature of 1,273 K and a current density of 3,000 kW and 3,000 rpm current of 3,000 amps per square meter.

of concentration voltage drop on fuel performance. As can be seen, by increasing the cell's operating temperature to 1,400 K, its overall efficiency will increase by about 75.5%.

Figure 6.10(a) shows the effect of cell working pressure on the electrical and thermal power generated in the proposed system. The results show an increase in both production capacities due to the increase of the working pressure of the cell. An important issue to consider in this case is the use of stronger air and fuel compressors in high-pressure ratios, which will reduce the system's electrical efficiency. Figure 6.10(b) shows the changes in the overall electrical, electrical, and thermal efficiency of the system relative to the working pressure of the cell. The results show that increasing the working pressure of the cell will reduce the electrical efficiency and increase the thermal efficiency and the whole system. As mentioned in the previous sections, the system's operating pressure has less effect on the concentration voltage drop than its operating temperature, which will cause this parameter to have less effect on changes in the system's electrical and thermal power output. According to the results, in the high-pressure ratio, the maximum overall efficiency of the system is estimated at 83.5%. Similar to the previous case, the thermal and electrical needs of the building can be an effective parameter in determining the pressure ratio.

Figure 6.11(a) shows the effects of changes in current cell density on the electrical and thermal power generated in the proposed system. As can be seen, increasing the current density in the cell, on the one hand, will reduce the electrical power produced in the system and, on the other hand, will increase the thermal power produced in it. In general, increasing the current density due to the effects of concentration drop voltage reduces the voltage of the battery at high densities, which will reduce the electrical power generated in the cell. On the other hand, increasing the current density will increase the heat produced in the cell, which will increase the temperature of the products leaving the cell and subsequently increase the heat obtained in the third retriever. Figure 6.11(b) shows the effects of this parameter on electrical, thermal, and system efficiency. As can be seen, with increasing current density, the electrical efficiency and the whole system will decrease, and its thermal efficiency will also increase.

Increasing the inlet fuel flow rate and keeping the fuel consumption coefficient constant means that more chemical energy in the cell is being converted into electrical energy. In this

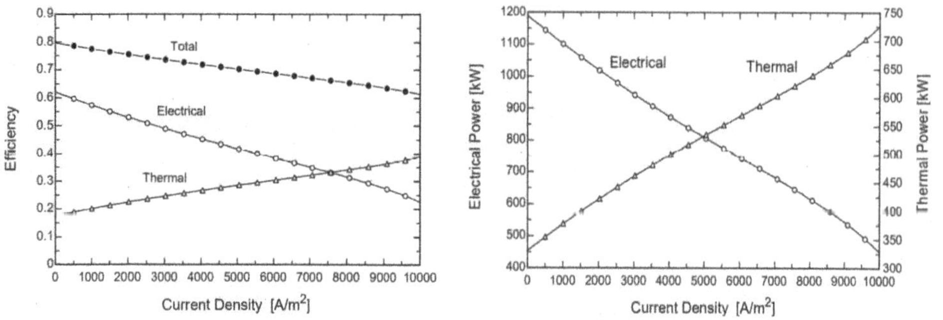

Figure 6.11 (a) Changes in electrical and thermal power of the system. (b) Changes in electrical efficiency and thermal conductivity to the current cell density at operating temperature. The proposed system to the cell flow density at 1,273 K and a working pressure of 1.25 bar. Operation 1,273 K and a working pressure of 1.25 bar.

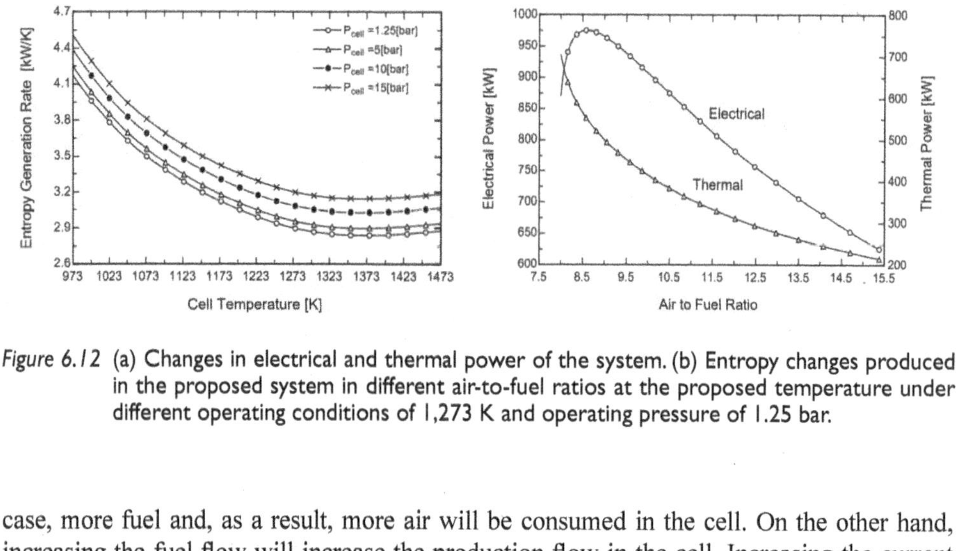

Figure 6.12 (a) Changes in electrical and thermal power of the system. (b) Entropy changes produced in the proposed system in different air-to-fuel ratios at the proposed temperature under different operating conditions of 1,273 K and operating pressure of 1.25 bar.

case, more fuel and, as a result, more air will be consumed in the cell. On the other hand, increasing the fuel flow will increase the production flow in the cell. Increasing the current increases the overvoltage of the cell and produces more heat in it, which increases the temperature of the cell. In general, increasing fuel flow does not have a favorable effect on system performance and greatly increases irreversibility. The positive effect of this parameter is to increase the system's power, while it will reduce the efficiency. The value of this parameter should be fixed in a suitable value. On the other hand, more than the flow of air entering the cell will increase the power output in the cell, which will reduce the heat produced in the cell and thus reduce the heat power required in the third retriever. According to the above, an optimal value should be obtained for the ratio of air to fuel entering the system. Figure 6.12(a) shows the changes in the system's electrical and thermal power output relative to the changes in this parameter; as can be seen in this figure, with increasing air to fuel ratio, the electrical power produced in the cell increases then decreases. Selecting a higher air-to-fuel ratio will significantly reduce the electric power and heat generated in the battery. Figure 6.12(b) shows the rate of entropy changes produced in the proposed system under its different operating conditions. As shown in this figure, increasing the cell temperature reduces the entropy production rate in the system in question, which is a constant rate at higher temperatures. Unlike increasing the temperature, increasing the system's operating pressure will increase the system's entropy production rate, which is not engineeringly desirable.

Figures 6.13(a) and 6.13(b) show the distribution of entropy production rates in different components of the proposed system in two different ratios of air to fuel. As can be seen, the highest entropy production rate is related to the combustion chamber and solid oxide fuel, and the lowest value belongs to the compressor and fuel recovery. As can be seen, increasing the air ratio to fuel entering the system will reduce the entropy rate produced in the system components.

In the final part, the proposed system is analyzed from an economic point of view, which is compared with a cycle of simultaneous production of gas microturbines. As mentioned in the fifth section, the purpose of the economic analysis is to determine the price of the proposed system equipment and calculate the cost of generating electricity during the economic life of the system (20 years). The results of Table 6.3 show that the proposed system has a low capacity in terms of the equipment purchase price but is less valuable considering the amount of electricity generated. Increasing the system's capacity causes the price of electricity generated based on the price of equipment and the price of electricity generated based on the economic life of the system to decrease, which is a useful parameter in choosing a system. According to the presented results, it can be stated that using the production system simultaneously with fuel is not economical for small residential units, but it can be used in residential complexes or small towns.

Table 6.4 also compares the proposed cogeneration system with a gas microturbine cogeneration system. As can be seen, at a constant capacity in electricity generation, the fuel cell system is more expensive than the microturbine system in terms of the price of electricity generated based on the price of equipment. Still, considering the economic life of the system, the price of electricity generated is lower.

6.5 SUMMARY

According to the contents of this chapter, the following can be presented as a summary of the discussion, increasing the operating temperature of the cell (up to a temperature of approximately 1,340 K) increases the electrical power and decreases the thermal power of the cogeneration system. Also, increasing the cell's operating temperature increases the overall electrical efficiency and decreases the thermal efficiency of the cogeneration system. The current density of the cell at the design point should be in the area of the drop due to the resistance and away from the area of concentration so that the system can generate its electrical power stably. According to the electrical and thermal needs of a building, the temperature range of the cell can be determined. If a building unit needs a high electrical load and a low thermal load, a high cell operating temperature should be considered, and if the opposite is the case, the use of solid oxide fuel cells with a low operating temperature should be developed. Increasing the working pressure of the cell will reduce the electrical efficiency and increase the thermal and overall efficiency of the system. As the current density increases, the electrical and overall efficiency of the system will decrease, and its thermal efficiency will also increase.

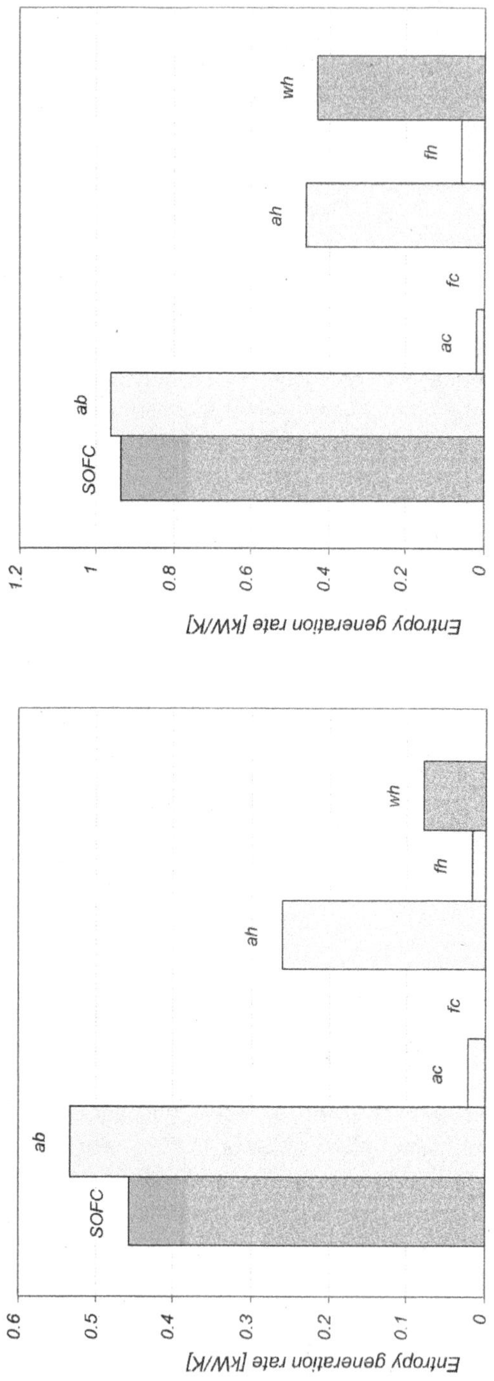

Figure 6.13 (a) Distribution of entropy production rate in different components of the proposed system in air-to-fuel ratio 9.4. (b) Distribution of entropy production rate in different components of the proposed system in air-to-fuel ratio 18.8.

Table 6.3 Comparison of different operating modes of the system from an economic point of view

Parameter/Cases	*6*	*5*	*4*	*3*	*2*	*1*
Air flow (kmol/h)	200	150	100	50	40	30
Fuel flow (kmol/h)	21.26	15.95	10.63	5.31	4.25	3.19
Cell temperature (C)	1,273	1,273	1,273	1,273	1,273	1,273
Cell working pressure(bar)	1.25	1.25	1.25	1.25	1.25	1.25
Number of tubercles	11,498	8,626	5,749	2,872	2,298	1,725
Electrical power generated by the system(kW)	1,889	1,417	944.4	471.9	377.7	283.4
Thermal power output of the system (kW)	1,366	1,025	682.9	341.2	273	204.9
System price ($/kW)	2,856,000	2,158,000	1,454,000	743,520	600,273	456,031
Price of electricity generated based on equipment price ($/kWh))	1,511	1,522	1,539	1,575	1,589	1,609
Price of electricity generated based on the economic life of the system (cents/kWh)	10.28	13.06	18.60	35.30	43.60	57.56

Table 6.4 Comparison of the operating mode of the proposed system with a gas turbine cycle

System	*Price of electricity generated based on the economic life of the system (cents per kilowatt-hour)*	*Price of electricity produced based on the price of equipment (dollars per kilowatt)*	*Electrical power generated by the system (kW)*
Simultaneous production system using microturbine	12	492	1,889
Simultaneous production system using fuel cell	10.28	1,511	1,889

To achieve high efficiency in the proposed cogeneration system, the ratio of air to fuel input to the system must be selected as an appropriate value. For the proposed system in this paper, and considering its maximum efficiency, the value of air to fuel ratio is 9.4. Increasing the temperature and decreasing the cell pressure will reduce the entropy production rate in the system. Therefore, the use of high operating temperatures and lower pressures will achieve an optimal system. The highest entropy production rate in the cogeneration system is proposed in the combustion chamber and fuel cell. Using this production system and fuel is not economical for small residential units, but it can be used in residential complexes or small towns. Also, as the results state at a constant capacity, the production system with fuel is more expensive than the gas microturbine system, but considering the price of electricity generated based on the economic life of the system is more appropriate. It also can be mentioned that this system helps the power plant designer and upcoming researcher to follow guidelines while selecting parameters for CHP modeling.

ABBREVIATIONS

E: Exergy rate (kW)
W: Power (kW)
F: Faraday constant (C/kmol)
h: Enthalpy (kJ/kg)
I: Electric current (A)
L: Length (m)
K: Constant equilibrium
m: Mass flow (kg / s)
LHV: Low calorific value (kJ / kg)
Q: Temperature rate (kJ/s)
R: Universal gas constant (kJ / (kmol.K))
P: Pressure (bar)
T: Temperature (K)
V: Voltage(volt)

Greek letters

ρ: Ohmic resistance per unit length (Ω/m)
η: Efficiency

i: Current density (A/cm^2)

i_o: Exchange current density (A/cm^2)
γ_c: Specific heat ratio
$i_{a,s}$: Anode limiting current density (A/cm^2)

a: Anode, air
AC: Air compressor
act: activation
AP: Air preheater
c: Cathode
ch: Chemical
Con: Concentration
e: Electrolyte
g: Gas, smoke
In: Input
Out: Output
Ohm: Ohmic
x: Concentration
GT: Gas turbine
V_o: Standard voltage (Volt)
P_{main}: Steam pressure
TIT: Temperature of combustion inlet products to gas turbine out (K)
T_{sat}: Vapor saturation temperature at corresponding pressure (K)
D: Losses
r_c: Compressor pressure ratio
N: Nernst

REFERENCES

Ahmadi, Rouhollah, Seyyed Muhammad Pourfatemi, & Saeed Ghaffari. 2017. "Exergoeconomic optimization of hybrid system of GT, SOFC and MED implementing genetic algorithm." *Desalination* 411: 76–88.

Alanne, Kari, Arto Saari, V. Ismet Ugursal, & Joel Good. 2006. "The financial viability of an SOFC cogeneration system in single-family dwellings." *Journal of Power Sources* 158(1): 403–416.

Aljundi, Isam H. 2009. "Energy and exergy analysis of a steam power plant in Jordan." *Applied Thermal Engineering* 29(2–3): 324–328.

Al-Sulaiman, Fahad A., Ibrahim Dincer, & Feridun Hamdullahpur. 2010a. "Exergy analysis of an integrated solid oxide fuel cell and organic Rankine cycle for cooling, heating and power production." *Journal of Power Sources* 195(8): 2346–2354.

Al-Sulaiman, Fahad A., Ibrahim Dincer, & Feridun Hamdullahpur. 2010b. "Energy analysis of a trigeneration plant based on solid oxide fuel cell and organic Rankine cycle." *International Journal of Hydrogen Energy* 35(10): 5104–5113.

Anastasiadis, Anestis G., Stavros A. Konstantinopoulos, Georgios P. Kondylis, Georgios A. Vokas, & Panagiotis Papageorgas. 2017. "Effect of fuel cell units in economic and environmental dispatch of a Microgrid with penetration of photovoltaic and micro turbine units." *International Journal of Hydrogen Energy* 42(5): 3479–3486.

Bao, Cheng, Ying Wang, Daili Feng, Zeyi Jiang, & Xinxin Zhang. 2018. "Macroscopic modeling of solid oxide fuel cell (SOFC) and model-based control of SOFC and gas turbine hybrid system." *Progress in Energy and Combustion Science* 66: 83–140.

Bejan, Adrian, George Tsatsaronis, & Michael Moran. 1996. *Thermal Design and Optimization*. New York: John Wiley and Sons.

Beyrami, Javid, Ata Chitsaz, Kiyan Parham, & Øystein Arild. 2019. "Optimum performance of a single effect desalination unit integrated with a SOFC system by multi-objective thermo-economic optimization based on genetic algorithm." *Energy* 186: 115811.

Choudhary, Tushar, & Mithilesh Kumar Sahu. 2017. "CFD modeling of SOFC cogeneration system for building application." *Energy Procedia* 109: 361–368.

Choudhary, Tushar, & Mithilesh Kumar Sahu. 2019. "Energy and exergy analysis of solid oxide fuel cell integrated with gas turbine cycle: A hybrid cycle." In Jayeeta Chattopadhyay, Rahul Singh, & Om Prakash (eds.), *Renewable Energy and its Innovative Technologies*, pp. 139–153. Singapore: Springer.

Choudhary, Tushar, Mithilesh Kumar Sahu, R. Sanjay, Anupam Kumari, & Alok Mohapatra. 2018. *Thermodynamic Modeling of Blade Cooled Turboprop Engine Integrated to Solid Oxide Fuel Cell: A Concept*. SAE Technical Paper, No. 2018-01-1308.

Cheddie, Denver F. 2011. "Thermo-economic optimization of an indirectly coupled solid oxide fuel cell/gas turbine hybrid power plant." *International Journal of Hydrogen Energy* 36(2): 1702–1709.

Chatrattanawet, Narissara, Dang Saebea, Suthida Authayanun, Amornchai Arpornwichanop, & Yaneeporn Patcharavorachot. 2018. "Performance and environmental study of a biogas-fuelled solid oxide fuel cell with different reforming approaches." *Energy* 146: 131–140.

Dusastre, Vincent, & John A. Kilner. 1999. "Optimisation of composite cathodes for intermediate temperature SOFC applications." *Solid State Ionics* 126(1–2): 163–174.

El-Emam, Rami S., Ibrahim Dincer, & Greg F. Naterer. 2012. "Energy and exergy analyses of an integrated SOFC and coal gasification system." *International Journal of Hydrogen Energy* 37(2): 1689–1697.

Eveloy, Valerie, Peter Rodgers, & Linyue Qiu. 2016. "Integration of an atmospheric solid oxide fuel cell-gas turbine system with reverse osmosis for distributed seawater desalination in a process facility." *Energy Conversion and Management* 126: 944–959.

Fani, Maryam, Nima Norouzi, & Molood Ramezani. 2020. "Energy, exergy, and exergoeconomic analysis of solar thermal power plant hybrid with designed PCM storage." *International Journal of Air-Conditioning and Refrigeration* 28(4): 2050030.

Ghirardo, Federico, Marco Santin, Alberto Traverso, & Aristide Massardo. 2011. "Heat recovery options for onboard fuel cell systems." *International Journal of Hydrogen Energy* 36(13): 8134–8142.

Hammond, Geoffrey P., & A. J. Stapleton. 2001. "Exergy analysis of the United Kingdom energy system." *Proceedings of the Institution of Mechanical Engineers, Part A: Journal of Power and Energy* 215(2): 141–162.

Hosseini, Mehdi, Ibrahim Dincer, Pouria Ahmadi, Hasan Barzegar Avval, & Masoud Ziaasharhagh. 2013. "Thermodynamic modelling of an integrated solid oxide fuel cell and micro gas turbine system for desalination purposes." *International Journal of Energy Research* 37(5): 426–434.

Ivers-Tiffée, Ellen, André Weber, & Dirk Herbstritt. 2001. "Materials and technologies for SOFC-components." *Journal of the European Ceramic Society* 21(10–11): 1805–1811.

Jehandideh, Sobhan, Hasan Hassanzade, & Seyyed Ehsan Shakib. 2020. "Environmental assessment of a hybrid system composed of solid oxide fuel cell, gas turbine and multiple effect evaporation desalination system." *Energy & Environment* 32(5): 874–901.

Khajehpour, Hossein, Nima Norouzi, & Maryam Fani. 2021. "An exergetic model for the ambient air temperature impacts on the combined power plants and its management using the genetic algorithm." *International Journal of Air-Conditioning and Refrigeration* 29(1): 2150008.

Khani, Leyla, S. Mohammad S. Mahmoudi, Ata Chitsaz, & Marc A. Rosen. 2016. "Energy and exergoeconomic evaluation of a new power/cooling cogeneration system based on a solid oxide fuel cell." *Energy* 94: 64–77.

Lisbona, Pilar, Javier Uche, & Luis M. Serra. 2005. "High-temperature fuel cells for fresh water production." *Desalination* 182(1–3): 471–482.

Matsuzaki, Yoshio, & Isamu Yasuda. 2000. "The poisoning effect of sulfur-containing impurity gas on a SOFC anode: Part I. Dependence on temperature, time, and impurity concentration." *Solid State Ionics* 132(3–4): 261–269.

Mortazaei, Mortaza, & Mostafa Rahimi. 2016. "A comparison between two methods of generating power, heat and refrigeration via biomass based solid oxide fuel cell: A thermodynamic and environmental analysis." *Energy Conversion and Management* 126: 132–141.

Najafi, Behzad, Ali Shirazi, Mehdi Aminyavari, Fabio Rinaldi, & Robert A. Taylor. 2014. "Exergetic, economic and environmental analyses and multi-objective optimization of an SOFC-gas turbine hybrid cycle coupled with an MSF desalination system." *Desalination* 334(1): 46–59.

Ni, Meng. 2012. "Modeling of SOFC running on partially pre-reformed gas mixture." *International Journal of Hydrogen Energy* 37(2): 1731–1745.

Norouzi, Nima. 2020. "4E Analysis and design of a combined cycle with a geothermal condensing system in Iranian Moghan diesel power plant." *International Journal of Air-Conditioning and Refrigeration* 28(3): 2050022.

Norouzi, Nima. 2021a. "4E Analysis of a fuel cell and gas turbine hybrid energy system." *Biointerface Research in Applied Chemistry* 11: 7568–7579.

Norouzi, Nima. 2021b. "The Pahlev Reliability Index: A measurement for the resilience of power generation technologies versus climate change." *Nuclear Engineering and Technology* 53(5): 1658–1663.

Norouzi, Nima. 2021c. "Thermodynamic and exergy analysis of cogeneration cycles of electricity and heat integrated with a solid oxide fuel cell unit." *Advanced Journal of Chemistry – Section A* 4(4): 244–257.

Norouzi, Nima, Saeedeh Choupanpiesheh, Saeed Talebi, & Hossein Khajehpour. 2021a. "Exergoenvironmental and exergoeconomic modelling and assessment in the complex energy systems." *Iranian Journal of Chemistry and Chemical Engineering (IJCCE).*

Norouzi, Nima, Hosseinpour Morteza, S. Talebi, & Maryam Fani. 2021b. "A 4E analysis of renewable formic acid synthesis from the electrochemical reduction of carbon dioxide and water: Studying impacts of the anolyte material on the performance of the process." *Journal of Cleaner Production* 293: 126149.

Norouzi, Nima, & Ghazal Kalantari. 2021. "An overview on sustainable hydrogen supply chain using the carbon dioxide utilization system of formic acid." *Asian Journal of Green Chemistry* 5(1): 71–90.

Norouzi, Nima, Ghazal Kalantari, & Saeed Talebi. 2020. "Combination of renewable energy in the refinery, with carbon emissions approach." *Biointerface Research in Applied Chemistry* 10(4): 5780–5786.

Norouzi, Nima, Maryam Fani, & Saeed Talebi. 2021. "Exergetic design and analysis of a nuclear SMR reactor tetrageneration (combined water, heat, power, and chemicals) with designed PCM energy storage and a CO_2 gas turbine inner cycle." *Nuclear Engineering and Technology* 53(2): 677–687.

Norouzi, Nima, & Saeed Talebi. 2020. "Exergy and energy analysis of effective utilization of carbon dioxide in the gas-to-methanol process." *Iranian Journal of Hydrogen & Fuel Cell* 7(1): 13–31.

Norouzi, Nima, Saeed Talebi, & Pouyan Najafi. 2021. "Thermal-hydraulic efficiency of a modular reactor power plant by using the second law of thermodynamic." *Annals of Nuclear Energy* 151: 107936.

Norouzi, Nima, Saeed Talebi, Maryam Fani, & Hossein Khajehpour. 2021c. "Exergy and exergoeconomic analysis of hydrogen and power cogeneration using an HTR plant." *Nuclear Engineering and Technology.*

Owebor, Kesiena, Chika Ogbonna Chima Oko, Ogheneruona E. Diemuodeke, & Oreva Joe Ogorure. 2019. "Thermo-environmental and economic analysis of an integrated municipal waste-to-energy solid oxide fuel cell, gas-, steam-, organic fluid-and absorption refrigeration cycle thermal power plants." *Applied Energy* 239: 1385–1401.

Padulles, Joel, Graham W. Ault, & Jim R. McDonald. 2000. "An integrated SOFC plant dynamic model for power systems simulation." *Journal of Power Sources* 86(1–2): 495–500.

Park, S. K., & T. S. Kim. 2006. "Comparison between pressurized design and ambient pressure design of hybrid solid oxide fuel cell–gas turbine systems." *Journal of Power Sources* 163(1): 490–499.

Pirkandi, Jamasb, Mostafa Mahmoodi, & Farhad Amanloo. 2015. "Thermodynamic modeling of an auxiliary power unit equipped with a tubular solid oxide fuel cell with application in aerospace power system." *Modares Mechanical Engineering* 15(6).

Radmanesh, Hamid, & Hamid Hadi. 2019. "Modeling and evaluation of technical, economical and environmental performance of molten carbonate fuel cell compared to micro turbine gas for the production of electricity and heat simultaneously." *Journal of Modeling in Engineering* 16(55): 411–426.

Rahimi, Mostafa 2018. "Exergy, economic, and environmental analysis of a trigeneration system based on a solid oxide fuel cell and gasifier." *Journal of Mechanical Engineering* 2(83): 317–328.

Ramezani, Fatemeh, Maryam Razmgir, Kiarash Tanha, Farinaz Nasirinezhad, Ali Neshasteriz, Amir Bahrami-Ahmadi, Michael R. Hamblin, & Atousa Janzadeh. 2020. "Photobiomodulation for spinal cord injury: A systematic review and meta-analysis." *Physiology & Behavior* 224: 112977.

Ranjbar, Faramarz, Ata Chitsaz, S. M. Seyed Mahmoudi, Shahram Khalilarya, & Marc A. Rosen. 2014. "Energy and exergy assessments of a novel trigeneration system based on a solid oxide fuel cell." *Energy Conversion and Management* 87: 318–327.

Reyhani, Hamed Akbarpour, Mousa Meratizaman, Armin Ebrahimi, Omid Pourali, & Majid Amidpour. 2016. "Thermodynamic and economic optimization of SOFC-GT and its cogeneration opportunities using generated syngas from heavy fuel oil gasification." *Energy* 107: 141–164.

Roy, Dibyendu, Samiran Samanta, & Sudip Ghosh. 2019. "Techno-economic and environmental analyses of a biomass based system employing solid oxide fuel cell, externally fired gas turbine and organic Rankine cycle." *Journal of Cleaner Production* 225: 36–57.

Shukla, Anoop Kumar, Sandeep Gupta, Shivam Pratap Singh, Meeta Sharma, & Gopal Nandan. 2018. Thermodynamic performance evaluation of SOFC based simple gas turbine cycle. *International Journal of Applied Engineering Research* 13(10): 7772–7778.

Singh, Devinder, Eduardo Hernández-Pacheco, Phillip N. Hutton, Nikhil Patel, & Michael D. Mann. 2005. "Carbon deposition in an SOFC fueled by tar-laden biomass gas: A thermodynamic analysis." *Journal of Power Sources* 142(1–2): 194–199.

Talebi, Saeed, & Nima Norouzi. 2020. "Entropy and exergy analysis and optimization of the VVER nuclear power plant with a capacity of 1000 MW using the firefly optimization algorithm." *Nuclear Engineering and Technology* 52(12): 2928–2938.

Tashakori-Miyanroudi, Mahsa, Kamran Rakhshan, Maral Ramez, Shghayegh Asgarian, Atousa Janzadeh, Yaser Azizi, Alexander Seifalian, & Fatemeh Ramezani. 2020. "Conductive carbon nanofibers incorporated into collagen bio-scaffold assists myocardial injury repair." *International Journal of Biological Macromolecules* 163: 1136–1146.

Tsatsaronis, George. 2007. "Definitions and nomenclature in exergy analysis and exergoeconomics." *Energy* 32(4): 249–253.

Vahed, Majid, Fatemeh Ramezani, Vida Tafakori, V. S. Mirbagheri, Akram Najafi, & Gholamreza Ahmadian. 2020. "Molecular dynamics simulation and experimental study of the surface-display of SPA protein via Lpp-OmpA system for screening of IgG." *AMB Express* 10(1): 1–9.

Valencia, Guillermo, Aldair Benavides, & Yulineth Cárdenas. 2019. "Economic and environmental multiobjective optimization of a wind-solar-fuel cell hybrid energy system in the Colombian Caribbean region." *Energies* 12(11): 2119.

Zhang, Xiuqin, Shanhe Su, Jincan Chen, Yingru Zhao, & Nigel Brandon. 2011. "A new analytical approach to evaluate and optimize the performance of an irreversible solid oxide fuel cell-gas turbine hybrid system." *International Journal of Hydrogen Energy* 36(23): 15304–15312.

Chapter 7

Fuel cell hybrid power system

Kriti Srivastava
Mechanical Engineering Department, IET Dr. R. M. L. Avadh University, Ayodhya, India

Abhinav Anand Sinha
PDPM Indian Institute of Information Technology, Design and Manufacturing, Jabalpur, India

7.1 INTRODUCTION

With ongoing challenges about emissions, energy stability, and potential oil supply, the international community is looking for petroleum free renewable fuels as well as more enhanced energy techniques to develop energy efficiency. A hydrogen-fueled fuel cell is one such system that improves efficiency while also generating environmentally friendly waste products (i.e., water). Hydrogen can be produced in numerous ways, including clean energy gasification with coal, steam reforming of methane and biomass, efficient solar hot water splitting, and so on (Semelsberger 2009).

Many researchers have worked individually on fuel cell, photovoltaic system, battery system, gas turbines, wind turbines, etc. Also some researchers try to integrate systems to each other and develop a hybrid model (Sharma et al. 2021). This kind of hybridization is for attaining better and improved performance of the system. Also, it is found that such hybrid systems are cost-effective. In this chapter, the hybridization of fuel cells with many other systems is discussed. The chapter is separated into two sections. First, the basics about fuel cells, solar panels, and batteries are discussed, because this is the additional system which can be installed with the existing conventional system to make it more efficient. Then in the second part, a hybrid system is considered.

7.2 FUEL CELL

A fuel cell (FC) is a system that uses chemical processes to produce electricity (Figure 7.1). Cathodes (positive electrodes) and anodes (negative electrodes) are located in a fuel cell. Chemical reactions occur at both the electrodes. A FC can produce electricity as long as there is supply of oxygen and hydrogen. An electrolyte is found in any cell. The electrically charged molecules are transported from one electrode to the other by this. In addition, catalysts are used to accelerate chemical processes to the electrodes. A little amount of direct current (DC) is generated by the fuel cell (Carrette et al. 2000).The various forms and properties of electrodes (Table 7.1) distinguish fuel cells primarily (Stambouli, Boudghene, & Traversa 2002).

7.3 SOLAR PANEL

As sunlight hits a photovoltaic (PV) cell, frequently called a solar cell (Figure 7.2), it can be absorbed, reflected, or transmitted through it. The solar cell is made of semiconductors,

DOI: 10.1201/9781003213741-7

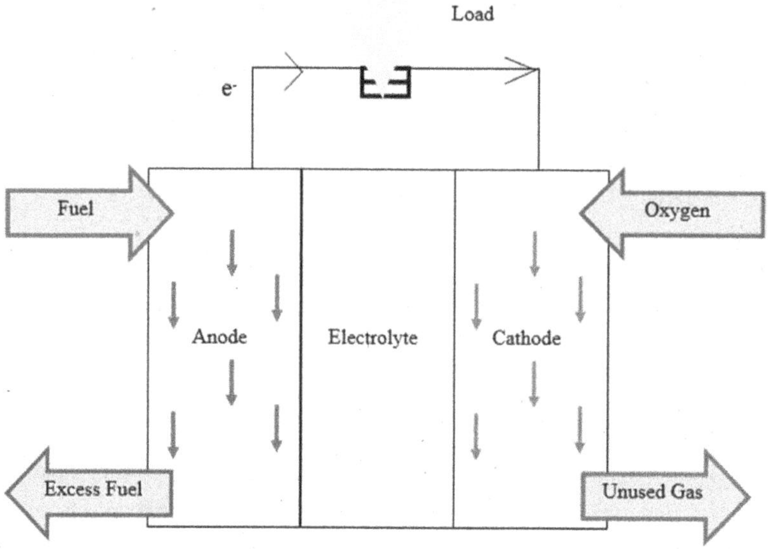

Figure 7.1 Fuel cell.

Table 7.1 Comparison of fuel cells

Characteristics	*AFC*	*PEMFC*	*DMFC*	*PAFC*	*MCFC*	*SOFC*
Operating temperature	90–100	50–100	60–200	150–200	600–700	600–1,000
System output	10–100	1–250	0.001–100	50–1,000	1–1,000	1–3,000
Electrical efficiency	>80	70–90	Approx. 80	>85	>80	<90
Cell voltage	1.0	1.1	0.2–0.4	1.1	0.7–1.0	0.8–1.0
Fuel	H_2	H_2	CH_3OH	H_2	H_2,CH_4, CO, other	H_2, CO, CH_4, other
Charge carrier	OH^-	H^+	H^+	H^+	CO_3^-	O^-
Reformer requirement	yes	Yes	-	yes	-	-
Disadvantage	Intolerance to CO_2	Slow process Poor heat and water management	Due to mixing poisons the cathode (Abdelkareem & Nakagawa 2006)	High cost due to PT catalyst Lower conductivity Long startup time (Hart & Hörmandinger 1998)	Corrosion Long startup time (Abdelkareem & Nakagawa 2006)	

Sources: Mekhilef, Saidur, and Safari (2012); Abdelkareem et al. (2021); Baroutaji et al. (2019).

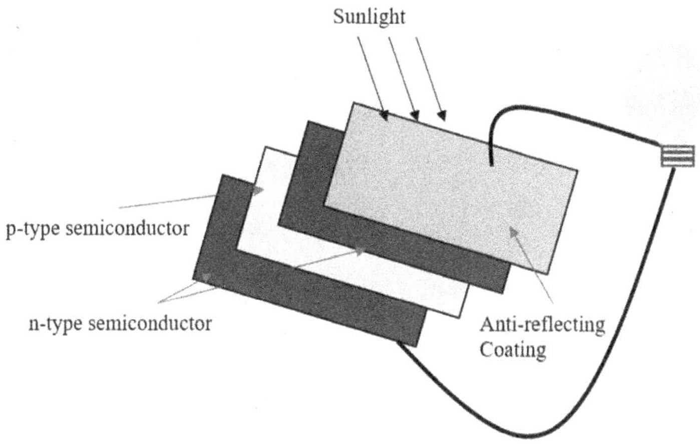

Figure 7.2 Solar cell.

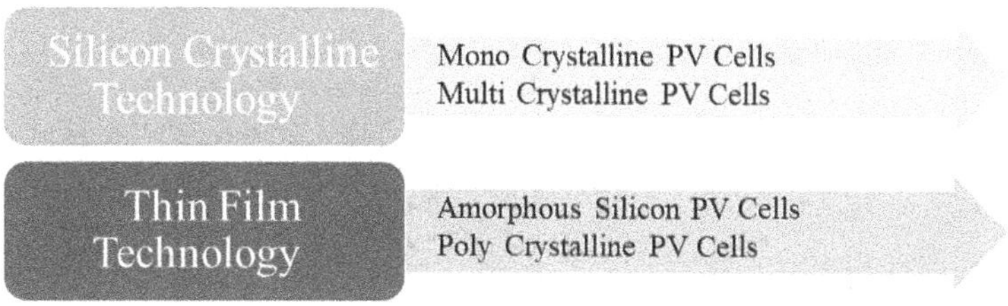

Figure 7.3 PV technology classification. (Source: Ramanujam & Singh 2017.)

which conduct electricity better than an insulating material but not as well as metal. When subjected to solar rays, the semiconductor absorbs the energy and transfers it to negatively charged particles, i.e., electrons, in the material. This extra energy makes the electrons pass as an electrical current through the material. This current is found by using conductive metal connections (the grid-like lines on photovoltaic panels) and would then be utilized to power the houses and supply the rest to the power station (Yadav et al. 2015).

Figure 7.3 shows the classification of the PV technology.

7.3.1 Battery

Regardless of whether or not it comprises one or more cells, the word *battery* has come to mean the substance that powers a computer. Electrochemical reactions occur at the electrodes of a cell, which transform chemical energy into electricity (Figure 7.4). For both types of batteries, the terms *anode* and *cathode* are used to refer to the positively and negatively charged electrodes, respectively (Pistoia 2005).

The battery capacity Q is calculated by multiplying the current and time. Its unit is ampere hour (Ah) or milliampere hour (mAh). It is given by

$$Q=It$$

The Watt hours of the battery is defined by the multiplication of the battery's potential difference and the rated ampere hours and is calculated as given by Sundén (2019).

Watt hours (Wh) = battery voltage Ah (Vt).

7.4 INTEGRATION OF FUEL CELL AND BATTERY

For the automobile sector, replacing fossil fuels with electricity and other non-conventional energy has been an unavoidable trend to decrease air pollution and energy usage. (Chen et al. 2020). Batteries provide high power density, a high throughput, and they don't pollute the atmosphere. They're popular in electric vehicles. On the other hand, batteries have certain drawbacks, such as less energy density, a prolonged waiting/charging period, and a limited life span (Li et al. 2018). Fuel cells, which have a huge efficiency and produce no emissions, are an excellent fossil fuel substitute. Hybrid vehicles have a lot of potential because they can merge the benefits of FC along with Li batteries (Lü et al. 2018; Gao et al. 2019).

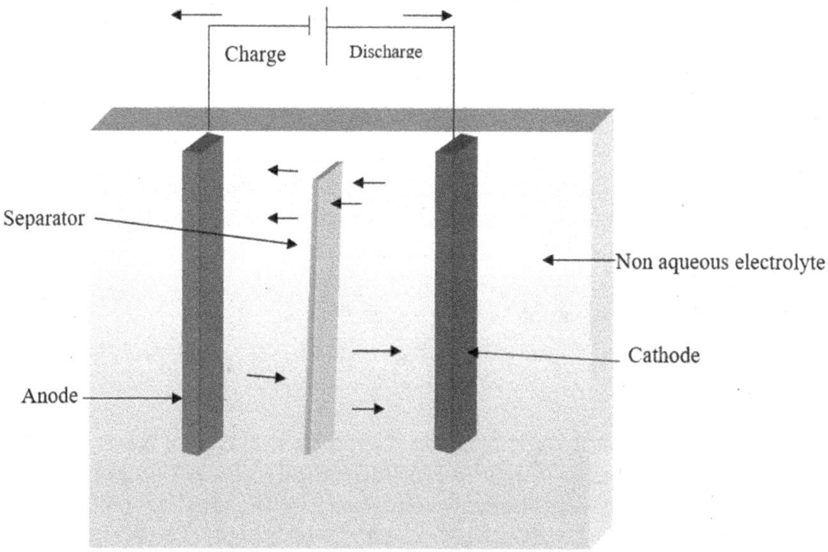

Figure 7.4 Working of battery. (Source: Balasubramaniam et al. 2020.)

Efficient energy management strategy (EMS) is the basis of hybrid vehicles' high performance (Sabri, Danapalasingam, & Rahmat, 2016). The series-parallel configuration is the most commonly known scheme for FC-battery power system, in which

a The FCs are not actually related to a mechanical transmission and gear drive train via a separate motor.
b The FCs link in sequence with the battery to run the motor, as well as straight in a parallel fashion to the motor in with the battery. The FCs are normally in charge of the main power source, with the battery acting as a backup.

The motor converts all of the electrical power from the fuel cells and battery into mechanical energy, which is then transmitted to the drive train device. For three driving conditions, the FCs and BAT are synchronized (Figure 7.5).

Case 1: The battery pack and fuel cell stacks work together to manage power peaks during acceleration.
Case 2: A portion of the energy produced during braking is transferred to the battery.
Case 3: When cruising, the fuel cell layer meets the key demand of power while also charging the battery box. The lifespan of fuel cells can be extended by using a BAT that can handle acceleration instantaneously (Ma et al. 2021).

Recently, several research organizations are keen on creating electric vehicles based on fuel cells (FCEVs) which are a potential solution to replace traditional ones. However, the FCEVs have some drawbacks in form of drive range, cost and energy consumption. Hybrid cars have more than one power source, which requires a management plan to assess the task profile of each source according to power and load (Zeng et al. 2018). An energy storage unit is a greater power polymer Li-ion battery store (Han, Charpentier, & Tang 2014).

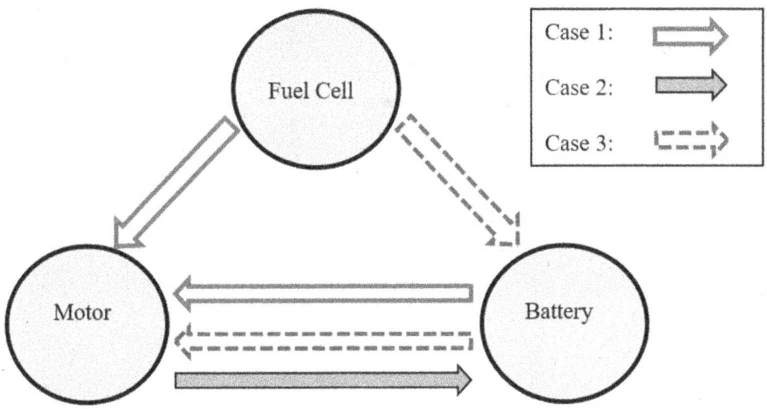

Figure 7.5 Energy flow diagram of the fuel cell-battery power system. (Source: Ma et al. 2021.)

7.5 INTEGRATION OF FUEL CELL AND PV CELLS

Solar panel power outcomes differ dramatically depending on solar radiation and air temperature in supplement with in active at night (Sharma, Jain, & Sharma 2015). When a large number of photovoltaic cells is attached to a distribution arrangement, the potential difference at load points will increase throughout the day due to backward power flow. As a consequence, a specific power storage system becomes required for taking out the maximum energy from the solar cells. While a low-cost lead battery has the potential to provide electric power for a few hours, its disposal has a significant environmental effect. Furthermore, since charge and discharge are often repeated, the lifespan will be short (Balog & Davoudi 2012). The chemical reaction of oxygen and hydrogen in a fuel cell will produce electricity (Shukla et al. 2018).

Different device topologies are formed by combining a photovoltaic cell, a fuel cell, and a solar inverter. As a consequence, the photovoltaics transform light into electricity directly. Since it aims to optimize the use of a renewable energy system (RES), the device ought to be an eco-sustainable solution. When solar auxiliary power is not required due to rain, a fuel cell is being used to retain the system stability at the same level as traditional ones while reducing the system's ecological consequences. When photovoltaic energy is insufficient, the fuel cell utilizes gases provided by an electrolyzer to meet user demand of load, acting as an auxiliary generator. The solar inverter is used to convert and distribute energy between the system's components.

Photovoltaic fuel cell hybrid power systems incorporate photovoltaic (PV), proton exchange membrane fuel cell (PEMFC), and electronic power conditioning units (PCU) to create a range of system topologies (Figures 7.6 and 7.7). As it aims to optimize the utilization of a RES, the device is supposed to be an environmentally sustainable approach. The user's demand load can be powered entirely by power generation if there is sufficient solar radiation.

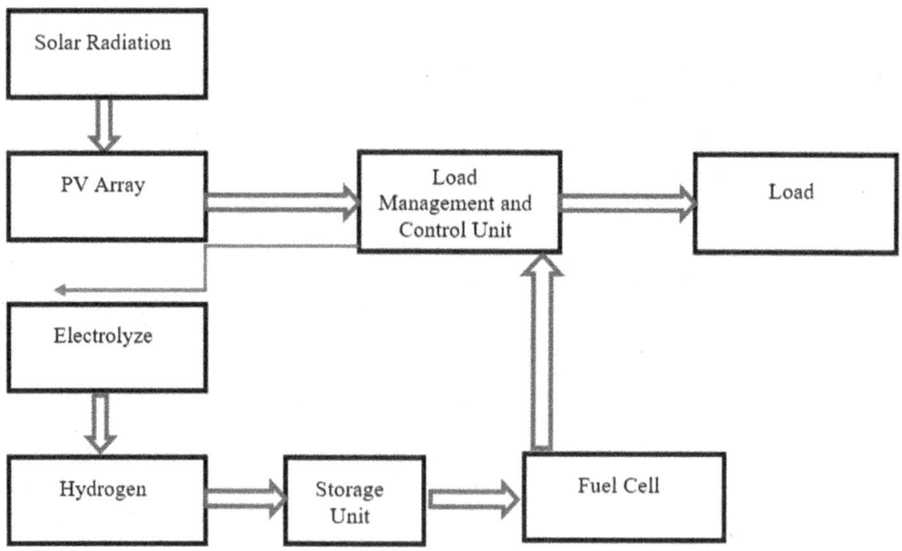

Figure 7.6 Integration of fuel cell and PV cells.

Auxiliary electricity is necessary during periods of low solar insolation (Rekioua, Bensmail, & Bettar 2014).

Integration of sustainable power systems with the power system is among the most effective strategies and effective ways for increasing renewable energy penetration in the energy mix at a low cost, reducing dependence on fossil fuels, and protecting the environment (reduction in greenhouse gas emissions). To meet a city's energy requirements, the fuel cell converter will effectively supplement the solar PV system's varying renewable resource (Rekioua, Bensmail, & Bettar 2014)

7.5.1 Integration of fuel cell, PV cell, and battery

There is an increment in the production and consumption cost of petroleum and, on the other hand, improvement in non-conventional energy sources technology. Due to which hybrid systems combining various renewable energy sources (HRES) are becoming common for rural locations generating electricity. These hybrid systems are gaining more and more attention for generation of power as they are economically considerable even for developing countries.

There is a requirement of continuous research and development in various RES like sun, geothermal, wind, and many other non-conventional energy technologies in order to improve efficiency, develop methods for correctly forecasting their production, and integrate them satisfactorily with other traditional energy sources (Shapiro et al. 2005).

Starting with the hydrogen container filled by the electrolyzer, which is supplied by surplus energy from solar insolation, energy can be stored and recovered. An intelligent hybrid model management system that can use solar cells and fuel cells as per system load while also saving photovoltaic and fuel cell energy is shown by Deshmukh and Deshmukh (2008). Due to changing environmental and operational conditions, the PV array's output characteristics differ significantly (Bigdeli 2015).The power output of a solar array deviates based on sun radiation and ambient temperature (Rezk, Kanagaraj, & Al-Dhaifallah 2020).

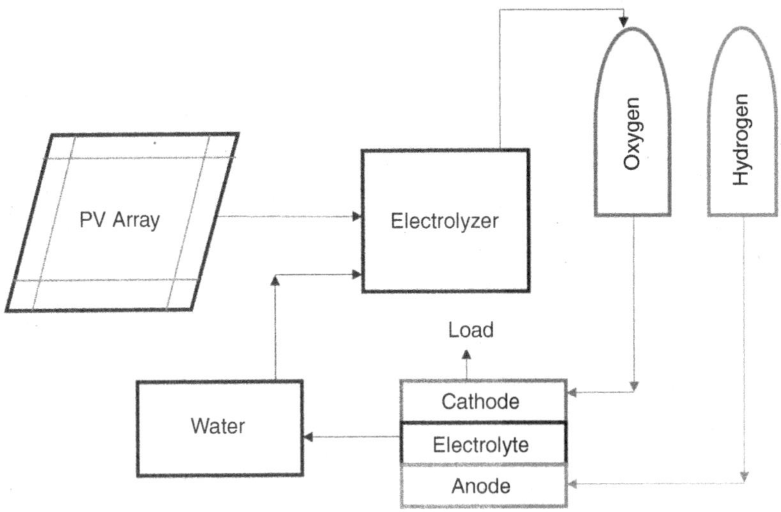

Figure 7.7 PV and fuel cell system. (Source: Ghenai & Bettayeb 2019.)

FC can be used in solar modules in two ways: as a standalone source of power or as component of a much larger storage system of energy for flexibility in season by using hydrogen from a specialized electrolyzer (Radziemska & Klugmann 2002; Agbossou et al. 2004).

A PV plate, PEMFC layers, and a lithium-ion battery make up the hybrid power system, which are all attached to the very same direct current, potential difference is by suitable converters and controls of DC-DC power. The hybrid power system's configuration is depicted in Figure 7.8. There are two primary energy sources: fuel cell stack and photovoltaic panel. Even though a battery is a power storage unit, it can also be utilized as a source of power when load needs more power. The solar cell supplies the load with as greater power as achievable. The FC's task is to provide the remaining power rating that solar panel can't provide to the load. The battery either provides momentary power to meet max out load demands or consumes residual power again from the main power supply. The battery, photovoltaic module, and fuel cell currents and voltages are all controlled, and the calculated potential difference and currents, each filtered correctly by a resistor capacitor filter, are supplied into the energy managing subsystem, that is utilized to organize the power conversion device. The charge controller arrangement (which includes power conversion system and related motor controller) regulates amount of power that flows from every unit and then supplies this power to charge the battery (Knaupp & Mundschau 2003).

The implementation of the proposed integrated power system yields satisfactory fiscal, energy, and environmental results. Enhanced ecological benefits may come at the expense of inferior monetary performance if the running strategy is changed. Not only can the addition of storage technology improve demand reaction, it can also react appropriately to time-changing energy cost by charging batteries (Ren et al. 2016).

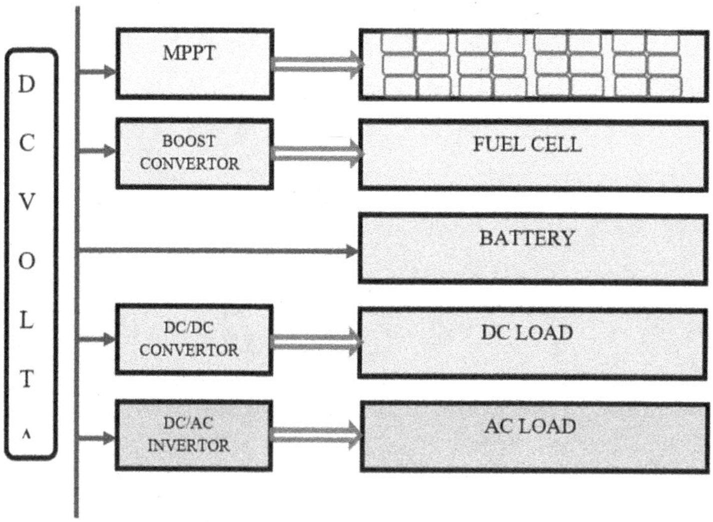

Figure 7.8 Schematic diagram of hybrid PV fuel cell power system. (Source: Jiang 2006.)

7.6 INTEGRATION OF FUEL CELL, PV CELL, AND WIND

It is suggested that combining a solar cell device as well as a wind generator with a FC is a viable option for ensuring electric energy output across the year. This integration aims to keep the battery bank's state of charge (SOC) as constant as possible to avoid blackouts and prolong battery life despite fluctuations in solar insolation and wind conditions. And when sunlight irradiance and wind speed differ, the control rule is to hold the sustainable voltage sources equal to the DC bus voltage. Also, to monitor the battery charging procedure depending on the charging current in order to increase the battery's life. A wind generator is a piece of apparatus that uses the wind to generate electricity (Bendary & Ismail 2019).

Samy, Mosaad, and Barakat (2021) perform cost analysis using HFA/HS and PSO optimization with the renewable sources. Using HFA/HS analysis PV, wind, and fuel cell cost was 0.0887 \$/kWh and using PSO, maximum LPSP is around 0.5%.

When employing HFA/HS, the cost of RES, like photovoltaic, wind, and FC, was 0.0887 \$/kWh, but it was 0.0888 \$/kWh when utilizing (particle swarm optimization) PSO at a loss of supply of power at probability of 0.5%. In addition, the convergence time for Hybrid Firefly and Harmony Search optimization methodology (HFA/HS) is 27.52 seconds and at the same LPSP max of 0.5%t, PSO took 112.40 seconds and PSO took 112.40 seconds. For the two introduced optimization strategies, the number of PV modules is reduced by 75% while the amount of wind turbines remains constant, the quantity of fuel cells is improved by 66.6%, and the amount of hydrogen (H_2) tanks and electrolyzers is increased twice (Samy, Mosaad, & Barakat 2021).

7.7 INTEGRATION OF FUEL CELL AND GAS TURBINE

For a long time, solid oxide FCs have been regarded as among the most successful energy changing techniques. The key advantage of this device is that it can work at high temperatures, resulting in extremely high electric power and heat efficiencies regardless of system size. Solid oxide FCs are particularly appealing for incorporation in a 70% efficient integrated system. As a result, both industry and academia have put in a huge amount of effort over the past few decades to develop and produce commercially active SOFC stacks. In reality, a variety of configurations were designed to account for almost all hybrid SOFC-GT plant configurations (Figures 7.9 and 7.11).

In real life application, in a traditional high pressure Brayton cycle, the combustor chamber can be exchanged with a SOFC array, which provides heat energy to the downstream to be extended in the turbine while still generating electric power (Zabihian & Fung 2009; McPhail et al. 2011; Zhang et al. 2010; Lim et al. 2008; Burbank, Witmer, & Holcomb 2009; Gengo et al. 2008). Sometimes the burner is placed at the exit of SOFC where the unburnt fuel of SOC can be further combusted in the afterburner.

In contrast, non-pressurized SOFC-GT cycles are not attached, also allowing for much easier control and activity. The configuration of a SOFC-GT plant (Table 7.2) is determined by some design characteristic such as:

- the temperature and pressure at which the performance of SOFC layer is depended;
- variety of fuel accessibility and particularities of the fuel producing subsystem (steam reforming, SR: either external or internal, indirect or direct; thermal decomposition, auto thermal reforming, etc.); and
- steam output needed for the process of reforming (negative electrode recirculation or heat recovery system).

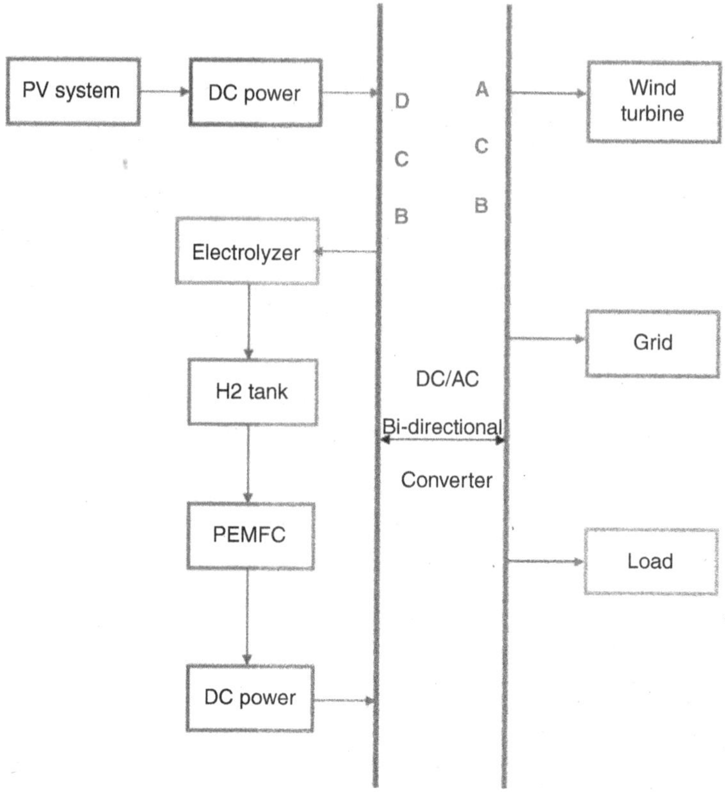

Figure 7.9 PV-SOFC-wind turbine layout. (Source: Samy, Mosaad, & Barakat 2021.)

Basic, intercooled, and/or reheated Brayton cycles and their combinations are the types of Brayton cycles. As a result, over the past few years, a number of researchers have built a variety of SOFC-GT configurations with the aim of increasing power conversion efficiency and/or lowering capital expenditure (Zabihian & Fung 2009; McPhail et al. 2011; Zhang et al. 2010).

The fuel processing system is an essential feature of the designing of a SOFC-GT power plant. In reality, SOFCs are particularly appealing because they can run on natural gas, which is much more economical and easier to handle for transportation than hydrogen fuel (e.g., utilized in PEM fuel cells). Steam for that process can be introduced via negative electrode recirculation or generated in a HRSG powered by exhaust gas of an SOFC stack. This choice has a huge effect on the overall machine architecture and performance. From an effective and economic standpoint, the anode recirculation structure worked best. However, a much more complicated management of the situation is needed (Henke et al. 2013).

Figure 7.10 shows a schematic diagram of a pressurized SOFC integrated with GT based hybrid device (a). An evaporator, pump, HE, SOFC, combustor, fuel processor, GT, and compressor make up the system. Steam and C_2H_5OH are injected into the evaporator, where they are vaporized. In the steam reactor, a mixture of C_2H_5OH and steam is heated to a recommended

Table 7.2 SOFC GT findings

Fuel cell – gas turbine hybrid system	*Findings*	*Findings*
SOFC-MGT	Pareto distribution of optimization, exergy efficiency is 60.7% and estimated cost of electrical Energy was calculated as 0.057 $\$kW^{-1}h^{-1}$	Sanaye & Katebi 2014
IRSOFC—GT	Fuel cell and combustion chamber is the main source of destruction of exergy	Bavarsad 2007

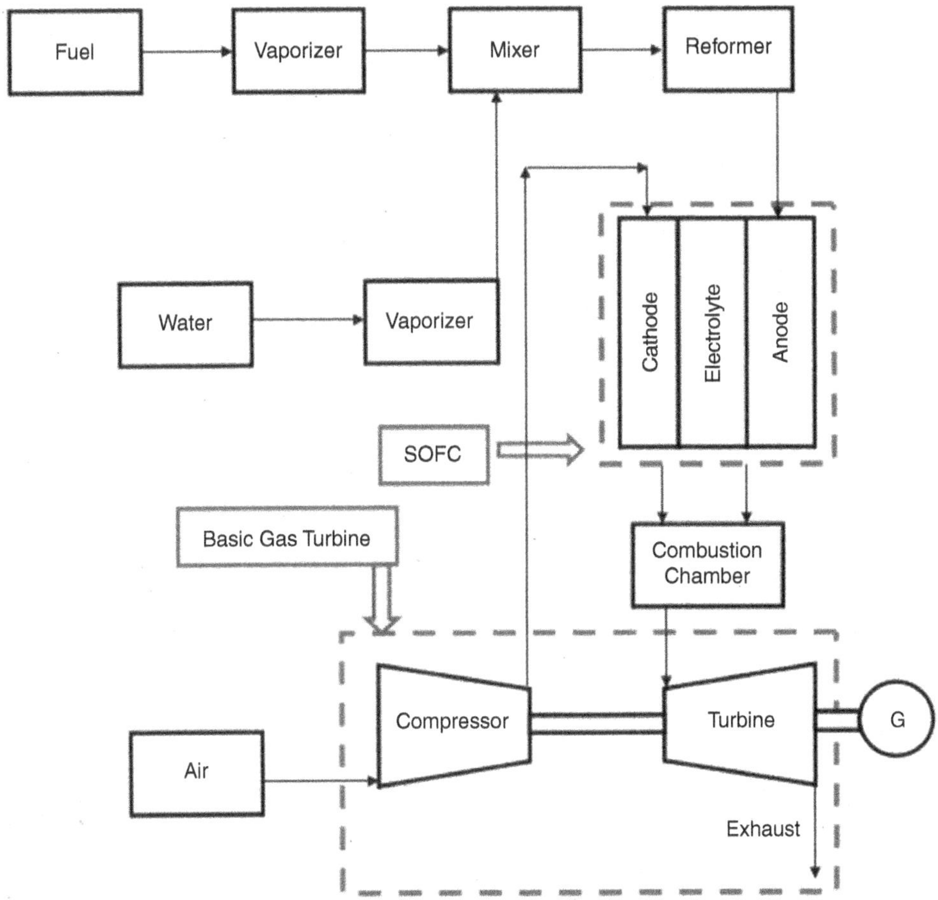

Figure 7.10 SOFC-GT.

value and converted into a syngas. After cleaning, the preheated syngas is fed into the SOFC stack. The electrochemical reaction produces electrical power and steam as H_2 in the synthesis gas combines with O_2 in pressurized gas. In general, SOFCs are unable to run at full fuel utilization. As a result, the SOFC downstream residue fuel is consumed in the combustion chamber, and the resulting downstream gas is supplied into the gas turbine to produce electric power. The GT outlet could take advantage of thermal source for other heat-requisite units owing to its large temperature. The SOFC system's heat management will help to reduce the need for heat generated. Separately, recycle ratio and operating pressure of a cathode exhaust gas, the output of the SOFC device in terms of electrical performance and thermal management is examined. The simulation results demonstrate that the pressurized SOFC system's optimum operating pressure is about 4–6 bar. The device can attain the maximum electrical efficiency in this state, whereas the recuperator heat is wasted (Nanaeda et al. 2010).

7.8 INTEGRATION OF FUEL CELL AND CHP

A heat retrieval model has been configured and incorporated into the SOFC in order to thermally combine the device with the heat input and monitor the thermal performance of the hybrid SOFC-CHP system. A duct burner, a high temperature heat exchanger, and a thermal storage container are the main contributors of the heat restoration physical model. Thermal energy generated by the SOFC is substituted by a burner to satisfy the consumer demand in the combined SOFC-CHP unit, with the TES serving as a heat storage reserve that separately sends hot water to the consumer. Water in a high temperature heat exchanger allows excess heat from the SOFC unit to improve the degree of hotness of water. If the heat from the SOFC is limited to assure the consumer's demand, the secondary flue gas duct burner heats it until it enters the water HE. If the surplus heat crosses the heating requirement, the excess heat is retained in the TES as hot water and released when required. In addition to the models of the system components the thermal storage facility was needed to satisfy the application's heating demand (Wang et al. 2021). The SOFC, which lies under HTFC, can operate typically from 700°C –1,000°C and is the better appealing for combined heat and power (CHP) applications among the different categories of fuel cells (Xu, Dang, & Bai 2013). The residential sector has recently seen a lot CHP plants used in different ways, which can simply turn into conventional heating systems. Due to the direct chemical energy conversion from the fuel into heat

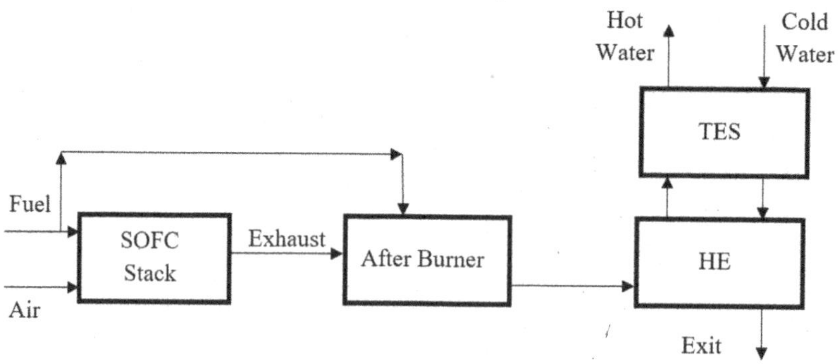

Figure 7.11 A layout of SOFC-CHP.

Table 7.3 Fuel cell – CHP hybrid system findings

Fuel cell – CHP hybrid system	*Findings*	*References*
Consider actual operating conditions	Gross and net electrical efficiency = 42.53% and 37.04% respectively.	Doherty, Reynolds, and Kennedy (2009)
Biogas-fed decentralized SOFC- (CHP)	The electrical efficiency is 55.6% and CHP efficiency is 85%.	Wang et al. (2021)
SOFC-CHP	Overall efficiency 73%	Pirkandi et al. (2012)
SOFC-CHP (commercial buildings)	Annual utility cost reduced by up to 14.5% and up to 62% CO_2 emissions reduced	Naimaster and Sleiti (2013)
SOFC-CHP	Efficiency 74%	Galvagno et al. (2016)
SOFC-mCHP	Electrical efficiency is approximately 40%. Overall efficiency obtained is 80%.	Kupeki (2015)
SOFC-mCHP	Electrical efficiency is approximately 40%. Overall efficiency obtained is 79%.	Braun, Klein, and Reindl (2006)

and power, FCs can be a more powerful, safer, cheaper, and easier way to generate energy (Doherty, Reynolds, & Kennedy 2009).

The biogas source is believed to be filtered by some cleaning methods that eliminate both sulfur and some other pollutants, and these methods are not accepted in this analysis. Until mixed with the steam generated by the evaporator, filtered biogas is supplied by a tank and preheated by adding heat taken from the exhaust of the post-burner and mixed with steam emitted by the evaporator. Before reaching the SOFC cathode, the compressed ambient air is heated by the high temperature exhaust. The fuel and O_2 from the surrounding air are electrochemically transformed inside the stack, resulting in the generation of electricity. The post-burner completely oxidizes the exhaust at anode of SOFC stack, which includes native CH_4, CO, and H_2. The pre-reformer, the HE preheating the biogas, the evaporator, and the air are all heated by the hot exhaust from the post-burner. And then for producing hot water, exhaust gas is used in the boiler. In this analysis, the impact of the biogas blower and pump on performance and efficiency are believed to be insignificant. Some of more findings of different researchers on SOFC-CHP hybrid systems are shown in Table 7.3. Efficiencies of some of the integrated systems have been shown in Table 7.4.

7.9 CONCLUSION AND FUTURE SCOPE

The combustion of fossil fuels results in greenhouse gas emissions which ultimately cause global warming. A long-term solution to this problem is to make use of non-conventional energy sources instead of conventional sources such as fossil fuels. Fuel cell prototypes of various types and sizes have been produced in the search for greater performance such as PEMFC, AFC, PAFC, MCFC, and SOFC. Only SOFC has capability to operate at high temperatures and allows the existing setup used for the fuel cell heating applications. Fuel cell hybridization has the following benefits:

- **Energy security:** Locally distributed and are more reliable. Improved efficiency and longer operating life. Easy for domestic application. Alternative of fossil fuel.
- **Economic impacts:** Fuel diversity makes it more attractive, and hence cost varies as per the type of fuel used. Efficiency of the system increases by reducing the cost of the system.

Table 7.4 Efficiency of some of integrated system

Reference	*Highlights*	*Results*
Zink, Lu, and Schaefer (2007)	A case analysis of the device is conducted, which includes a commercial LiBr absorption system and pre-commercial SOFC system. The combined system provides major benefits in both technological and environmental terms.	The overall efficiencies of system in various modes are expected to >87%.
De Souza et al. (2021)	In a biodiesel plant, a steam reforming system utilizing CHP glycerol was modeled.	The utilization of SOFCs for electricity generation was found to be unsuitable, owing to their high capital cost, which range from 41.8% to 65.1% of the overall expenditures depending on the expenditure scenario.
Ghorbani et al. (2020)	The ORC working fluids R407C and R404A are used to generate the extra fuel. There have been exergetic, exergoeconomic, and exergo-environmental analyses. MATLAB is used to carry out the integrated system.	The proposed optimum device has overall energy and exergy efficiencies of 49.42% and 46.83%, respectively
Guo et al. (2021)	For flight electric generation, an integrated turbofan engine/ SOFC/Al-H_2O reactor system is used.	It is observed that the specific fuel consumption of the SOFC-GT engine can be decreased by 10.30%, 14.41%, 18.98%, and 24.08% compared with engine utilizing turbofan when the electric power percentages are 15%, 20%, 25%, and 30% respectively.
Uchman (2021)	The economics of integrated systems is extensively discussed. Sensitivity analysis was carried out with the aim of achieving a fair price for micro-CHP.	Different approaches of the integrated prosumer device were examined in terms of energy demand self-coverage, with degree of self ranging from 28% to 98%.
Roushenas et al. (2020)	Energy, exergy analysis and CO_2 emissions of the planned system. Lowering emissions by combining fuel cells and energy storage units.	78% and 58% energetic and exergetic round trip efficiencies respectively. Obtaining electrical efficiencies of 43% in charge and 64% in discharge periods.

- **Environmental impacts:** Land waste is utilized for production of fuel (syngas) so it will reduce the wastage from the environment. Use of hydrogen or other renewable fuels makes system emission free or negligible emissions.

Since a variety of fuel is supplied to the fuel cell, the material selection for the anode is more challenging and is another research area. For high-temperature fuel cells such as MCFC and SOFC, the material of the system needs to be selected carefully to withstand that high operating temperature.

One major consideration is wastage of fuel; that is, after completing the fuel cell circuit, some of the fuel remains unburnt at the exit of fuel and has higher temperature than at entrance, so it can be further utilized for various other purposes. Fuel cell hybrid systems can be used in different modes of transportation like roadways, airways, space programs, etc. Many hybrid systems don't have any experimental setups due to their high initial cost.

ABBREVIATIONS

BAT	Battery	LTFC	Low temperature fuel cell
CHP	Combined heat and power	MCFC	Molten carbonate fuel cell
DC	Direct current	NG	Natural gas
EMS	Energy management strategy	ORC	Organic Rankine cycle
EV	Electric vehicle	PAFC	Phosphoric acid fuel cell
FC	Fuel cell	PCU	Power conditioning unit
FCEV	Fuel cell electric vehicle	PEMFC	Proton exchange membrane fuel cell
G	Generator	PSO	Particle swarm optimization
HE	Heat exchanger	PV	Photovoltaic
LPSP	Loss of power supply probability	RC filter	Resistor capacitor filter
HFA/HS	Hybrid Firefly and Harmony Search optimization technique	RES	Renewable energy system
HRES	Hybrid renewable energy system	SOC	State of charge
HRSG	Heat recovery steam generator	SOFC	Solid oxide fuel cell
HTFC	High temperature fuel cell	SOFC-GT	Solid oxide fuel cell-gas turbine
ITFC	Intermediate temperature fuel cell	TES	Thermal energy system

REFERENCES

Abdelkareem, Mohammad Ali, Khaled Elsaid, Tabbi Wilberforce, Mohammed Kamil, Enas Taha Sayed, & A. Olabi. 2021. "Environmental aspects of fuel cells: A review." *Science of the Total Environment* 752: 141803.

Abdelkareem, Mohammad Ali, & Nobuyoshi Nakagawa. 2006. "DMFC employing a porous plate for an efficient operation at high methanol concentrations." *Journal of Power Sources* 162(1): 114–123.

Agbossou, Kodjo, Mohanlal Kolhe, Jean Hamelin, & Tapan K. Bose. 2004. "Performance of a stand-alone renewable energy system based on energy storage as hydrogen." *IEEE Transactions on energy Conversion* 19(3): 633–640.

Balasubramaniam, Bhuvaneshwari, Narendra Singh, Swati Verma, & Raju K. Gupta. 2020. "Recycling of lithium from li-ion batteries." In Saleem Hashmi & Imtiaz Ahmad Choudhury, Eds., *Encyclopedia of Renewable and Sustainable Materials*, pp. 546–554. New York: Elsevier.

Balog, Robert S., & Ali Davoudi. 2012. "Batteries battery management and battery charging technology." In Robert A. Meyers, Ed., *Encyclopedia of Sustainability Science and Technology*, pp. 671–706. New York: Springer.

Baroutaji, Ahmad, Tabbi Wilberforce, Mohamad Ramadan, & Abdul Ghani Olabi. 2019. "Comprehensive investigation on hydrogen and fuel cell technology in the aviation and aerospace sectors." *Renewable and Sustainable Energy Reviews* 106: 31–40.

Bavarsad, Pegah Ghanbari. 2007. "Energy and exergy analysis of internal reforming solid oxide fuel cell–gas turbine hybrid system." *International Journal of Hydrogen Energy* 32(17): 4591–4599.

Bendary, Ahmed F., & Mohamed M. Ismail. 2019. "Battery charge management for hybrid PV/wind/fuel cell with storage battery." *Energy Procedia* 162: 107–116.

Bigdeli, Nooshin. 2015. "Optimal management of hybrid PV/fuel cell/battery power system: a comparison of optimal hybrid approaches." *Renewable and Sustainable Energy Reviews* 42: 377–393.

Braun, R. J., S. A. Klein, & D. T. Reindl. 2006. "Evaluation of system configurations for solid oxide fuel cell-based micro-combined heat and power generators in residential applications." *Journal of Power Sources* 158(2): 1290–1305.

Buonomano, Annamaria, Francesco Calise, Massimo Dentice d'Accadia, Adolfo Palombo, & Maria Vicidomini. 2015. "Hybrid solid oxide fuel cells–gas turbine systems for combined heat and power: a review." *Applied Energy* 156: 32–85.

Burbank Jr, Winston, Dennis Witmer, & Frank Holcomb. 2009. "Model of a novel pressurized solid oxide fuel cell gas turbine hybrid engine." *Journal of Power Sources* 193(2): 656–664.

Carrette, Linda, K. Andreas Friedrich, & Ulrich Stimming. 2000. "Fuel cells: Principles, types, fuels, and applications." *ChemPhysChem* 1(4): 162–193.

Chen, Xu, Guangdi Hu, Feng Guo, Mengqi Ye, & Jingyuan Huang. 2020. "Switched energy management strategy for fuel cell hybrid vehicle based on switch network." *Energies* 13(1): 247.

Deshmukh, Mukund K., & Sandip S. Deshmukh. 2008. "Modeling of hybrid renewable energy systems." *Renewable and Sustainable Energy Reviews* 12(1): 235–249.

De Souza, Túlio A. Z., Christian J. R. Coronado, José Luz Silveira, & Gabriel M. Pinto. 2021. "Economic assessment of hydrogen and electricity cogeneration through steam reforming-SOFC system in the Brazilian biodiesel industry." *Journal of Cleaner Production* 279: 123814.

Doherty, Wayne, Anthony Reynolds, & David Kennedy. 2009. "Modelling and simulation of a biomass gasification-solid oxide fuel cell combined heat and power plant using Aspen Plus." www.researchgate.net/publication/229014271_Modelling_and_simulation_of_a_biomass_gasification-solid_oxide_fuel_cell_combined_heat_and_power_plant_using_Aspen_Plus.

Galvagno, Antonio, Mauro Prestipino, Giovanni Zafarana, & Vitaliano Chiodo. 2016. "Analysis of an integrated agro-waste gasification and 120 kW SOFC CHP system: modeling and experimental investigation." *Energy Procedia* 101: 528–535.

Gao, Jinwu, Meng Li, Yunfeng Hu, Hong Chen, & Yan Ma. 2019. "Challenges and developments of automotive fuel cell hybrid power system and control." *Science China Information Sciences* 62(5): 1–25.

Gengo, Tadashi, Yoshimasa Ando, Tatsuo Kabata, Yoshinori Kobayashi, Nagao Hisatome, & Kenichiro Kosaka 2008. "Development of 200 kW class SOFC combined cycle system and future view." *Technical Review Mitsubishi Heavy Industries, Ltd* 45.

Ghenai, Chouki, & Maamar Bettayeb. 2019. "Grid-tied solar PV/fuel cell hybrid power system for university building." *Energy Procedia* 159: 96–103.

Ghorbani, S., Mohammad H. Khoshgoftar-Manesh, M. Nourpour, & Ana María Blanco-Marigorta. 2020. "Exergoeconomic and exergoenvironmental analyses of an integrated SOFC-GT-ORC hybrid system." *Energy* 206: 118151.

Guo, Fafu, Jiang Qin, Zhixing Ji, He Liu, Kunlin Cheng, & Silong Zhang. 2021. "Performance analysis of a turbofan engine integrated with solid oxide fuel cells based on Al-H2O hydrogen production for more electric long-endurance UAVs." *Energy Conversion and Management* 235: 113999.

Han, Jingang, Jean-Frederic Charpentier, & Tianhao Tang. 2014. "An energy management system of a fuel cell/battery hybrid boat." *Energies* 7(5): 2799–2820.

Hart, David, & Günter Hörmandinger. 1998. "Environmental benefits of transport and stationary fuel cells." *Journal of Power Sources* 71(1–2): 348–353.

Henke, Moritz, Caroline Willich, Mike Steilen, Josef Kallo, & K. Andreas Friedrich. 2013. "Solid oxide fuel cell–gas turbine hybrid power plant." *ECS Transactions* 57(1): 67.

Jiang, Zhenhua. 2006. "Power management of hybrid photovoltaic-fuel cell power systems." In *2006 IEEE Power Engineering Society General Meeting*. New Jersey: IEEE.

Knaupp, Werner, & Eva Mundschau. 2003. "Photovoltaic-hydrogen energy systems for stratospheric platforms." In *Proceedings of 3rd World Conference on Photovoltaic Energy Conversion*, vol. 3, pp. 2143–2147. New Jersey: IEEE.

Kupecki, Jakub. 2015. "Off-design analysis of a micro-CHP unit with solid oxide fuel cells fed by DME." *International Journal of Hydrogen Energy* 40(35): 12009–12022.

Li, Huan, Alexandre Ravey, Abdoul N'Diaye, & Abdesslem Djerdir. 2018. "A novel equivalent consumption minimization strategy for hybrid electric vehicle powered by fuel cell, battery and supercapacitor." *Journal of Power Sources* 395: 262–270.

Lim, Tak-Hyoung, Rak-Hyun Song, Dong-Ryul Shin, Jung-Il Yang, Heon Jung, Izaak C. Vinke, & Soo-Seok Yang. 2008. "Operating characteristics of a 5 kW class anode-supported planar SOFC stack for a fuel cell/gas turbine hybrid system." *International Journal of Hydrogen Energy* 33(3): 1076–1083.

Lü, Xueqin, Yan Qu, Yudong Wang, Chao Qin, & Gang Liu. 2018. "A comprehensive review on hybrid power system for PEMFC-HEV: Issues and strategies." *Energy Conversion and Management* 171: 1273–1291.

Ma, Shuai, Meng Lin, Tzu-En Lin, Tian Lan, Xun Liao, François Maréchal, Yongping Yang, Changqing Dong, & Ligang Wang. 2021. "Fuel cell-battery hybrid systems for mobility and off-grid applications: A review." *Renewable and Sustainable Energy Reviews* 135: 110119.

McPhail, Stephen J., Anja Aarva, Hary Devianto, Roberto Bove, & Angelo Moreno. 2011. "SOFC and MCFC: Commonalities and opportunities for integrated research." *International Journal of Hydrogen Energy* 36(16): 10337–10345.

Mekhilef, Saad, Rahman Saidur, & Azadeh Safari. 2012. "Comparative study of different fuel cell technologies." *Renewable and Sustainable Energy Reviews* 16(1): 981–989.

Naimaster IV, Edward J., & Ahmad K. Sleiti. 2013. "Potential of SOFC CHP systems for energy-efficient commercial buildings." *Energy and Buildings* 61: 153–160.

Nanaeda, Kimihiro, Fabian Mueller, Jacob Brouwer, & Scott Samuelsen. 2010. "Dynamic modeling and evaluation of solid oxide fuel cell–combined heat and power system operating strategies." *Journal of Power Sources* 195(10): 3176–3185.

Pirkandi, Jamasb, Majid Ghassemi, Mohammad Hossein Hamedi, & Rafat Mohammadi. 2012. "Electrochemical and thermodynamic modeling of a CHP system using tubular solid oxide fuel cell (SOFC-CHP)." *Journal of Cleaner Production* 29: 151–162.

Pistoia, Gianfranco. 2005. *Batteries for Portable Devices*. New York: Elsevier.

Popel', O. S., Tarasenko, A. B., & S. P. Filippov. 2018. "Fuel cell based power-generating installations: State of the art and future prospects." *Thermal Engineering* 65(12): 859–874.

Radziemska, E., & Eugeniusz Klugmann. 2002. "Thermally affected parameters of the current–voltage characteristics of silicon photocell." *Energy Conversion and Management* 43(14): 1889–1900.

Ramanujam, Jeyakumar, & Udai P. Singh. 2017. "Copper indium gallium selenide based solar cells: A review." *Energy & Environmental Science* 10(6): 1306–1319.

Rekioua, Djamila, Samia Bensmail, & Nabila Bettar. 2014. "Development of hybrid photovoltaic-fuel cell system for stand-alone application." *International Journal of Hydrogen Energy* 39(3): 1604–1611.

Ren, Hongbo, Qiong Wu, Weijun Gao, & Weisheng Zhou. 2016. "Optimal operation of a grid-connected hybrid PV/fuel cell/battery energy system for residential applications." *Energy* 113: 702–712.

Rezk, Hegazy, N. Kanagaraj, & Mujahed Al-Dhaifallah. 2020. "Design and sensitivity analysis of hybrid photovoltaic-fuel-cell-battery system to supply a small community at Saudi NEOM City." *Sustainability* 12(8): 3341.

Roushenas, Ramin, Amir Reza Razmi, Madjid Soltani, Morteza Torabi, Maurice B. Dusseault, & Jatin Nathwani. 2020. "Thermo-environmental analysis of a novel cogeneration system based on solid oxide fuel cell (SOFC) and compressed air energy storage (CAES) coupled with turbocharger." *Applied Thermal Engineering* 181: 115978.

Sabri, Mohd. Faizul M., Kumeresan A. Danapalasingam, & Mohd Fuaad Rahmat. 2016. "A review on hybrid electric vehicles architecture and energy management strategies." *Renewable and Sustainable Energy Reviews* 53: 1433–1442.

Saebea, Dang, Yaneeporn Patcharavorachot, Suttichai Assabumrungrat, & Amornchai Arpornwichanop. 2013. "Analysis of a pressurized solid oxide fuel cell–gas turbine hybrid power system with cathode gas recirculation." *International Journal of Hydrogen Energy* 38(11): 4748–4759.

Samy, Mohamed Mahmoud, Mohamed I. Mosaad, & Shimaa Barakat. 2021. "Optimal economic study of hybrid PV-wind-fuel cell system integrated to unreliable electric utility using hybrid search optimization technique." *International Journal of Hydrogen Energy* 46(20): 11217–11231.

Sanaye, Sepehr, & Arash Katebi. 2014. "4E analysis and multi objective optimization of a micro gas turbine and solid oxide fuel cell hybrid combined heat and power system." *Journal of Power Sources* 247: 294–306.

Semelsberger, Troy A. 2009. "Fuels – hydrogen storage: Chemical carriers." In *Encyclopedia of Electrochemical Power Sources*, pp. 504–518. New York: Elsevier.

Shapiro, Daniel, John Duffy, Michael Kimble, & Michael Pien. 2005. "Solar-powered regenerative PEM electrolyzer/fuel cell system." *Solar Energy* 79(5): 544–550.

Sharma, Achintya, Anoop Kumar Shukla, Onkar Singh, & Meeta Sharma. 2021. "Recent advances in gas/steam power cycles for concentrating solar power." *International Journal of Ambient Energy*. doi:10.1080/01430750.2021.1919552.

Sharma, Shruti, Kamlesh Kumar Jain, & Ashutosh Sharma. 2015. "Solar cells: In research and applications – a review." *Materials Sciences and Applications* 6(12): 1145.

Shukla, Anoop Kumar, Sandeep Gupta, Shivam Pratap Singh, Meeta Sharma, & Gopal Nandan. 2018. "Thermodynamic performance evaluation of SOFC based simple gas turbine cycle." *International Journal of Applied Engineering Research* 13(10): 7772–7778. www.ripublication.com/ijaer18/ijaerv13n10_69.pdf.

Stambouli, A. Boudghene, & Enrico Traversa. 2002. "Fuel cells, an alternative to standard sources of energy." *Renewable and Sustainable Energy Reviews* 6(3): 295–304.

Sundén, Bengt. 2019. *Hydrogen, Batteries and Fuel Cells*. Cambridge, MA: Academic Press.

Uchman, Wojciech. 2021. "The cost of increasing prosumer self-sufficiency." *Applied Thermal Engineering* 186: 116361

Wang, Yuqing, Lukas Wehrle, Aayan Banerjee, Yixiang Shi, & Olaf Deutschmann. 2021. "Analysis of a biogas-fed SOFC CHP system based on multi-scale hierarchical modeling." *Renewable Energy* 163: 78–87.

Xu, Han, Zheng Dang, & Bo-Feng Bai. 2013. "Analysis of a 1 kW residential combined heating and power system based on solid oxide fuel cell." *Applied Thermal Engineering* 50(1): 1101–1110.

Yadav, Akash, Gajendra Singh, Reza Nekovei, & Ramanujam Jeyakumar. 2015. "c-Si solar cells formed from spin-on phosphoric acid and boric acid." *Renewable Energy* 80: 80–84.

Zabihian, Farshid, & Alan Fung. 2009. "A review on modeling of hybrid solid oxide fuel cell systems." *International Journal of Engineering* 3(2): 85–119.

Zeng, Tao, Caizhi Zhang, Minghui Hu, Yan Chen, Changrong Yuan, Jingrui Chen, & Anjian Zhou, 2018. "Modelling and predicting energy consumption of a range extender fuel cell hybrid vehicle." *Energy* 165: 187–197.

Zhang, Xiongwen, Siewhwa H. Chan, Guojun Li, Hiang Kwee Ho, Jun Li, & Zhenping Feng. 2010. "A review of integration strategies for solid oxide fuel cells." *Journal of Power Sources* 195(3): 685–702.

Zink, Florian, Yixin Lu, & Laura Schaefer. 2007. "A solid oxide fuel cell system for buildings." *Energy Conversion and Management* 48(3): 809–818.

Chapter 8

Solid oxide fuel cell integrated blade cooled gas turbine hybrid power cycle

Tushar Choudhary
PDPM Indian Institute of Information Technology, Design and Manufacturing, Jabalpur, India

Sanjay
National Institute of Technology, Jamshedpur, India

Mithilesh Kumar Sahu
Gayatri Vidya Parishad College of Engineering (A), Visakhapatnam, India

Tikendra Nath Verma
Maulana Azad National Institute of Technology, Bhopal, India

8.1 INTRODUCTION

In any scenario, energy is a critical asset on a global scale as well as a crucial factor for sustainable growth and development (Nami, Anvari-Moghaddam, & Arabkoohsar 2020). The rising world population and growing demand for electricity accelerate urbanization and industrialization across the world, whereas around 70% of energy consumption correlates with cities. In the future, communities will face severe challenges, as around 80–86% of the global energy requirement depends on fossil fuels, and they are depleting rapidly (Zeynalian et al. 2020). From the conventional energy sources, greenhouse gas (GHS) emissions are regarded as a significant global challenge as they are the major stakeholder for depleting the ozone layer and global warming (Sadeghi & Askari 2019). Taking into account the various scenarios for potential global warming under four different representative pathways, by the end of this century, the global average temperature is expected to increase by 0.3°C to 4.8°C (Roushenas et al. 2020; Shukla et al. 2020). Therefore, implementation of several protocols and actions such as Kyoto and Montreal are required worldwide to crosscut the impact of fossil fuel usage and chlorofluorocarbons (CFCs) (Herrera-Orozco, Valencia-Ochoa, & Duarte-Forero 2021).

Compared with other systems like gas turbines, steam turbines, and internal combustion engines, fuel cells are durable, highly powerful, and noiseless, and their performance is not limited to the Carnot cycle (Larminie, Dicks, & McDonald 2003). Fuel cells' high-temperature exhaust gases would be utilized for space heating or other processes. Therefore, such fuel cells are highly applicable for CHP purposes (Roushenas et al. 2020). Moreover, a fuel cell can be integrated with the gas turbine cycle for distributed energy generation markets. For this a high-temperature fuel cell would be required, i.e., solid oxide fuel cell (SOFC). There are various options for the effective integration for SOFC with gas turbine such as in regenerative gas turbine cycle, where hot gases from the turbine would be utilized to power

DOI: 10.1201/9781003213741-8

up the fuel cell, and the hot gases from the fuel cell would also be introduced to gas turbine to produce clean and perfect means for waste heat utilization (Chen et al. 2017; Sghaier, Khir, & Ben Brahim 2018; Shukla et al. 2018). SOFCs can also be integrated with a cycle in which they act as a heat source in low-temperature cycles such as the Rankine cycle (Sharifzadeh, Meghdari, & Rashtchian 2017), Kalina cycle (Chitgar et al. 2019), organic Rankine cycle (OCR) (Karimi et al. 2020), and Stirling engine (Rokni 2014).

For the production of highly efficient clean power, SOFC is the most viable and almost zero-emission technology. SOFCs offer fuel flexibility and handle various fuel gases such hydrogen, natural gas, syngas, and carbon monoxide to produce power through electrochemical reactions. Generally, SOFC consists of a thick electrolyte material that is sandwiched between two porous electrodes. As an electrolyte, yttria-stabilized zirconium oxide is used, and for electrodes, porous nickel and porous perovskite for anodes and cathodes. SOFCs are available in tubular as well as in plated forms (Zhang et al. 2018). In tubular SOFC, electrodes are widely spaced and have distant current transfer, resulting in high internal resistance and low power density; however, TSOFC has better durability and no sealing issues. In plate SOFC has higher power density but has sealing issues, and it further withstands the temperature range of 600–1,000°C depending upon the support structure, i.e., anode, electrolyte, and metal. In metal-supported SOFCs, cerium gadolinium oxide is used as the electrolyte and has great mechanical robustness with no loss of volumetric power density (Xue et al. 2019).

SOFCs provide a viable alternative to conventional coal-fired power plants, as they can operate using coal-fired syngas (mainly include CO and H_2) while also delivering high energy efficiency and a low level of GHS emissions (Adams et al. 2012; Nease & Adams 2014). Figure 8.1 shows a simple SOFC cell pattern that utilizes CO and H_2 as a fuel source (Choudhary, Sanjay, & Murty 2015). Despite several advantages, SOFCs are vulnerable to deterioration, which can shorten their lives by up to 1.5 years if the continuous output of power is sustained (baseload power production mode). Current density and fuel flow rate are the two most important factors that influence the life of a fuel cell and the degree of degradation (Tucker, Abreu-Sepulveda, & Harun 2014).

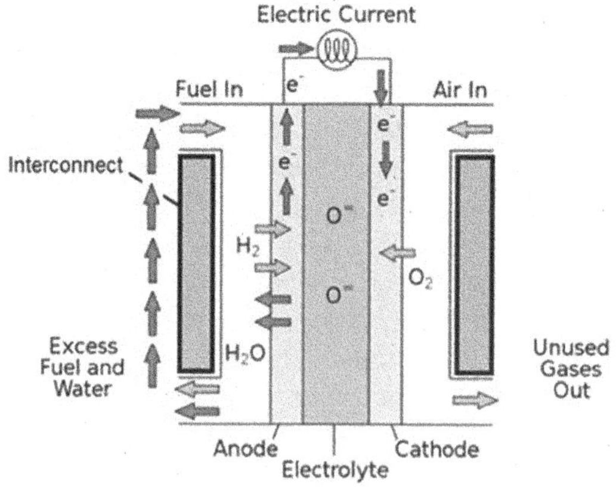

Figure 8.1 A generalized solid oxide fuel cell. (Source: Choudhary, Sanjay, & Murty 2015.)

Due to the high usage of fossil fuels in various industrial applications (Amador et al. 2017), many researchers are exploring other ways to generate clean energy using waste heat recovery technologies (Seyedkavoosi, Javan, & Kota 2017; Su & Shedd 2018). To achieve this, existing technology can be upgraded to hybrid by combining different power systems. For example, combining fuel cells and gas turbines. Choudhary et al. (Choudhary & Sanjay 2017; Choudhary et al. 2018) combined a blade-cooled SOFC with a gas turbine power cycle and achieved maximum efficiency and utilized the maximum available unused heat from the gas turbine outlet.

Various studies have been conducted for several SOFC-based hybrid systems and analyzed them with respect to the first thermodynamic law (energy), the second thermodynamic law (exergy), and lastly with economic aspect. All these are briefly discussed here.

When developing a hybrid energy conversion cycle, the SOFC's integration with the gas turbine (GT) is most promising. The SOFC operating temperature is close to the gas turbine exhaust temperature, allowing for near perfect thermal integration of the two energy conversion devices. This synergetic hybrid cycle should have an overall efficiency of over 65% (Calise et al. 2018).

Large power systems based on SOFC-GT have come a long way since the 1990s. Several studies (Zhang et al. 2010) show that the SOFC-GT hybrid cycle can achieve over 70% efficiency. Because SOFCs operate at high temperatures, they can directly use fuels like syngas, natural gas, and biofuels. Cooling of gas turbine blades, steam induction in combustion chamber, and cooling system of air inlet are all used to increase performance (Zhang et al. 2010). Different SOFC-GT hybrid cycle configurations, analyses and applications were undertaken over recent years in search of an optimal hybrid cycle for efficiency, energy and economy (Calise et al. 2006). Moreover, evaluating the presence of irreversibility and minimizing it is the key to system optimization.

Calise et al. (2006) have optimized the SOFC-GT plant using a single level approach. The obtained results of polarization are compared with data from Siemens-Westinghouse and finally the developed capital cost function in terms of a thermo-economic model for each component has been presented. Akkaya, Sahin, and Huseyin Erdem (2008) have carried out exergetic analysis of an SOFC/GT CHP system. The obtained results have been compared with the data from Siemens-Westinghouse. Their thermodynamic modeling of SOFC and GT is oversimplified and has not considered gas turbine blade cooling and component-level entropy generation.

Zhao, Shah, and Brandon (2011) compare the performance of SOFC-GT hybrid systems under two optimization strategies. In addition, they propose a methodology for selecting optimal synthesis/design parameters. Zhao et al. (2011) have presented optimal integration strategies for SOFC-GT systems fueled by syngas. In their work, numerical analysis has been carried out to evaluate the optimum system behavior. However, they have not accounted for the entropy minimization and turbine blade cooling. Zhang et al. (2011) have established a generic model for a SOFC-GT hybrid system in which multiple irreversibility that exists in a real hybrid system has been examined. They have considered an analytically derived expression of efficiencies and power outputs for both subsystems and hybrid system. Moreover, they also discussed the effect of irreversibility existing in the fuel cell and gas turbine on the performance of the hybrid systems. But they have ignored entropy generation minimization.

Najafi et al. (2014) have performed four E analyses (energy, exergy, economic, and environmental) for SOFC-GT system integrated to multi-stage flash desalination unit. They have considered exergetic efficiency and the total cost rate of the system as the objective function for optimization and reported minimization of total cost rate and maximization of exergetic efficiency. Meratizaman, Monadizadeh, and Amidpour (2014) have introduced an SOFC-GT based hybrid system for residential application fueled by natural gas and consequently performed techno-economic analysis. They observed that in Iran maximum power consumption takes place during hot and humid climatic conditions. Bakalis and Stamatis (2014) have reported a methodology for optimizing turbomachines for a SOFC-GT hybrid system. In their proposed procedure, optimum system performance has been observed without any alteration in the SOFC system which delivers optimum geometries of turbine and compressor using proposed optimization technique and turbomachinery design codes.

Mehrpooya et al. (2014) have proposed the optimum design and modeling of a SOFC-GT hybrid in which the model of SOFC has been analyzed by finite difference approach. The model of SOFC has been divided into five control volumes in order to evaluate the mass and heat transfer, electrochemical process, and temperature profile of anode, cathode, and interconnect of co-flow and counter-flow configuration. Facchinetti, Favrat, and Marechal (2014) have designed and optimized the SOFC-GT hybrid cycle for small-scale power generation for residential application. In their work for optimal configuration, heat exchanger network design was performed and the proposed design has exergetic efficiency higher than 65%.

Zhang et al. (2014) have dealt with performance optimization of SOFC-GT hybrid cycle on a theoretical basis. In their model, internal reforming of fuel occurs, and the combustor (GT) uses the unutilized fuel to heat up the working fluid of cycle until it reaches turbine inlet temperature (TIT). Moreover, they have also evaluated the optimal fuel flow rate for SOFC for achieving maximum efficiency. Ozcan and Dincer (2015) have examined the performance characteristics of an SOFC-based tri-generation system fueled by multi-gases derived from biomass gasification. They have observed that heating option from organic Rankine cycle enhances the cogeneration efficiency and utilizing exhaust gas also improves the efficiency by 54%.

Zhang et al. (2015) have examined the performance characteristics and optimum design parameters for syngas fueled molten carbonate fuel cell with GT hybrid system. The influence of fuel consumption and molar fraction of oxygen in oxidant on the hybrid cycle performance has been investigated. It has been observed that at 78% fuel utilization, maximum power output can be achieved. Khani et al. (2016) have conducted a multi-objective optimization approach to determine optimal design parameters for the indirect integrated SOFC with the gas turbine cogeneration system, in which maximum exergetic efficiency has been attained and the sum of the unit costs of products is minimum, i.e., 55.11% and 170.5 $/GJ respectively.

For the long-term degradation effect, Lai et al. (2021) designed and analyzed a SOFC-GT hybrid system. In their work, they compared standalone SOFC performance of fuel cell-gas turbine hybrid system; for steam bottoming cycle SOFC-GT hybrid system, 44.6% efficiency is achieved. Moreover, compared to SOFC operations, the hybrid cycle is less sensitive to fuel cell price.

Huang and Turan (2019) proposed a novel SOFC-GT system model that incorporates mechanical equilibrium. The proposed model tested under and off-design conditions and found that SOFC and hybrid system performance improved by 26.1% and 13.1%, respectively, from 0.13 to 0.25. But it has only a minor impact on the system's operation. Huan and Turan (2019) optimize SOFC-GT performance by varying parameters in energetic and exergetic performance. They also did a fuel sensitivity analysis and found that using CH_4, H_2, and CO gives the best efficiency of 56.1%, 60.7%, and 54.3%.

Wang, Lv, and Weng (2020) studied a SOFC-GT hybrid system using biogas and anode-combustor exhaust gas recirculation loops. Injector technology is used in the recirculation loop. An anode recycling loop improves the hybrid system's electrical performance and maintains the SOFC temperature gradient. Using an anode recirculation looping system also prevented carbon deposition in the reformer and thermal crack in the fuel cell. The combustor exhaust recirculation loop, on the other hand, allows for a wide temperature range. The hybrid system achieves 62.21% efficiency with anode and combustor recirculation of 0.4 and 0.425, respectively.

Kumar and Singh (2019) have conducted thermo-economic investigation for integrated gas turbine (GT) cycle, vapor absorption refrigeration (VAR) system and organic Rankine cycle (ORC) with SOFC system for power and cooling application. The results at 1,473 K and the cycle pressure of 16 maximum efficiency of 68.9% have been achieved for the GT-VAR-ORC-SOFC system. This is a 16.3% improvement compared to the SOFC-GT combined system.

Research on combining a fuel cell and a gas turbine cycle to optimize energy parameters or maximize the overall energy production through fuel cell integration has been outlined in the first scenario. However, there is a lack of literature available on the application of fuel cell integration with blade-cooled gas-turbine cycle.

Whereas in the area of cooled gas turbine, Sanjay et al. (Sanjay, Singh, & Prasad 2008, 2009; Kumari & Sanjay 2015; Sahu & Sanjay 2017; Mohapatra & Sanjay 2018) have done significant work, including energy, exergy, emission and thermo-economic analysis. Moreover, they also have integrated various thermal systems such as vapor absorption inlet cooling system, intercooled triple pressure reheat steam cycle, cogeneration, and steam cooled reheated gas-steam cycle. All these are briefly discussed here.

An analysis by Sanjay et al. (Sanjay & Prasad 2013; Sanjay, Singh, & Prasad 2009) compares the effects of seven different blade cooling techniques. There is an open gas turbine topping cycle and a triple pressure HRSG bottoming cycle. The analysis shows that the closed loop steam cooling technique achieves maximum plant efficiency and specific work output. Closed-loop steam cooling and air convection cooling also meet the minimum and maximum coolant flow requirements respectively. So the compressor pressure ratio has a minor impact on cooling flow requirements.

Kumari and Sanjay (2015) explores the main variables that influence the exergy and emission efficiency of basic and intercooled gas turbine (IcGT) cycles. They observed that IcGT has a higher rational efficiency as compared to the basic one, whereas the level of exergy destruction is quite low within the components of IcGT cycle such as compressor, gas turbine, and stack. Moreover, the emission level of CO and NOx make the IcGT more environmentally friendly.

Sahu and Sanjay (2017) compared the exergo-economics of BGT and CCGT. Exergo-economic analysis shows that CCGT is more efficient and produces more power than BGT at a cost that is 13% higher. Mohapatra and Sanjay (2018) investigated the exergy destruction of a triple pressure, reheat combined cycle. The inlet air cooling system improved the cycle's energetic and exergetic performance. A study by Sanjay and Prasad (2013) compares intercooled triple pressure reheat steam cycle to basic gas triple pressure reheat steam combined cycle. They concluded that an intercooled one outperforms both energetically and exergetically.

Various studies have examined the performance of the SOFC-GT system alone or in combination with other energy systems. No work has been reported on integrating SOFC with a blade cooled gas turbine. This chapter fills that gap by demonstrating SOFC integration with a blade cooled gas turbine cycle. The proposed hybrid configuration has been subjected to a thorough thermodynamic evaluation. Following are some key benefits of the proposed hybrid configuration.

- Thermodynamic model for effective integration of high-temperature fuel cell (SOFC) with gas turbine.
- Integrating a fuel cell to a traditional gas turbine loop improves its thermodynamic performance significantly.
- The integrated system has enough potential to supply simultaneous power for mid-scale distributed generation applications.
- The integrated system provides adequate heat and power from the fuel cell and also enhances the overall power generation.

Employing a high-temperature fuel cell as a link between the recuperator and combustor recovers the waste heat of high-temperature exhaust gases from a gas turbine and applies it to heat up the released air and fuel. From this, the fuel consumption gets reduced to some extent, and the waste heat would be used to reach the desired turbine inlet temperature. To model the system, a code was developed in MATLAB software for:

- Quantifying energy production, sensitive response, and waste heat utilization.
- Analyzing the impact of key variables on hybrid system performance.
- Exclusive comparison between the two power producing units, i.e., fuel cell (SOFC) and gas turbine.

8.2 SYSTEM DESCRIPTION

The hybrid cycle SOFC-intercooled recuperated blade cooled GT scheme is depicted in Figure 8.2. Intercooler, pump, compressor, recuperator, combustion chamber, fuel cell (SOFC), and gas turbine are among the six components of the proposed SOFC-GT hybrid cycle. Sysgas has been used to power the hybrid cycle's efficiency. The working fluid in this case is air, and its nature in terms of temperature, enthalpy, and enthalpy is shown in Figure 8.3. At state 1 the air enters the low pressure compressor and gets pressurized and again passes to the high-pressure compressor after getting intercooled through the intercooler, the pressurized air after the high-pressure compressor enters the recuperator and gets preheated before entering the fuel cell. However, the recuperator has been charged by utilizing the high grade of available waste heat at the gas turbine outlet. Consequently, the performance of hybrid cycle can be improved by recuperation process. In a fuel cell, pressurized fuel and air enters at the cathode and anode channel, where an electrochemical reaction takes place and reaction specifics are described in the modeling section of fuel cells. The fuel cell by-product consists of unused syngas, heat and electricity (DC). In the combustion chamber, the unutilized syngas is completely burnt to reach desired turbine inlet temperature (TIT). In the gas turbine, the blades of a turbine are exposed to high temperature and flue gases strain, resulting in a critically high-temperature oxidation and creeping blade failure. Therefore, the air/blade cooling system was used to maintain the temperature of the gas turbine blade within the permitted limit. The compressed air was bled from the compressor at various times in order to cool the turbine blade. Though the working fluid has enough potential and is allowed to expand within the gas turbine to generate useful work.

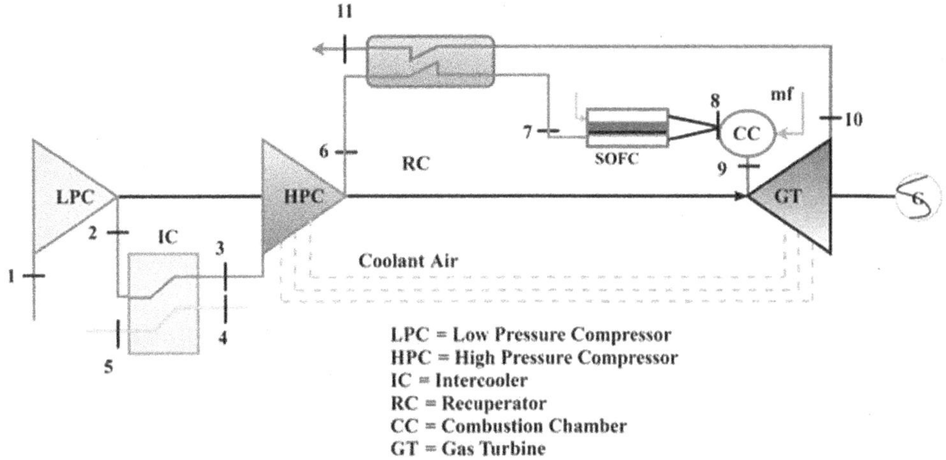

Figure 8.2 SOFC-intercooled recuperated gas turbine (blade cooled gas turbine) hybrid cycle schematic configuration.

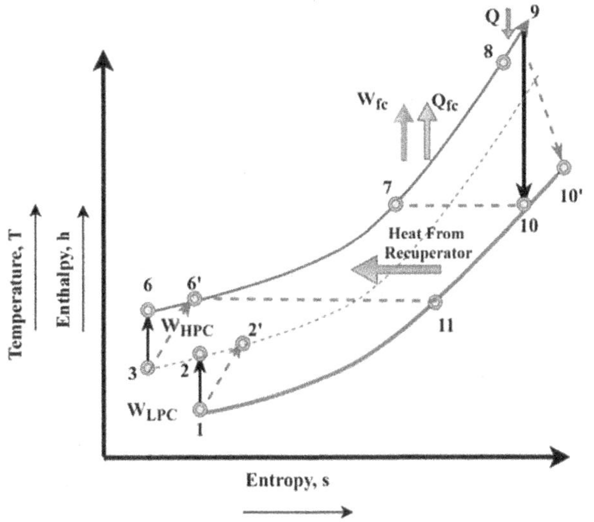

Figure 8.3 SOFC-GT hybrid cycle T–s diagram.

8.3 MODELING AND SIMULATION

The thermodynamic analysis of the SOFC-GT hybrid cycle was carried out using the cycle component equation. The hybrid cycle parameters and operational parameters used to measure thermodynamic properties at different cycle stages are listed in Table 8.1 (pressure, temperature and enthalpy). For the acquisition of all cycle state point properties, the following thermodynamic modeling of the cycle components was used.

Table 8.1 The SOFC-GT hybrid plant's primary operating parameters

Parameters	*Values*
Compressor efficiency, η_{comp}	92%
Turbine efficiency, η_{GT}	92%
Power turbine efficiency, η_{PT}	89%
Recuperator effectiveness, ε	80%
Combustor efficiency, η_{comb}	99.5%
Mechanical efficiency, η_{gen}	98.5%
Alternator efficiency	98.5%
Air inlet temperature,	288K
SOFC parameters	
Wall condition	Adiabatic
Fuel used	Syngas
Composition: Syngas $\{H_2=40, CH_4=21, CO=20, CO_2=18, N_2=1\}$	%
Air composition: $\{N_2=79, O_2=21\}$	%
Active surface area (A)	10,000 mm^2
Fuel utilization ratio (U_F)	0.85
Anode exchange current density ($i_{cd,a}$)	0.65 A/cm^2
Cathode exchange current density ($i_{cd,c}$)	0.25 A/cm^2
Effective gaseous diffusivity through the anode (D_{aeff})	0.2 cm^2/s
Effective gaseous diffusivity through the cathode (D_{ceff})	0.05 cm^2/s
Thickness of anode (τ_a)	500 μm
Thickness of electrolyte (τ_e)	10 μm
Thickness of cathode (τ_c)	50 μm
Electrolyte thermal conductivity	2 W/mK
Pressure losses	
Compressor inlet plenum loss = 0.5% of entry pressure	
Recuperator and intercooler gas/air side	1%
Gas turbine exhaust hood loss	3%
Fuel cell stack	3%
Afterburner	5%

Sources: Choudhary and Sanjay (2017); Choudhary et al. (2018); Buonomano et al. (2015).

$$h = \int_{T_a}^{T} c_p(T)\,dT \tag{8.1}$$

$$\varnothing = \int_{T_a}^{T} c_p(T)\frac{dT}{T} \tag{8.2}$$

$$s = \varnothing - R \cdot ln\left|\frac{P}{P_a}\right| \tag{8.3}$$

8.3.1 Compressor

The isentropic compressor has been considered in this research. For the hybrid power cycle shown, the blade cooled gas turbine model was used. To prevent back flow in the gas turbine unit, coolant (air) is extracted from the compressor at a sufficient pressure level to cool the gas turbine blades, as shown in Figure 8.1. In terms of state point and compressor work efficiency, the governing equations are as follows:

$$\eta_{comp} = \frac{W_{comp,ideal}}{W_{comp,actual}} = \frac{h_2 - h_1}{h_2' - h_1} \tag{8.4}$$

$$\frac{T_2}{T_1} = \left(\frac{P_2}{P_1}\right)^{\left(\frac{\gamma-1}{\gamma}\right)} \tag{8.5}$$

$$\frac{T_2}{T_1} = \left(r_{pc}\right)^{\left(\frac{\gamma-1}{\gamma}\right)} \tag{8.6}$$

The following work is needed for the compressor to generate a compression ratio of r_{pc}, according to the mass and energy balance for the system,

$$-\dot{W}_{comp} = -\dot{m}_e \cdot h_e + \sum \dot{m}_{c,j} \cdot h_{c,j} - \dot{m}_{c,in} \cdot h_{c,in} \tag{8.7}$$

8.3.2 Intercooler

Intercooling between multi-stage air compression is widely used to reduce compression work input. Intercooling reduces the temperature of the air before compression, reducing overall compression work and increasing the net work output of the gas turbine cycle proportional to the amount of compression work saved. As shown in Figure 8.4, the intercooler uses ambient water to cool. Intercooling occurs after partial compression at a pressure ratio determined by the total compressor pressure ratio.

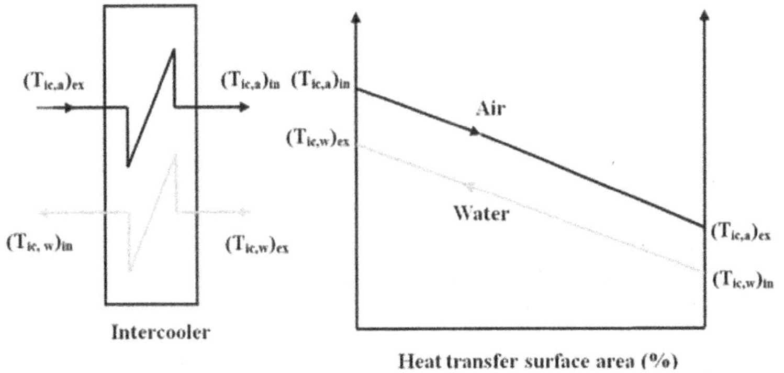

Figure 8.4 A counter-flow surface type intercooler schematic and temperature-heat transfer surface area diagram.

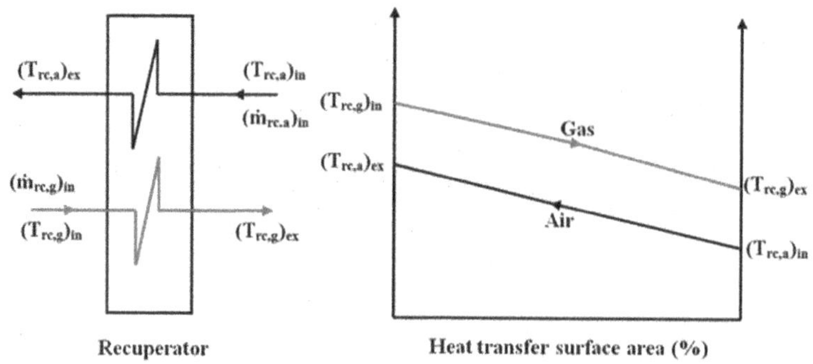

Figure 8.5 A counter-flow surface type recuperator schematic and temperature-heat transfer surface area diagram.

The effectiveness (ψ) of the intercooler is given by:

$$\psi = \frac{\left(T_{ic,a}\right)_{in} - \left(T_{ic,a}\right)_{ex}}{\left(T_{ic,a}\right)_{in} - \left(T_{rc,w}\right)_{in}} \tag{8.8}$$

The energy balance of the intercooler's control volume is calculated as follows:

$$m_{ic,g} c_{pw} \cdot \psi \cdot \left(T_{w,in} - T_{w,ex}\right) = m_{ic,a} c_{pa} \cdot \left(T_{a,in} - T_{a,ex}\right) \tag{8.9}$$

8.3.3 Recuperator

A recuperator is a gas-to-gas heat exchanger that can be integrated into the gas turbine cycle to utilize the exhaust's thermal energy. The recuperator is usually installed between the compressor and the combustor of a gas turbine cycle to raise the temperature of compressed air exiting the compressor, thus saving fuel (Figure 8.5). It has been used to charge the operating fluid and fuel using the gas turbine's waste heat. The recuperator's effectiveness (ε) is calculated as follows (Choudhary & Sanjay 2017):

$$\varepsilon = \frac{\left(T_{rc,a}\right)_{e} - \left(T_{rc,a}\right)_{in}}{\left(T_{rc,g}\right)_{in} - \left(T_{rc,a}\right)_{e}} \tag{8.10}$$

In the case of a recuperator, the mass and energy balance equation are:

$$\dot{m}_{rc,g} c_{pg} \cdot \psi\left(T_{g,in} - T_{g,e}\right) = \dot{m}_{rc,a} c_{pa} \cdot \psi\left(T_{g,in} - T_{g,e}\right) \tag{8.11}$$

8.3.4 Fuel cell (SOFC)

To build the SOFC-GT hybrid system, a high-temperature internal reformed SOFC fueled by syngas has been integrated with blade cooled gas turbine cycle. The SOFC model took into account the following major assumptions: adiabatic cell; equivalent temperature anode-cathode exit streams; cathode channel composed of O_2 and N_2; electrical flow-responsible H_2 ion transport; anode channel composed of CO, CO_2, CH_4, H_2, H_2O; flow-stream pressure decreases by 3%

The following are the electrochemical reactions that take place within the fuel cell anode and cathode:

$$H_2 \rightarrow 2H^+ + 2e^- \text{ at anode electrolyte inteface} \tag{8.12}$$

$$2H^+ + O^{2-} \rightarrow H_2O \text{ at anode electrolyte inteface} \tag{8.13}$$

$$CH_4 + H_2O \rightleftharpoons CO + 3H_2 \tag{8.14}$$

$$CO + H_2O \rightleftharpoons H_2 + CO_2 \tag{8.15}$$

Overall fuel cell reaction:

$$H_2 + 0.5O_2 \rightarrow H_2O \tag{8.16}$$

An electrochemical reaction describes the H_2O formation from hydrogen gas during an oxidation reaction that releases two electrons (Janardhanan & Deutschmann 2006; Zhu et al. 2005). These reactions appear in the three-phase boundary region (TPB), which is found in Ni or LSM electronic conducting materials and is where the ionic (YSZ) and gas-phase interact. Oxygen ions travel from the cathode to the anode via the electrolyte layer. The hydrogen electrons (Ni anode, LSM cathode) go through the electronic phase. Mass transport moves gas particles through the electrode open voids, as shown in Figure 8.6. The steam is recirculated, increasing hydrogen production and facilitating the exothermic water gas change reaction. The electrochemical potential of the reforming reaction is higher than that of the water gas shift reaction.

The basic solutions for the conservation of fuel cell mass and energy equations include the calculation of the voltage and current produced in the cell. The polarization (loss/irreversibility) inside the SOFC is induced by three sources: activation polarization (Vact), ohms (Vohm), and concentration polarization (Vconc). Different governing equations are used to measure the individual output characteristics of SOFC (Equations (8.17–8.26). In Choudhary and Sanjay (2016) the full procedures for computing cell voltage and loss were detailed.

$$E_{Nernst} = -\frac{\Delta G_T^0}{n_e F} + \frac{RT}{n_e F} ln\, ln\left(\frac{X_{H_2} X_{O_2}^{0.5}}{X_{H_2O}}\right) + \frac{1}{2}\frac{RT}{n_e F} ln\, ln\left(\frac{P}{P^o}\right) \tag{8.17}$$

E_{Nernst} varies from 0.99 V to 1.01 V.

The kinetics of electrochemical reactions generates activation polarization (Hajimolana et al. 2011). When the current is minimal, it becomes a significant loss because to initiate the electrochemical reactions at the electrodes–electrolyte interface, the reactants must cross the energy barrier referred to the activation energy, which results in polarization. The activation barrier is the product of several complex electrochemical reaction phases, with polarization

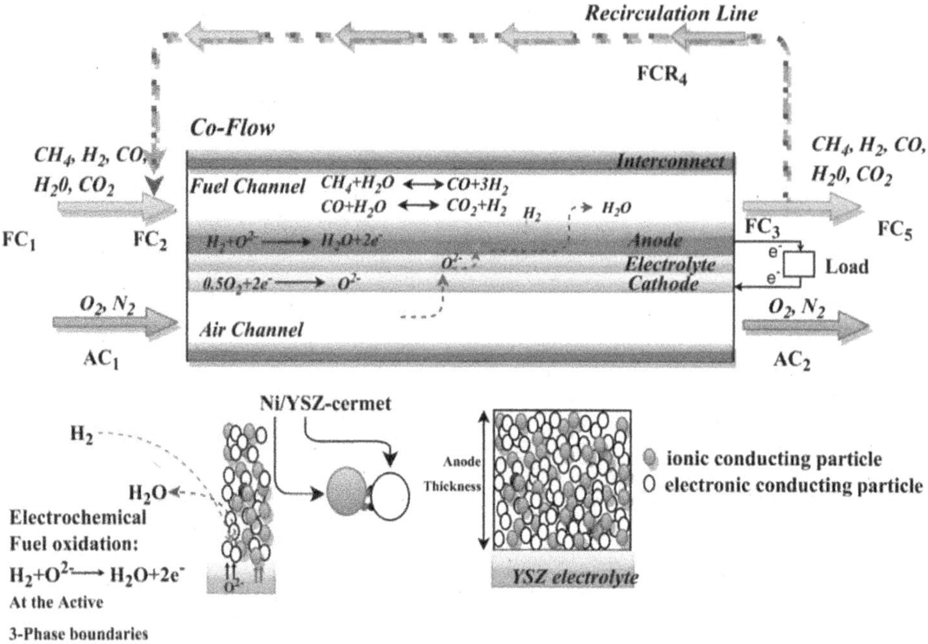

Figure 8.6 Schematic of the co-flow planar SOFC cell. (Source: Choudhary & Sanjay 2016.)

being the rate-limiting step in most cases. For accounting anode and cathode activation polarization, the Butler-Volmer equation can be used and it is given by:

$$E_{act} = E_{act,a} + E_{act,c} = \frac{RT}{F} \cdot \left(\frac{i}{2i_{cd,a}} \right) + \frac{RT}{F} \cdot \left(\frac{i}{2i_{cd,c}} \right) \tag{8.18}$$

Ohmic polarization is the presence of resistance for exchange of ions across the electrolyte, electrodes and current collector. This polarization generally associates with the contact resistance which presents in the components of fuel cell. It can be calculated as follows:

$$E_{ohm} = R_{ohm} \cdot i = \left(\frac{\tau_a}{\sigma_a} + \frac{\tau_{elec}}{\sigma_{elec}} + \frac{\tau_c}{\sigma_c} \right) = \left(R_{contact} + \sum\nolimits_k^n \rho_k \cdot L_k \right) i \tag{8.19}$$

Concentration polarization is caused by changes in the concentration of essential species as a result of mass transport processes.

Both diffusion between bulk flows and fuel channel surfaces and reactant and product transport through electrodes are common sources of mass loss in transport. It is thus dependent on the gases used and their spread distance (Hajimolana et al. 2011).

$$E_{conc} = E^{a}_{conc} + E^{c}_{conc} = \left[\frac{-RT}{n_e F} ln\, ln\left(1 - \frac{i}{i_{as}}\right) + \frac{RT}{n_e F} ln\, ln\left(1 + \frac{X_{H_2}}{X_{H_2O}}\frac{i}{i_{as}} \cdot \frac{P}{P^o}\right)\right] + \left[\frac{-RT}{n_e F} ln\, ln\left(1 - \frac{i}{i_{cs}}\right)\right] \tag{8.20}$$

Where, $$i_{as} = \frac{n_e F \cdot X_{H_2} \cdot D_{aeff} \cdot P}{RT \cdot \tau_a} \tag{8.20a}$$

$$i_{cs} = \frac{n_e F \cdot X_{O_2} \cdot D_{ceff} \cdot P}{\left(1 - \frac{X_{O_2} P}{X_{O_2} P^o}\right) RT \cdot \tau_a} \tag{8.20b}$$

SOFC's actual cell voltage is calculated by Equation (8.21):

$$\Delta E_{loss} = E_{act} + E_{ohm} + E_{conc} \tag{8.21}$$

$$E = E_{Nerst} - E_{loss} \tag{8.22}$$

The current produced by the fuel cell can be calculated as follows:

$$I = i \cdot A = 2F \cdot c = \frac{\dot{m}_f^{H_2}}{1 - r + r \cdot U_F} \tag{8.23}$$

In this case, r′ stands for the recirculation ratio, which helps keep the steam to carbon ratio of fuel entering the fuel channel low enough to avoid carbon deposition within the cell.

$$r = \frac{\dot{n}_{fuel,untilised}}{\dot{n}_{fuel,untilised}} \tag{8.23a}$$

U_F stands for the fuel consumption ratio, which is the proportion of hydrogen reacting electrochemically with hydrogen entering the system.

$$U_F = \frac{\dot{n}_{H_2,utilised}}{\dot{n}_{H_2,inlet}} \tag{8.23b}$$

Due to the presence of the irreversibilities described earlier, some heat generation occurs inside the cell stack. To calculate the rate at which heat is produced inside the cell stack, use the following equation:

$$Q_{gen,fc} = I \cdot \Delta E_{loss} [kW] \tag{8.24}$$

The SOFC's inlet and outlet flows are shown schematically in Figure 8.7.

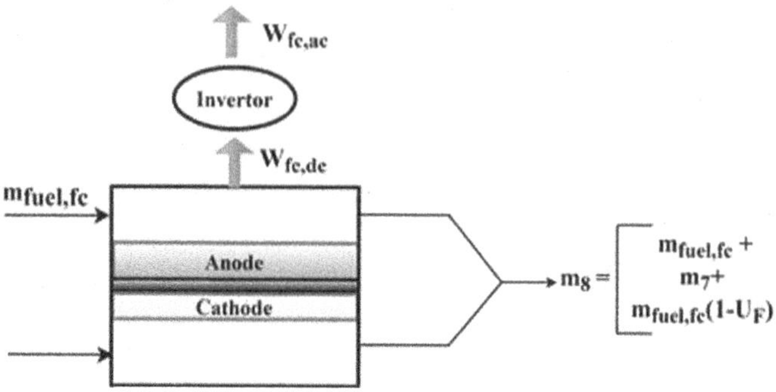

Figure 8.7 SOFC inlet and outlet flow. (Source: Haseli, Dincer, & Naterer 2008.)

The work produced in the fuel cell is computed by:

$$\dot{W}_{fc} = I \cdot E = I.E \tag{8.25}$$

At last, the fuel cell stack's electrical efficiency (ηcell) is calculated using:

$$\eta_{cell} = \underline{n} \cdot \frac{\dot{W}_{fc}}{\dot{m}_f \cdot LHV} \tag{8.26}$$

8.3.5 Blade cooled gas turbine

Combustion products can expand inside a gas turbine rotor. If the flue gases are not cooled, high temperature oxidation and creep cause blade progressive collapse. Thus, various cooling methods are used to keep turbine blade temperatures within acceptable limits (Sanjay, Singh, and Prasad 2008, 2009). This work used air film cooling (Figure 8.8). The blade surface temperature must be lower than the blade material temperature (Tb = 1,123 K) for blade cooling.

The coolant air mass flow rate (ς) is given by (Sanjay, Singh, and Prasad 2008, 2009):

$$\zeta = \frac{\dot{m}_c}{\dot{m}_g} = \left(1 - \eta_{iso,air}\right) . \frac{St_{in} \cdot S_g}{\varepsilon_c \cdot t \cdot cos\alpha} \times \frac{c_{pg}\left(T_{g,in} - T_b\right)}{c_{p,c}\left(T_b - T_{c,in}\right)} \times F_{sa} \tag{8.27}$$

$$\dot{W}_{gt} = \sum\nolimits_{row} \dot{m}_{g,in} \cdot \left(h_{g,a_1} - h_{g,b_1}\right)_{cooled} + \sum\nolimits_{row} \dot{m}_{g,in} \cdot \left(h_{g,in} - h_{g,e}\right)_{uncooled} \tag{8.28}$$

$$\eta_{hybrid} = \frac{\dot{W}_{plant}}{\dot{Q}_{total}} \tag{8.29}$$

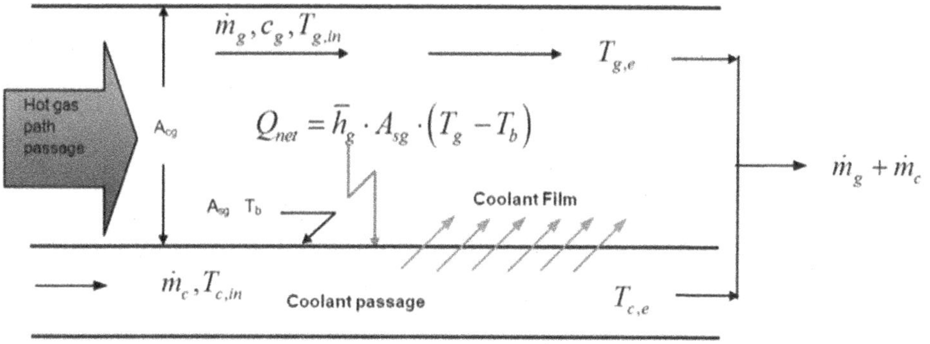

Figure 8.8 Air film cooling model.

$$\dot{W}_{plant} = \left(\dot{W}_{fc,ac} + \dot{W}_{gt,net}\right) \cdot \eta_{gen} \tag{8.30}$$

$$\dot{W}_{gt,net} = \dot{W}_{gt} - \frac{\left|\dot{W}_{comp}\right|}{\eta_{mech}} \tag{8.31}$$

$$\dot{W}_{fc,ac} = W_{fc} \cdot \eta_{invert} \tag{8.32}$$

$$\dot{Q}_{total} = \dot{m}_{fuel,fc} \times U_F \times LHV_{syn-gas} + Q_{comb} \tag{8.33}$$

8.3.6 Combustion chamber

The products from the fuel cell, i.e., unreacted fuel, air, steam and operating cycle fluids are further heated inside the combustion chamber (Figure 8.9) to reach the required turbine inlet temperatures.

The balance of mass and energy in the combustion chamber can be expressed as:

$$m_f \cdot LHV \cdot \eta_{comb} = m_{comb,e} h_{comb,e} - m_{comb,e} h_{comb,e} \tag{8.34}$$

$$\left(\dot{m}_8 + \dot{m}_{fuel,FC} \cdot U_F\right) h_8 + \dot{Q}_{comb} = \dot{m}_9 h_9 + \dot{Q}_{loss} \tag{8.35}$$

$$\dot{Q}_{comb} = \left[\dot{m}_{fuel,FC} \times (1 - U_F) + \dot{m}_{fuel,comb}\right] \times LHV \tag{8.35a}$$

$$\dot{Q}_{loss} = \left[\dot{m}_{fuel,FC} \times (1 - U_F) + \dot{m}_{fuel,comb}\right] \times (1 - \eta_{comb}) \times LHV \tag{8.35b}$$

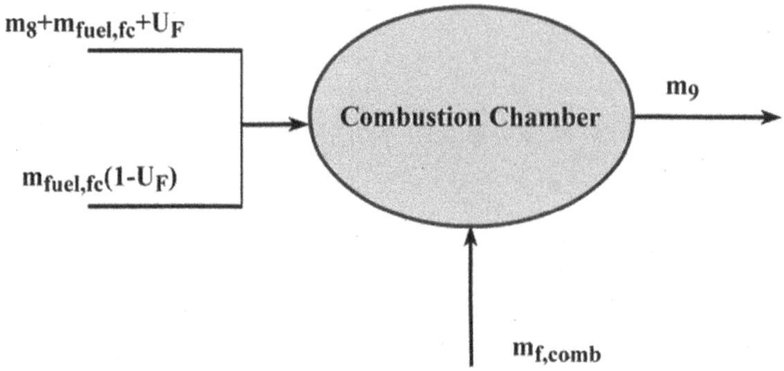

Figure 8.9 Combustor flow diagram with inlet and outlet. (Source: Haseli, Dincer, & Naterer 2008.)

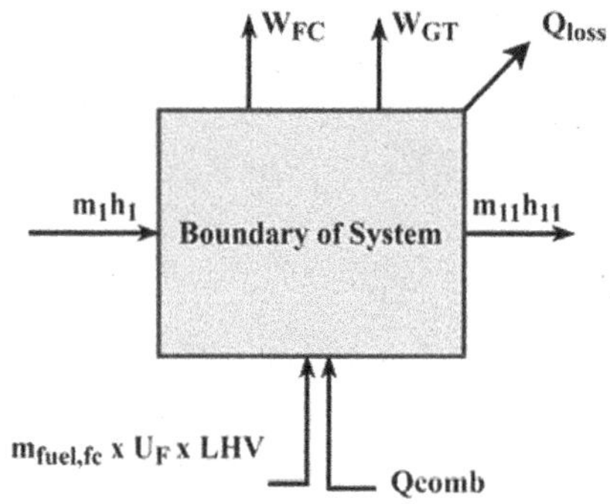

Figure 8.10 Energy balance of entire hybrid system.

8.4 RESULT AND DISCUSSION

8.4.1 Validation

A MATLAB code was developed for each component, based on the thermodynamic modeling as discussed, to simulate SOFC-GT hybrid cycle thermodynamic efficiency characteristics. In order to validate the performance of the integrated SOFC, Massardo and Lubelli (2000) have validated its performance with previously documented work. The details of validation of SOFC are tabulated in Table 8.2.

8.4.2 Influence of TIT on blade coolant requirement

Figure 8.11 shows the pattern of blade coolant mass requirements at various compression ratios and inlet temperature of the turbine. It has shown that the need for blade coolant increases as the TIT and compression ratio increase. It can be inferred from the graph that the variation of TIT on the requirement for blade coolant is more remarkable compared to the variation of the compression ratio. Due to mixing and cooling losses, the benefits of

Table 8.2 Validation of the performance characteristics of proposed hybrid plants

Output parameters	*Massardo and Lubelli 2000*	*Choudhary and Sanjay 2017 (SOFC- blade cooled GT)*	*Present (SOFC-ICGT)*
Hybrid plant specific work	1700 kJ/kg, at $r_{p,c}$ 20	1839 kJ/kg at $r_{p,c}$ 20	2218 kJ/kg at $r_{p,c}$ 20
Plant efficiency, %	72.56%, at $r_{p,c}$ 20	73.46%, at $r_{p,c}$ 20	74.13%, at $r_{p,c}$ 20
Power produced by GT, %	37%	33%	34%
Power produced by SOFC, %	63%	67%	66%

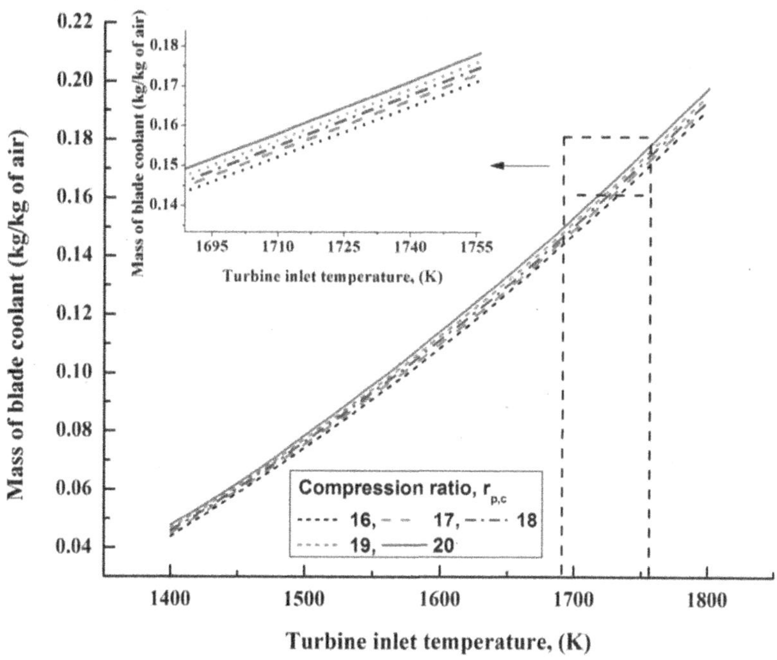

Figure 8.11 Variation of turbine inlet temperature w.r.t to mass of blade coolant.

reduced compressor work, increased fuel mass flow rate, and increased plant specific work are offset.

8.4.3 Sensitivity analysis

Figure 8.12(a) reveals the SOFC-GT hybrid cycle sensitivity analysis. The influence of change in compression ratio (r_{pc}) on thermodynamic performance has been analyzed. The result shows that there is a decrease in hybrid net specific work and SOFC specific work by 0.6223% and 0.299%, respectively when r_{pc} increases from 19 to 20. Whereas, decrease in r_{pc} from 19 to 18, hybrid net specific work, SOFC specific work, and SOFC efficiency increase by 0.415%, 1.121%, and 0.3398, respectively.

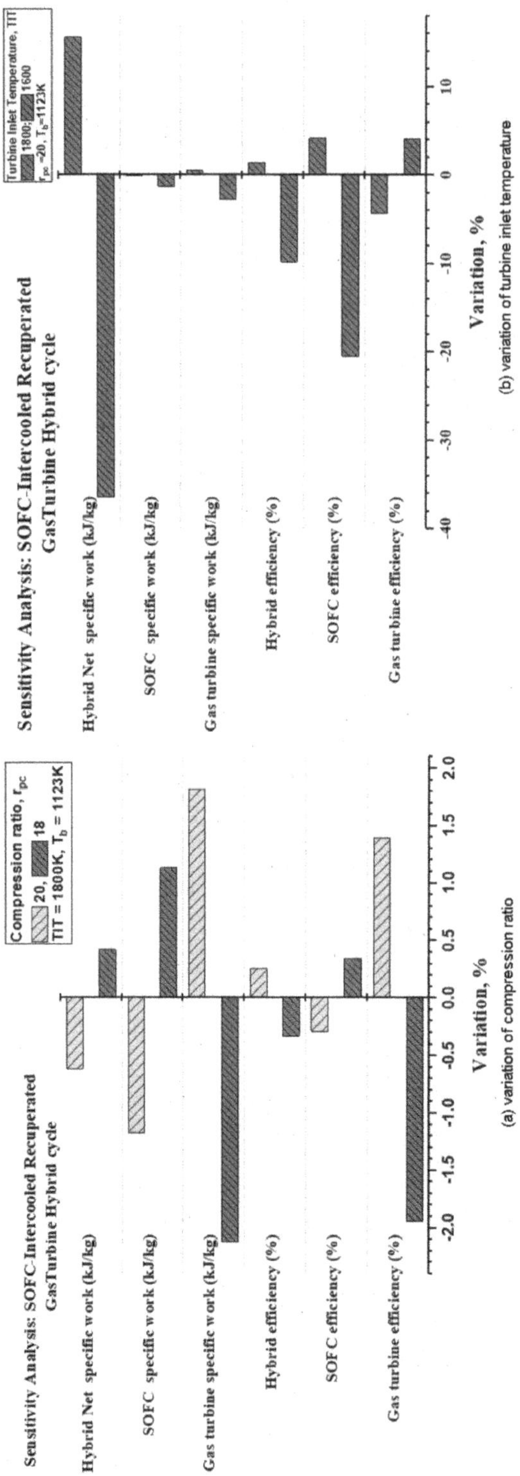

Figure 8.12 Sensitivity analysis.

Figure 8.12(b) illustrates the sensitivity analysis of SOFC-GT hybrid cycle. The influence of change in turbine inlet temperature (TIT) on thermodynamic performance has been analyzed. The result shows that for the decrease in TIT from 1,700 K to 1,600 K there is a decrease in hybrid efficiency, SOFC efficiency by 9.867% and –20.56%. Whereas the gas turbine efficiency increases by 4.044% respectively, whereas, increase in TIT from 1,700 K to 1,800 K, SOFC efficiency, gas turbine specific work, and hybrid net specific work increase by 4.099%, 0.4560%, and 15.486%, respectively.

From the sensitivity analysis, it can be revealed that the proposed hybrid system is more sensitive towards the TIT operation, as the SOFC is highly responsive towards the variation in TIT as it gets power up through recuperation.

8.4.4 Effect of fuel utilization ratio and recirculation ratio

Figure 8.13 shows the voltage and power patterns of different fuel utilization ratios. Because of the high degree of activation polarization across the anode area, the power density and voltage decrease with increasing fuel utilization ratio. The cathode ohmic polarization remains negligible. Thus, fuel composition affects fuel cell performance. Different fuels have different electrochemical reactions and activation polarizations.

Recirculation of fuel and air ensures uniform temperature distribution inside the fuel cell. Figure 8.14 shows the voltage and power density performance curves for various recirculation ratios. A higher recirculation ratio reduces fuel cell performance. The fuel channel reduces the

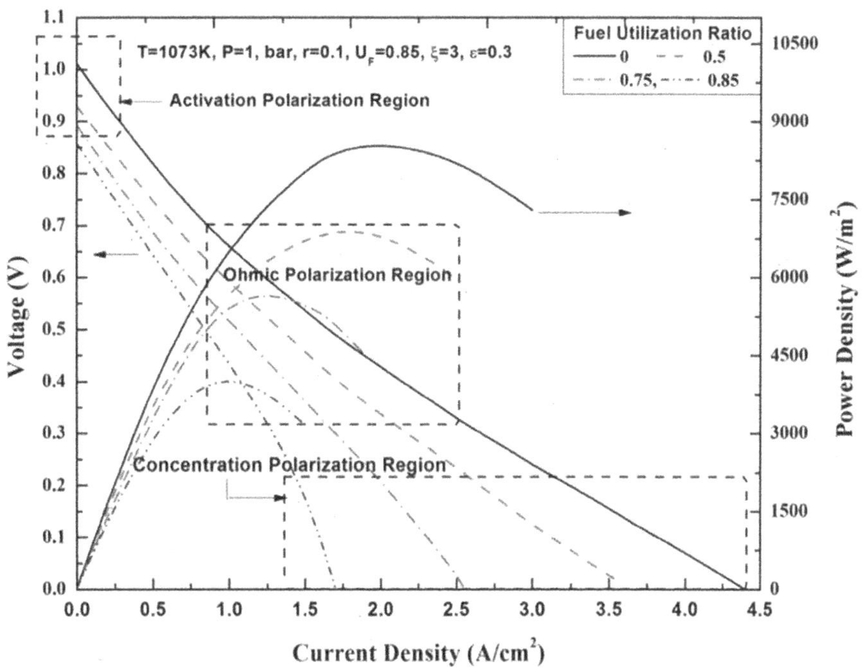

Figure 8.13 Effect of fuel utilization ratio on fuel cell performance.

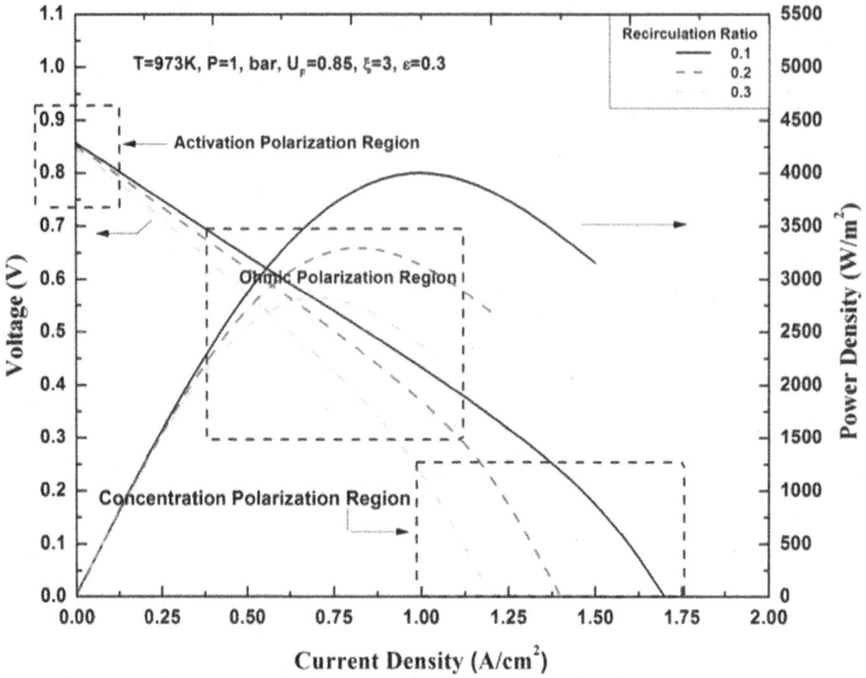

Figure 8.14 Effect of recirculation ratio on fuel cell performance.

molar concentration of H_2 and CO. Recirculation can achieve uniform temperature within the fuel cell, but dilutes the fuel stream, lowering performance.

8.4.5 Effect of fuel utilization ratio and recirculation ratio on fuel cell performance

Figure 8.15 shows the impact of compression ratio on hybrid plant-specific work. The hybrid plant's specific work decreases linearly with compression ratio. The level of plant-specific effort reduces as the compression ratio rises while TIT remains constant. Also, as TIT increases, the plant-specific work increases at constant $r_{p,c}$. As TIT increases, the fuel cell specific work increases faster than gas turbine specific work, causing the slope of the line in Figure 8.15 to decrease. The slope is justified because the fuel cell recovers unused heat from the gas turbine exhaust, so the operating temperature of the fuel cell rises with TIT.

8.4.6 Influence of compression ratio ($r_{p,c}$)

Figure 8.16 depicts hybrid plant efficiency and compression ratio. With increasing compression ratio, hybrid plant efficiency for TIT 1,550–1,650 K decreases. The performance of hybrid plants improves steadily as TIT rises above 1,650 K. Fuel cell output is low at lower TIT and $r_{p,c}$ because the GT exhaust temperature is far below the fuel cell's activation thermal energy. Fuel cell power output increases as TIT and $r_{p,c}$ increase. As a result, increasing TIT and $r_{p,c}$ improves cell efficiency.

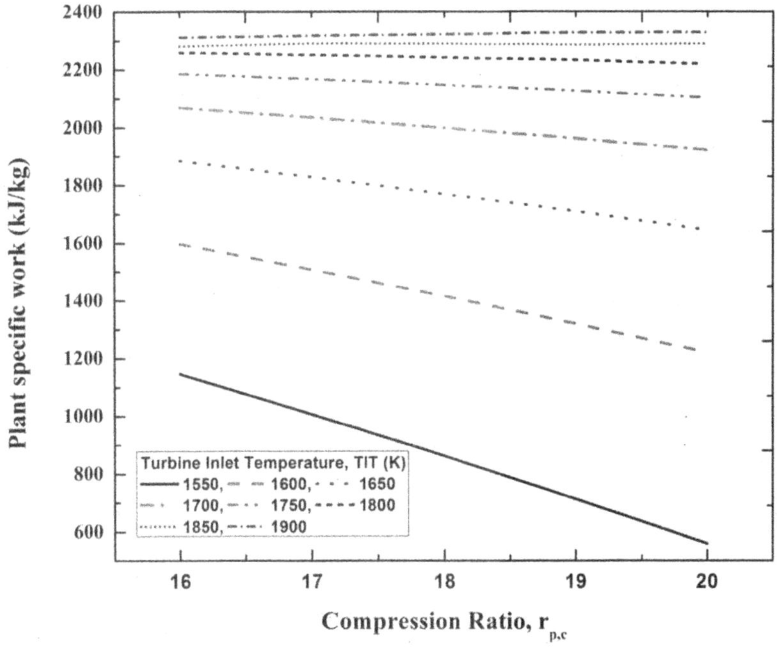

Figure 8.15 Influence of compression ratio on hybrid plant-specific work for various TIT.

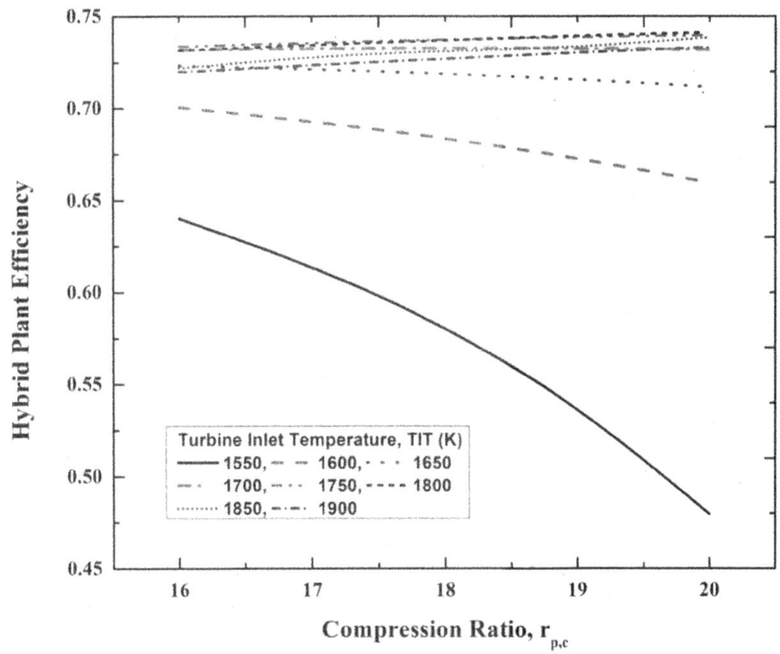

Figure 8.16 Variation of hybrid plant efficiency with compression ratio for various TIT.

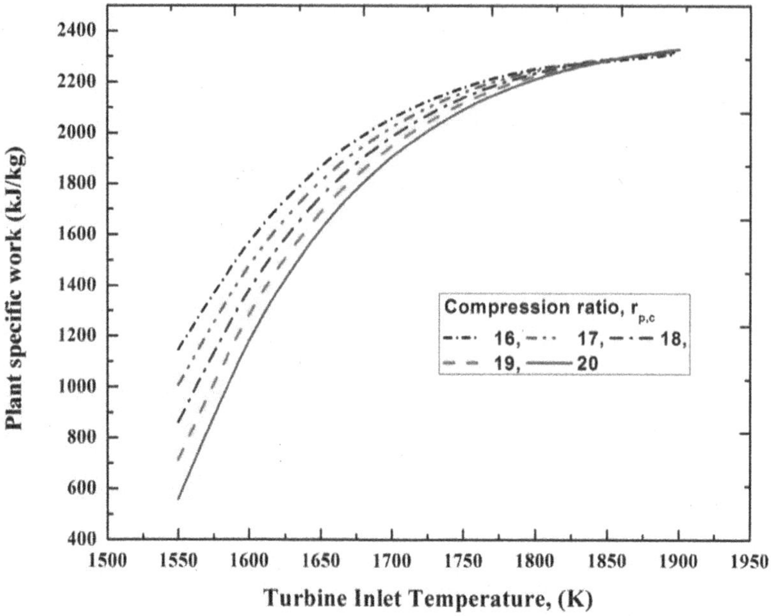

Figure 8.17 Variation of hybrid plant specific work with TIT.

8.4.7 Influence of turbine inlet temperature (TIT) on plant specific work

The relationship between the hybrid plant specific work with TIT is illustrated in Figure 8.17. The hybrid plant specific work has been observed to increase along a parabolic path as TIT increases. Furthermore, it has also been noted that as TIT increases, higher compression ratios have less impact and the different $r_{p,c}$ lines converge. This is because as TIT rises, fuel cell work output rises as well. The results show that the performance of fuel cells is significantly higher than that of gas turbines. Over the entire plotted range, an increase in TIT and compression ratio increases the fuel cell performance, while an increase in TIT tends to decrease gas turbine output (refer to Figure 8.20 for the nature of this variation). The improvement in fuel cell performance as TIT and compression ratio increase is attributed to a higher gas turbine exhaust temperature, which helps to boost the fuel cell operating temperature due to the recuperation process.

8.4.8 Influence of turbine inlet temperature (TIT) on hybrid efficiency

The impact of turbine inlet temperature (TIT) on hybrid plant performance is depicted from Figure 8.18. When TIT increases, hybrid plant efficiency improves until it reaches a maximum (optimum) and then begins to decline. This is because as TIT rises, the blade cooling demand rises, and as a result, additional coolant mass reduces turbine performance. This is due to coolant mixing losses and losses as a result of the main gas's thermal energy dropping due to cooling, canceling out the benefits of lower compression work, higher fuel flow rate, and

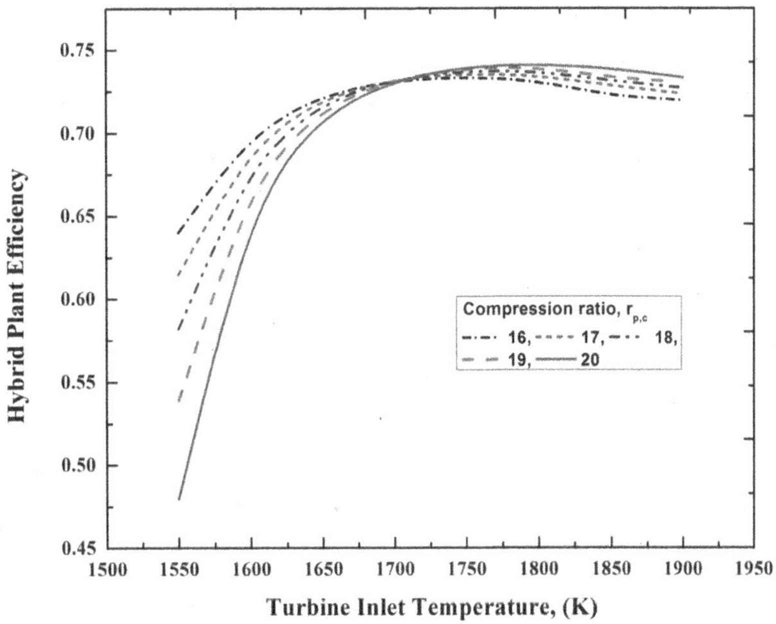

Figure 8.18 Influence of TIT on the hybrid plant efficiency.

greater GT work. When TIT is 1,670–1,850 K, plant efficiencies plummet. Figure 8.20 shows the differences in fuel cell and gas turbine performance.

8.4.9 Comparative analysis of power-generating units

An exclusive comparison of power generating units, such as SOFC and GT, at different TIT is shown in Figure 8.19. The performance of a gas turbine decreases dramatically as TIT increases, while the efficiency of a hybrid plant appears to increase as TIT increases (Figure 8.18). When $r_{p,c}$=20, and TIT=1,550 K are used, the maximum GT efficiency is observed. Similarly, with the TIT range of 1,850–1,900 K, the highest electrical efficiency for fuel cells can be found. At a TIT of around 1,850 K, the collection of curves reflecting fuel cell efficiency for different values of $r_{p,c}$ is observed to be twisted, which may be considered the maximum value of TIT for the SOFC-ICGT hybrid cycle.

8.4.10 Specific fuel consumption within SOFC-ICGT hybrid cycle

Figure 8.20 illustrates the specific fuel consumption within the SOFC-ICGT hybrid cycle. Specific fuel consumption is an important performance parameter for the utility industry as it can be used to compare the performance of various power plant utilities with direct reference to the fuel consumed per kWh of power produced. A performance map has been plotted between specific work and specific fuel consumption of the proposed SOFC-ICGT hybrid

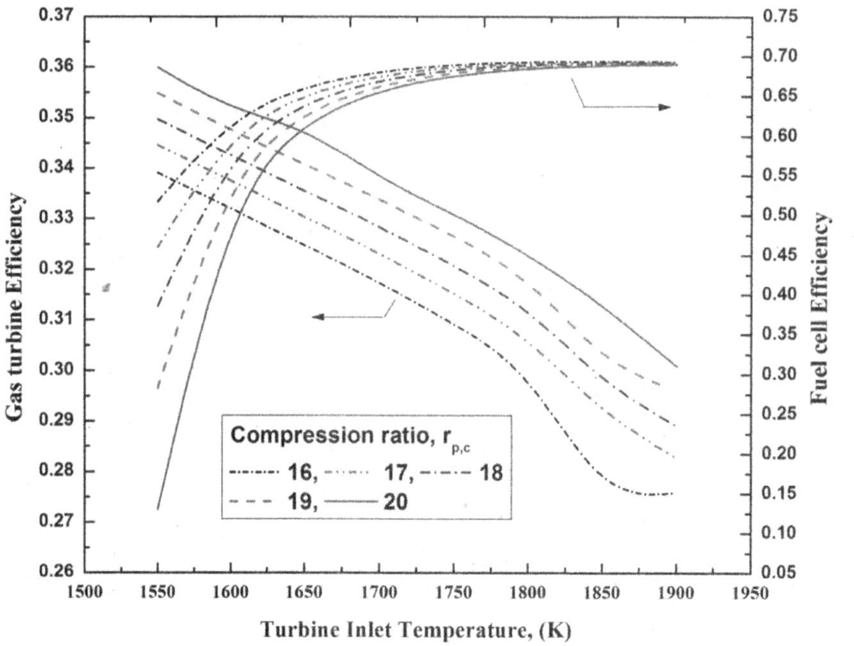

Figure 8.19 Comparative analysis of power generating unit at different TIT.

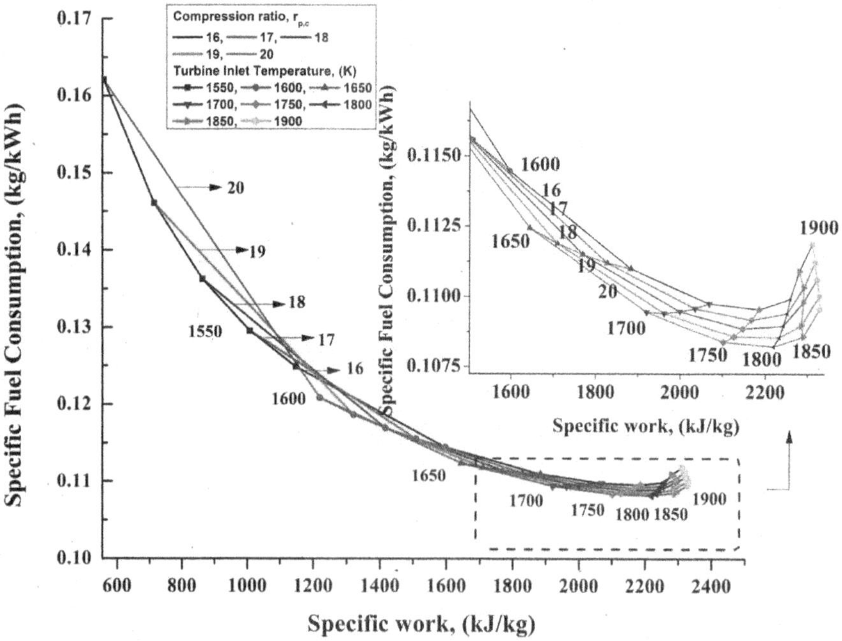

Figure 8.20 Specific fuel consumption within SOFC-ICGT hybrid cycle.

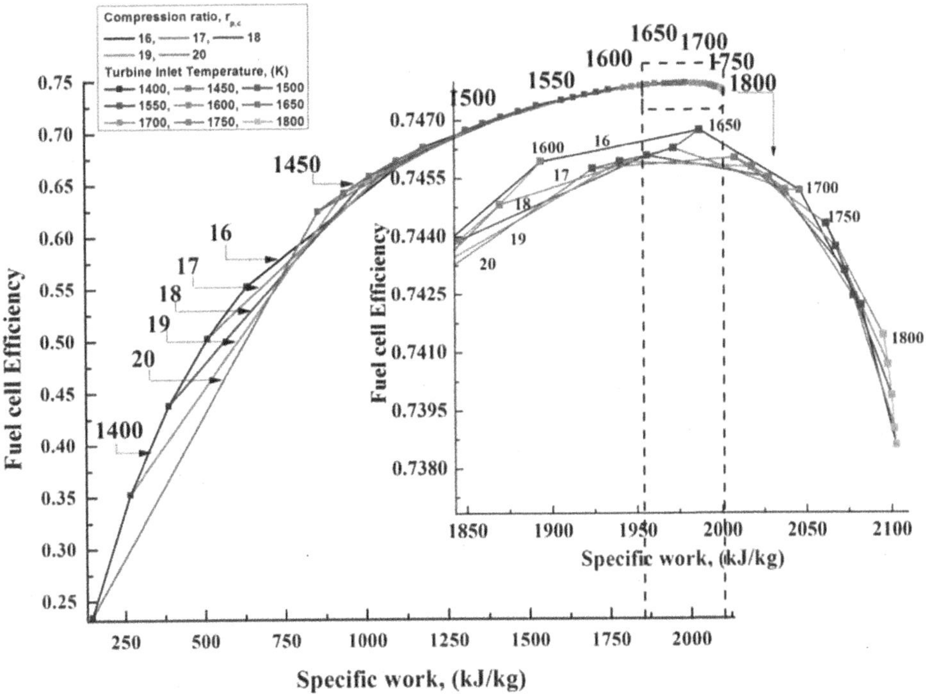

Figure 8.21 Fuel cell performance map.

plant. Optimum (minimum) specific fuel consumption of 0.1082 kg/kWh at the optimum performance of 74.65% for the hybrid cycle has been observed at TIT 1,700 K and $r_{p,c}$ 20 (see Figure 8.22). This unique performance map for specific fuel consumption and specific work for the SOFC-ICGT hybrid cycle can be used by power utility designers and researchers alike to identify the most suitable cycle operating parameters while designing similar systems.

8.4.11 Performance map

An exclusive performance map has been plotted from the parametric analysis for fuel cell and hybrid cycle as shown in Figures 8.21 and 8.22. Fuel cell performance and the efficiency of the hybrid plant can be seen by rising as the compression ratio and TIT to an acceptable level and decreases further on increase in compression ratio and TIT. This is due to an improvement in the degree of regeneration, improving fuel cell efficiency as fuel cell polarization decreases.

It can also be seen from the graph that the trend in the efficiency of hybrid plants and the specific work is the combined effect of gas turbines and fuel cells. The trend is similar to the fuel cell performance map, as the fuel cell is the major power producer as compared to the gas turbine.

Based on Figure 8.21, maximum fuel cell efficiency of (about 74.65%) occurred at TIT 1,700 K and $r_{p,c}$=20. Similarly, from Figure 8.22 maximum efficiency of 74.13% has been achieved for SOFC-GT hybrid cycle at $r_{p,c}$=20 and TIT=1,800 K.

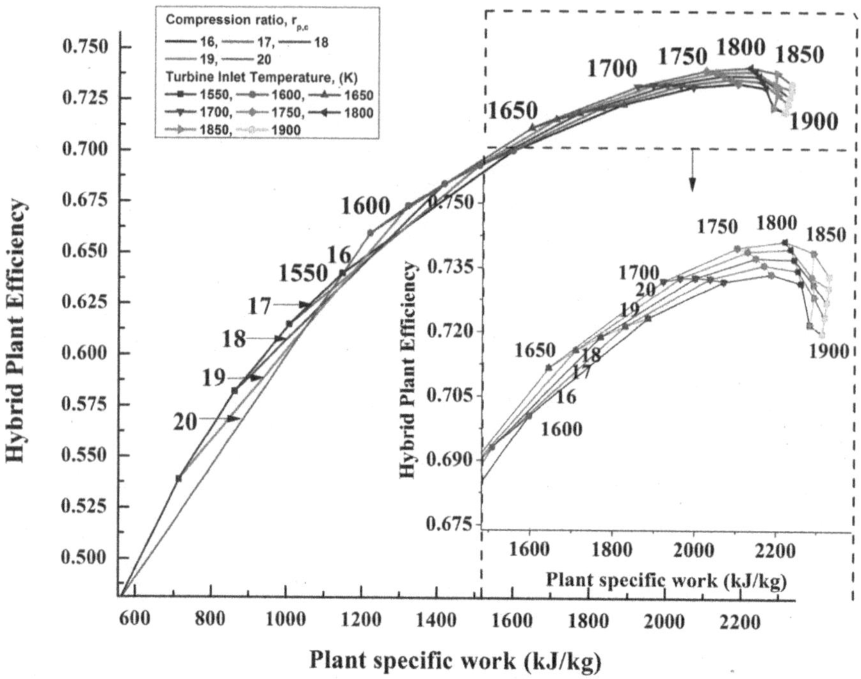

Figure 8.22 SOFC-GT hybrid cycle performance map.

8.5 SUMMARY

In this chapter, an intercooled recuperated blade cooled gas turbine has successfully integrated with high temperature internal reformed SOFC. Using the first law of thermodynamics, thermodynamic analysis has been carried out, and a MATLAB program has been developed. Several conclusions were drawn based on the results obtained.

- The mass of the blade coolant increases with a higher TIT and compression ratio. However, the TIT variation is noteworthy.
- From the sensitivity analysis, the proposed hybrid system is more sensitive towards the TIT operation.
- The efficiency of the fuel cell decreases as the recirculation ratio increases. Whereas, an increase in TIT increases the degree of recuperation and improves the efficiency of fuel cell.
- The maximum hybrid output of 74.13% was at $r_{p,c}$=20 and TIT=1,800 K.
- Based on parametric thermodynamic analysis, a novel efficiency map gas was plotted for the fuel cell and hybrid cycle.

The result presented for the proposed hybrid cycle opens a new research line for the researchers and power plant designers whose work focuses on developing alternative energy conversion technologies. Consequently, this would assist them in developing a more feasible integration option for the SOFC-GT hybrid power cycle.

NOMENCLATURE

A	Active surface area, cm^2
$Daeff$	Effective gaseous diffusivity through the anode, cm^2/s
$Dceff$	Effective gaseous diffusivity through the cathode, cm^2/s
E	Voltage, V
F	Faraday constant, C
h	Enthalpy, kJ/kg
$\underline{h}$	Specific molar enthalpy, J/mol
H	Enthalpy flow rate, W
icd,a	Exchange current density of anode, A/cm^2
i	Current density, A/cm^2
icd,c	Exchange current density of cathode, A/cm^2
ias	Anode-limiting current density, A/cm^2
ics	Cathode-limiting current density, A/cm^2
I	Current, A
K	Equilibrium constant
s	Entropy
T	Thickness of a cell component, µm
LHV	Lower heating value, J/mol
$\dot{m}$	Mass flow rate, kg/s
M	Molecular weight, g/mol
n_e	Number of Electrons
r	Recirculation ratio
P	Pressure, bar
R	Universal gas constant, J/molK
T	Temperature, K
TIT	Turbine inlet temperature, K
U_F	Fuel utilization ratio
U_a	Air utilization ratio
W_{fc}	Power output of cell, W
$\dot{W}$	Power, kW
X	Molar concentration

Greek letters

ρ_k	Electrical resistivity of the cell components
η	Cell efficiency
$\Delta\bar{G}$	Change in specific molar gibbs free energy j/mol

Subscripts

a	Air/ambient
an	Anode
act	Activation
b	Blade
c	Coolant
cat	Cathode
ch	Channel
$conc$	Concentration
$comb$	Combustor
$comp$	Compressor
$elec$	Electrolyte
f	Fuel
fc	Fuel cell
g	Gas
e	Exit
in	Inlet
j	Coolant stream
k	components
net	Difference
ohm	Ohmic
w	Water
rc	Recuperator

Acronyms

AEI	Anode electrolyte interface
GT	Gas turbine
PEN	Positive electrode/ electrolyte/ negative-electrode
SOFC	Solid oxide fuel cell
TIT	Turbine inlet temperature

REFERENCES

Adams II, Thomas A., Jake Nease, David Tucker, & Paul I. Barton. 2012. "Energy conversion with solid oxide fuel cell systems: A review of concepts and outlooks for the short- and long-term." *Industrial and Engineering Chemistry Research* 52(9): 3089–3111. doi:10.1021/IE300996R.

Akkaya, Ali Volkan, Bahri Sahin, & Hasan Huseyin Erdem. 2008. "An analysis of SOFC/GT CHP system based on exergetic performance criteria." *International Journal of Hydrogen Energy* 33(10): 2566–2577. doi:10.1016/J.IJHYDENE.2008.03.013.

Amador, German, Jorge Duarte Forero, Adriana Rincon, Armando Fontalvo, Antonio Bula, Ricardo Vasquez Padilla, & Wilman Orozco. 2017. "Characteristics of auto-ignition in internal combustion engines operated with gaseous fuels of variable methane number." *Journal of Energy Resources Technology* 139(4). doi:10.1115/1.4036044.

Bakalis, Diamantis P., & Anastassios G. Stamatis. 2014. "Optimization methodology of turbomachines for hybrid SOFC–GT applications." *Energy* 70: 86–94. doi:10.1016/J.ENERGY.2014.03.093.

Buonomano, Annamaria, Francesco Calise, Massimo Dentice d'Accadia, Adolfo Palombo, & Maria Vicidomini. 2015. "Hybrid solid oxide fuel cells–gas turbine systems for combined heat and power: A review." *Applied Energy* 156: 32–85. doi:10.1016/J.APENERGY.2015.06.027.

Calise, Francesco, Massimo Dentice d' Accadia, Laura Vanoli, & Michael R. von Spakovsky. 2006. "Single-level optimization of a hybrid SOFC–GT power plant." *Journal of Power Sources* 159(2): 1169–1185. doi:10.1016/J.JPOWSOUR.2005.11.108.

Calise, Francesco, Massimo Dentice d'Accadia, Luigi Libertini, & Maria Vicidomini. 2018. "Thermoeconomic analysis of an integrated solar combined cycle power plant." *Energy Conversion and Management* 171: 1038–1051. doi:10.1016/j.enconman.2018.06.005.

Chen, Jinwei, Maozong Liang, Huisheng Zhang, & Shilie Weng. 2017. "Study on control strategy for a SOFC-GT hybrid system with anode and cathode recirculation loops." *International Journal of Hydrogen Energy* 42(49): 29422–29432. doi:10.1016/J.IJHYDENE.2017.09.165.

Chitgar, Nazanin, Mohammad Ali Emadi, Ata Chitsaz, & Marc A. Rosen. 2019. "Investigation of a novel multigeneration system driven by a SOFC for electricity and fresh water production." *Energy Conversion and Management* 196: 296–310. doi:10.1016/J.ENCONMAN.2019.06.006.

Choudhary, Tushar, Mithilesh Kumar Sahu, R. Sanjay, Anupam Kumari, & Alok Mohapatra. 2018. "Thermodynamic modeling of blade cooled turboprop engine integrated to solid oxide fuel cell: A concept." *SAE Technical Papers*. doi:10.4271/2018-01-1308.

Choudhary, Tushar, & Sanjay. 2016. "Computational analysis of IR-SOFC: Thermodynamic, electrochemical process and flow configuration dependency." *International Journal of Hydrogen Energy* 41(2): 1259–1271. doi:10.1016/J.IJHYDENE.2015.10.098.

Choudhary, Tushar, & Sanjay. 2017. "Thermodynamic assessment of advanced SOFC-blade cooled gas turbine hybrid cycle." *International Journal of Hydrogen Energy* 42(15): 10248–10263. doi:10.1016/J.IJHYDENE.2017.02.178.

Choudhary, Tushar, Sanjay, & Pilaka Murty. 2015. "Parametric analysis of syn-gas fueled SOFC with internal reforming." *SAE Technical Papers*. doi:10.4271/2015-01-1176.

Facchinetti, E., D. Favrat, & F. Marechal. 2014. "Design and optimization of an innovative solid oxide fuel cell–gas turbine hybrid cycle for small scale distributed generation." *Fuel Cells* 14 (4): 595–606. doi:10.1002/FUCE.201300196.

Hajimolana, S. Ahmad, M. Azlan Hussain, W. M. Ashri Wan Daud, Masoud Soroush, & Ahmad Shamiri. 2011. "Mathematical modeling of solid oxide fuel cells: A review." *Renewable and Sustainable Energy Reviews* 15(4): 1893–1917. doi:10.1016/J.RSER.2010.12.011.

Haseli, Yousef, Ibrahim Dincer, & Greg F. Naterer. 2008. "Thermodynamic modeling of a gas turbine cycle combined with a solid oxide fuel cell." *International Journal of Hydrogen Energy* 33(20): 5811–5822. doi:10.1016/J.IJHYDENE.2008.05.036.

Herrera-Orozco, Israel, Guillermo Valencia-Ochoa, & Jorge Duarte-Forero. 2021. "Exergo-environmental assessment and multi-objective optimization of waste heat recovery systems based on organic Rankine cycle configurations." *Journal of Cleaner Production* 288: 125679. www.sciencedirect.com/science/article/pii/S0959652620357255.

Huang, Yu, & Ali Turan. 2019. "Fuel sensitivity and parametric optimization of SOFC – GT hybrid system operational characteristics." *Thermal Science and Engineering Progress* 14: 100407. doi:10.1016/J.TSEP.2019.100407.

Janardhanan, Vinod M., & Olaf Deutschmann. 2006. "CFD analysis of a solid oxide fuel cell with internal reforming: coupled interactions of transport, heterogeneous catalysis and electrochemical processes." *Journal of Power Sources* 162(2): 1192–1202. doi:10.1016/J.JPOWSOUR.2006.08.017.

Karimi, Mohammad Hossein, Nazanin Chitgar, Mohammad Ali Emadi, Pouria Ahmadi, & Marc A. Rosen. 2020. "Performance assessment and optimization of a biomass-based solid oxide fuel cell and micro gas turbine system integrated with an organic Rankine cycle." *International Journal of Hydrogen Energy* 45(11): 6262–6277. doi:10.1016/J.IJHYDENE.2019.12.143.

Khani, Leyla, Ali Saberi Mehr, Mortaza Yari, & S. M. Seyed Mahmoudi. 2016. "Multi-objective optimization of an indirectly integrated solid oxide fuel cell-gas turbine cogeneration system." *International Journal of Hydrogen Energy* 41(46): 21470–21488. doi:10.1016/J.IJHYDENE.2016.09.023.

Kumar, Pranjal, & Onkar Singh. 2019. "Thermoeconomic analysis of SOFC-GT-VARS-ORC combined power and cooling system." *International Journal of Hydrogen Energy* 44(50): 27575–27586. doi:10.1016/J.IJHYDENE.2019.08.198.

Kumari, Anupam, & Sanjay. 2015. "Investigation of parameters affecting exergy and emission performance of basic and intercooled gas turbine cycles." *Energy* 90: 525–536. doi:10.1016/J.ENERGY.2015.07.084.

Lai, Haoxiang, Nor Farida Harun, David Tucker, & Thomas A. Adams. 2021. "Design and eco-technoeconomic analyses of SOFC/GT hybrid systems accounting for long-term degradation effects." *International Journal of Hydrogen Energy* 46(7): 5612–5629. doi:10.1016/J.IJHYDENE.2020.11.032.

Larminie, James, Andrew Dicks, & Maurice S. McDonald. 2003. *Fuel Cell Systems Explained.* Chichester: John Wiley.

Massardo, Aristide F., & F. Lubelli. 2000. "Internal reforming solid oxide fuel cell-gas turbine combined cycles (IRSOFC-GT): Part A. Cell model and cycle thermodynamic analysis." *Journal of Engineering for Gas Turbines and Power* 122(1): 27–35. doi:10.1115/1.483187.

Mehrpooya, Mehdi, Sepide Akbarpour, Ali Vatani, & Marc A. Rosen. 2014. "Modeling and optimum design of hybrid solid oxide fuel cell-gas turbine power plants." *International Journal of Hydrogen Energy* 39(36): 21196–21214. doi:10.1016/J.IJHYDENE.2014.10.077.

Meratizaman, Mousa, Sina Monadizadeh, & Majid Amidpour. 2014. "Techno-economic assessment of high efficient energy production (SOFC-GT) for residential application from natural Gas." *Journal of Natural Gas Science and Engineering* 21: 118–133. doi:10.1016/J.JNGSE.2014.07.033.

Mohapatra, Alok K., & Sanjay. 2018. "Exergetic evaluation of gas-turbine based combined cycle system with vapor absorption inlet cooling." *Applied Thermal Engineering* 136: 431–443. doi:10.1016/J.APPLTHERMALENG.2018.03.023.

Najafi, Behzad, Ali Shirazi, Mehdi Aminyavari, Fabio Rinaldi, & Robert A. Taylor. 2014. "Exergetic, economic and environmental analyses and multi-objective optimization of an SOFC-gas turbine hybrid cycle coupled with an MSF desalination system." *Desalination* 334(1): 46–59. doi:10.1016/J.DESAL.2013.11.039.

Nami, Hossein, Amjad Anvari-Moghaddam, & Ahmad Arabkoohsar. 2020. "Application of CCHPs in a centralized domestic heating, cooling and power network: Thermodynamic and economic implications." *Sustainable Cities and Society* 60: 102151. doi:10.1016/j.scs.2020.102151.

Nease, Jake, & Thomas A. Adams. 2014. "Coal-fuelled systems for peaking power with 100% CO_2 capture through integration of solid oxide fuel cells with compressed air energy storage." *Journal of Power Sources* 251: 92–107. doi:10.1016/J.JPOWSOUR.2013.11.040.

Ozcan, Hasan, & Ibrahim Dincer. 2015. "Performance evaluation of an SOFC based trigeneration system using various gaseous fuels from biomass gasification." *International Journal of Hydrogen Energy* 40(24): 7798–7807. doi:10.1016/J.IJHYDENE.2014.11.109.

Rokni, Masoud. 2014. "Biomass gasification integrated with a solid oxide fuel cell and Stirling engine." *Energy* 77: 6–18. doi:10.1016/J.ENERGY.2014.01.078.

Roushenas, Ramin, Amir Reza Razmi, Madjid Soltani, Morteza Torabi, Maurice B. Dusseault, & Jatin Nathwani. 2020. "Thermo-environmental analysis of a novel cogeneration system based on solid oxide fuel cell (SOFC) and compressed air energy storage (CAES) coupled with turbocharger." *Applied Thermal Engineering* 181: 115978. doi:10.1016/J.APPLTHERMALENG.2020.115978.

Sadeghi, Saber, & Ighball Baniasad Askari. 2019. "Prefeasibility techno-economic assessment of a hybrid power plant with photovoltaic, fuel cell and compressed air energy storage (CAES)." *Energy* 168: 409–424. doi:10.1016/J.ENERGY.2018.11.108.

Sahu, Mithilesh Kumar, & Sanjay. 2017. "Comparative exergoeconomics of power utilities: air-cooled gas turbine cycle and combined cycle configurations." *Energy* 139: 42–51. doi:10.1016/J.ENERGY.2017.07.131.

Sanjay, Onkar Singh, & Bishwa N. Prasad. 2008. "Influence of different means of turbine blade cooling on the thermodynamic performance of combined cycle." *Applied Thermal Engineering* 28 (17–18): 2315–2326. doi:10.1016/J.APPLTHERMALENG.2008.01.022.

Sanjay, Onkar Singh, & Bishwa N. Prasad. 2009. "Comparative performance analysis of cogeneration gas turbine cycle for different blade cooling means." *International Journal of Thermal Sciences* 48 (7): 1432–1440. doi:10.1016/J.IJTHERMALSCI.2008.11.016.

Sanjay, & Bishwa N. Prasad. 2013. "Energy and exergy analysis of intercooled combustion-turbine based combined cycle power plant." *Energy* 59: 277–284. doi:10.1016/J.ENERGY.2013.06.051.

Seyedkavoosi, Seyedali, Saeed Javan, & Krishna Kota. 2017. "Exergy-based optimization of an organic Rankine cycle (ORC) for waste heat recovery from an internal combustion engine (ICE)." *Applied Thermal Engineering* 126: 447–457. doi:10.1016/J.APPLTHERMALENG.2017.07.124.

Sghaier, Salha Faleh, Tahar Khir, & Ammar Ben Brahim. 2018. "Energetic and exergetic parametric study of a SOFC-GT hybrid power plant." *International Journal of Hydrogen Energy* 43(6): 3542–3554. doi:10.1016/J.IJHYDENE.2017.08.216.

Sharifzadeh, Mahdi, Mojtaba Meghdari, & Davood Rashtchian. 2017. "Multi-objective design and operation of solid oxide fuel cell (SOFC) triple combined-cycle power generation systems: integrating energy efficiency and operational safety." *Applied Energy* 185: 345–361. doi:10.1016/J.APENERGY.2016.11.010.

Shukla, Anoop Kumar, Sandeep Gupta, Shivam Pratap Singh, Meeta Sharma, & Gopal Nandan. 2018. "Thermodynamic performance evaluation of SOFC based simple gas turbine cycle." *International Journal of Applied Engineering Research* 13(10): 7772–7778. www.ripublication.com/ijaer18/ijaerv13n10_69.pdf.

Shukla, Anoop Kumar, Zoheb Ahmad, Meeta Sharma, Gaurav Dwivedi, Tikendra Nath Verma, Siddharth Jain, Puneet Verma, & Ali Zare. 2020. "Advances of carbon capture and storage in coal-based power generating units in an Indian context." *Energies* 13(6): 1–17. doi:10.3390/en13164124.

Su, Xingyuan, & Timothy A. Shedd. 2018. "Towards working fluid properties and selection of Rankine cycle based waste heat recovery (WHR) systems for internal combustion engines – a fundamental analysis." *Applied Thermal Engineering* 142: 502–510. doi:10.1016/J.APPLTHERMALENG.2018.07.036.

Tucker, David, Maria Abreu-Sepulveda, & Nor Farida Harun. 2014. "SOFC lifetime assessment in gas turbine hybrid power systems." *Journal of Fuel Cell Science and Technology* 11(5). doi:10.1115/1.4028158.

Wang, Xusheng, Xiaojing Lv, & Yiwu Weng. 2020. "Performance analysis of a biogas-fueled SOFC/GT hybrid system integrated with anode-combustor exhaust gas recirculation loops." *Energy* 197: 117213. doi:10.1016/J.ENERGY.2020.117213.

Xue, Yejian, Changrong He, Man Liu, Jinliang Yuan, & Weiguo Wang. 2019. "Study on the fracture behavior of the planar-type solid oxide fuel cells." *Journal of Alloys and Compounds* 782: 355–362. doi:10.1016/J.JALLCOM.2018.12.203.

Zeynalian, Mirhadi, Amir Hossein Hajialirezaei, Amir Reza Razmi, & Morteza Torabi. 2020. "Carbon dioxide capture from compressed air energy storage system." *Applied Thermal Engineering* 178: 115593. doi:10.1016/J.APPLTHERMALENG.2020.115593.

Zhang, Ding, Shujun Mu, Chingchuen C. Chan, & George You Zhou. 2018. "Optimization control of SOFC based on bond graph model." *Energy Procedia* 152: 174–179. doi:10.1016/J.EGYPRO.2018.09.077.

Zhang, Xiongwen, Siewhwa H. Chan, Guojun Li, Hiang Kwee Ho, Jun Li, & Zhenping Feng. 2010. "A review of integration strategies for solid oxide fuel cells." *Journal of Power Sources* 195(3): 685–702. doi:10.1016/J.JPOWSOUR.2009.07.045.

Zhang, Xiuqin, Huiying Liu, Meng Ni, & Jincan Chen. 2015. "Performance evaluation and parametric optimum design of a syngas molten carbonate fuel cell and gas turbine hybrid system." *Renewable Energy* 80: 407–414. doi:10.1016/J.RENENE.2015.02.035.

Zhang, Xiuqin, Shanhe Su, Jincan Chen, Yingru Zhao, & Nigel Brandon. 2011. "A new analytical approach to evaluate and optimize the performance of an irreversible solid oxide fuel cell-gas turbine hybrid system." *International Journal of Hydrogen Energy* 36(23): 15304–15312. doi:10.1016/J.IJHYDENE.2011.09.004.

Zhang, Xiuqin, Yuan Wang, Tie Liu, & Jincan Chen. 2014. "Theoretical basis and performance optimization analysis of a solid oxide fuel cell–gas turbine hybrid system with fuel reforming." *Energy Conversion and Management* 86: 1102–1109. doi:10.1016/J.ENCONMAN.2014.06.068.

Zhao, Yingru, Jhuma Sadhukhan, Andrea Lanzini, Nigel Brandon, & Nilay Shah. 2011. "Optimal integration strategies for a syngas fuelled SOFC and gas turbine hybrid." *Journal of Power Sources* 196 (22): 9516–9527. doi:10.1016/J.JPOWSOUR.2011.07.044.

Zhao, Yingru, Nilay Shah, & Nigel Brandon. 2011. "Comparison between two optimization strategies for solid oxide fuel cell–gas turbine hybrid cycles." *International Journal of Hydrogen Energy* 36(16): 10235–10246. doi:10.1016/J.IJHYDENE.2010.11.015.

Zhu, Huayang, Robert J. Kee, Vinod M. Janardhanan, Olaf Deutschmann, & David G. Goodwin. 2005. "Modeling elementary heterogeneous chemistry and electrochemistry in solid-oxide fuel cells." *Journal of the Electrochemical Society* 152(12): A2427. doi:10.1149/1.2116607.

Chapter 9

Municipal solid waste-fueled plants

Ahmad Arabkoohsar and Amirmohammad Behzadi
Department of Energy Technology, Aalborg University, Denmark

9.1 INTRODUCTION

According to the latest energy reports, the average global energy consumption has doubled compared to 2010 and has hit a record high (Behzadi, Arabkoohsar, & Perić 2021). This trend will undoubtedly continue because of population growth and increase in the quality of life, and this will have a significant direct and indirect influence on the energy demand (Khan et al. 2021). An increase in the global rate of greenhouse gas emissions and fossil fuel depletion are catastrophic results of increased energy demand. These concerns must be addressed to hinder the economic problems and climate changes in the coming year. The replacement of fossil fuels by renewable energies is the central solution for policymakers and scientists to deal with global warming and meet the unprecedented energy demand growth (Vojdani, Fakhari, & Ahmadi 2021). Of all renewable fuels, the biogenic portion of municipal solid waste (MSW) as an abundant, carbon-neutral, and controllable resource is a favorable option to make a big step toward the world's green transition. In this regard, most countries have applied waste-to-energy plants to convert the waste into precious energy and reduce the need for landfilling. Efficient renewable integration and flue gas condensation process are other energy improvement and cost mitigation solutions to cope with the worldwide energy problem. The organic Rankine cycle (ORC) is a promising power generation technology to exploit the waste heat of a low-temperature heat source like flue gas because of low maintenance cost and reliability (Hussain, Sharma, & Shukla 2021).

The feasibility of exploiting the waste heat of MSW-fired plants through different technologies such as ORC has been studied in the literature from techno-economic and environmental perspectives. Eboh, Andersson, and Richards (2019) investigated and compared various configurations of MSW-based power plants. They showed that the highest exergy efficiency corresponds to the system driven by gasification technology with a flue gas condensation process. Exergoeconomic and environmental evaluation of a waste-fired plant combined with an ORC unit using different working fluids were studied by Lu et al. (2020). They reported that the energy and exergy efficiencies would rise to 37.66% and 35.65% resulting from the flue condensation process via ORC using butane. Holik et al. (2021) studied and compared the waste heat utilization of a waste-fired power plant applying the Rankine cycle against ORC from techno-economic facets. They showed that although the Rankine system is economically favorable, ORC is superior from an exergy standpoint. A waste-fired power plant integrated with an ORC unit and gasifier was assessed techno-economically by Georgousopoulos et al. (2021), concluding that the plant efficiency increases about 2.81% due to the flue gas condensation. In another study, Arabkoohsar and Nami (2019)

DOI: 10.1201/9781003213741-9

proposed an innovative waste-driven power plant integrated with ORC. They demonstrated that exploiting the waste heat of flue gas leads to the higher energy efficiency of 20%. The feasibility of integrating a waste-driven power plant with an ORC unit was investigated by Maria et al. (2016), concluding that 250 kW higher electricity is generated by the flue gas condensation process. Yatsunthea and Chaiyat (2020) studied a small-scale waste-fired power plant combined with an ORC unit using R-245fa as the working fluid. They revealed that compared to the waste-to-energy power plant alone, higher energy and exergy efficiencies of 0.91% and 0.89% and lower levelized cost of electricity of 0.048 $/kWh are achieved.

The idea of developing a thermoelectric generator (TEG) has been developed to improve the power share of combined heating and power (CHP) conventional systems via direct conversion of thermal energy to power (Behzadi, Arabkoohsar, & Gholamian 2020). This technology has recently been prevalent due to its low purchase cost, silent operating, and carbon dioxide-free performance. Various researchers have studied the performance assessment and optimization of different TEGs (Sharma, Dwivedi, & Pandit 2014; Biswas et al. 2018; Arora & Arora 2018). The combination of TEG with conventional power generation systems to increase power production is also investigated in the literature from various aspects. An innovative geothermal-driven ORC system combined with TEG was studied by Aliahmadi, Moosavi, and Sadrhosseini (2021). They concluded that the replacement of the condenser with the TEG unit leads to higher performance efficiencies and a lower payback period. Musharavati et al. (2021) proposed and evaluated techno-economic aspects of an innovative system comprising ORC and TEG to obtain higher output electricity and a lower total cost rate. Various positions for adding TEG unit in a biomass-fired CHP system combined with gasifier, ORC unit, and the water heater were examined and compared by Khanmohammadi et al. (2019). According to their results, the highest exergy efficiency of 17.93% is obtained when the TEG substitutes the condenser. Ziapour et al. (2017) studied the TEG unit's potential in exploiting the waste heat of a salinity gradient solar pond system combined with an ORC unit for power generation. They revealed that when the TEG replaces the condenser, higher performance efficiencies will be attained. An innovative power and hydrogen production system integrated with ORC and TEG was introduced by Gholamian, Habibollahzade, and Zare (2018). They indicated that the system is economically superior to the conventional ORC because of a lower product unit cost. Zare and Palideh (2018) investigated the combination of TEG with the Kalina cycle and concluded that compared to the conventional Kalina cycle, 7.3% increase in performance efficiency is obtained using TEG.

In addition to efficient renewable integration and applying performance enhancement approaches like waste heat recovery, optimizing the energy system is of great significance for achieving sustainability, improved performance efficiency, and cost-effectiveness (Arabkoohsar, Behzadi, & Alsagri 2021). Özahi, Tozlu, and Abuşoğlu (2018) optimized a waste-fired power plant combined with an ORC unit using various working fluids and concluded that the net produced power increases considerably while the total cost decreases. Pan et al. (2020) applied multi-objective optimization to a waste-fired CHP plant. According to their results, the payback period of 0.48 years and investment cost of 325.94 $/GJ is obtained at the best optimum point. Behzadi et al. (2019) studied and optimized an innovative system comprising a TEG unit for higher power production. They indicated that the optimum exergy efficiency is 12.01%, and the optimum total cost rate is 0.176 $/h. Wang et al. (2020) carried out tri-objective optimization of a power system equipped with ORC, contemplating thermal and exergy efficiencies to be maximized and product cost to be minimized. Exergo-economic assessment and multi-objective optimization of a power cycle equipped with an ORC unit recovering a diesel engine's waste heat was investigated by Fang et al. (2019), finding out that the optimum evaporator and condenser temperatures should be kept at their upper and lower bound, respectively. Yang et al. (2017) applied multi-objective optimization using a genetic algorithm approach to a dual loop ORC system driven by the compressed natural gas (CNG) engine's waste heat. They revealed that a favorable techno-economic condition is obtained when the evaporation pressure is kept above 2.5 MPa.

Based on the above-discussed literature, it can be concluded that there are three critical issues in existing conventional power plants comprising waste-to-energy power plants. First, increasing the power productivity of such power plants should be prioritized on top of every other output, such as heat. This goal is in line with the critical objective of further electrification of future energy systems. Second, the existing power plants need to become more environmentally friendly to stay alive in the green transition pathway. Third, innovative passive and active energy improvement and cost reduction approaches should be applied to mitigate the plant's energy consumption and enhance the energy conversion quality. If such measures are not taken, such power plants will indeed be taken out of the global energy system.

An example of this is so many waste-driven CHP plants in Denmark that will be closing down by 2025 even though they have many years before the end of their useful lifetime. This work introduces an innovative configuration of a gasifier-driven waste-to-energy power plant to obtain higher power generation, cleaner production, and lower product cost. First, the plant's waste heat is recovered through the flue gas condensation process to run an ORC unit. Second, the condensers are substituted by a thermoelectric generator as clean, efficient, and reliable equipment directly converting the heat to power. The proposed novel plant is then investigated and compared against the conventional plant from performance, exergoeconomic, sustainability, and environmental aspects. A parametric investigation is performed to examine the influence of main decision parameters on each system. Subsequently, multi-criteria optimization is implemented to the superior plant to maximize the exergy efficiency while minimizing the levelized cost of power and levelized carbon dioxide emission. Finally, the optimal variables and objectives are presented through scatter distribution and Pareto frontier diagrams.

9.2 SYSTEM DESCRIPTION AND ASSUMPTIONS

The schematic diagram of the conventional and the proposed waste-driven power plant is demonstrated in Figure 9.1. Both plants comprise three main parts: the waste gasifier unit integrated with combustion chamber for heat generation, Rankine cycle for power production, and organic Rankine cycle exploiting the waste heat of flue gases for power generation. In the proposed innovative system, the Rankine cycle and ORC unit's condensers are substituted by TEGs as a secondary power production unit. As depicted, the municipal solid waste and air enter the gasifier to start the thermochemical reactions generating high enthalpy syngas. The produced gas then goes into the combustion chamber to be mixed with secondary air. The combustion reaction occurs, and the high temperature-produced gases move toward the steam generator to drive the Rankine cycle. Afterward, the working fluid, water, is cooled in the condenser through the water and air. The steam generator outlet exhaust gases are high enthalpy enough to run a low-temperature power system. So, the heat of exhaust gases is exploited in the ORC unit's evaporator through the flue gas condensation process.

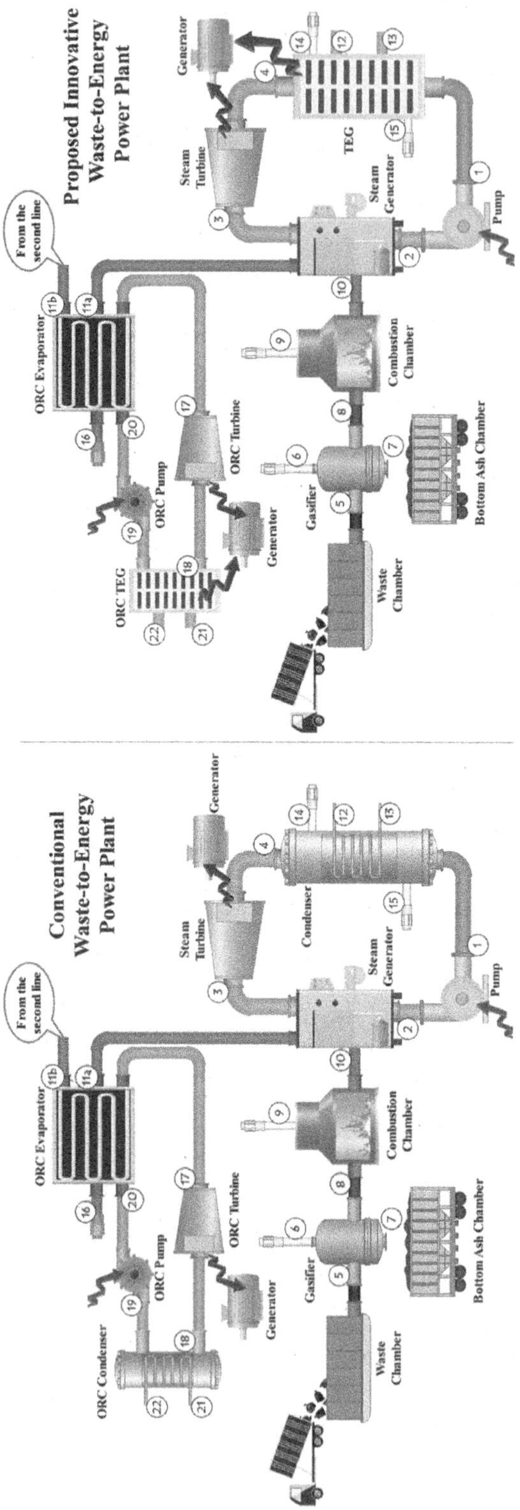

Figure 9.1 Schematic diagram of each system.

Table 9.1 Input parameters of each subsystem

Parameter	*Value*	*Parameter*	*Value*
Gasifier		*Rankine*	
MC	0.35	*TIP* (kPa)	2,000
$T_{Gasification}$ (°C)	980	CIP (kPa)	40
T_{Ash} (°C)	680	ΔT_{SH}	198
Air composition (mole %)	N_2:79 O_2:21	ΔT_{PP}	222
$P_{Gasification}$ (kPa)	100	$\Delta T_{Cooling\ air}$	25
$\dot{m}_{MSW}$ (ton/day) per line	100	$\Delta T_{Cooling\ water}$	10
Combustion		$\eta_{is,pu}$ (%)	85
$T_{Combustion}$ (°C)	1,100	$\eta_{is,ST}$(%)	80
$P_{Combustion}$ (kPa)	100	**Organic Rankine cycle**	
Thermoelectric generator		ORC *TIP* (kPa)	2,000
ZT_m	0.8	ORC ΔT_{SH}	10
TEG $\Delta T_{Cooling\ water}$	10	ORC ΔT_{PP}	5
Ambient condition		ORC $\Delta T_{Cooling\ water}$	5
$T_{Ambient}$ (°C)	25	ORC $\eta_{is,pu}$ (%)	90
$P_{Ambient}$ (kPa)	100	ORC $\eta_{is,ST}$(%)	80

R123 is selected as the ORC working fluid due to the low ozone depletion potential and techno-economical superiority compared to other fluids (Behzadi et al. 2018). In the ORC unit, the evaporator outlet fluid (state 17) enters the ORC turbine to produce power. While in the conventional system the ORC working fluid goes into the condenser (state 18), in the proposed innovative system, it passes through the TEG unit to generate the extra power. The TEG unit operates like a heat engine converting heat directly to electricity via thermoelectric effect. In the condenser and TEG units, the coolant fluid is water. It is noted that the proposed waste-to-energy plant has two same parallel lines integrated with only one ORC unit.

Each subsystem's input parameters, including gasifier and combustion units, waste-to-energy power plant, ORC unit, and TEG, are listed in Table 9.1.

9.3 MODELING

Engineering equation software (EES) is applied to model each system (conventional and proposed waste-driven power plant) and develop mathematical equations of each component. Afterward, multi-criteria optimization applying the genetic algorithm method in MATLAB is accomplished contemplating techno-economic-environmental aspects of the proposed innovative system to be optimized simultaneously.

9.3.1 Thermodynamic evaluation

Every component is contemplated as a lumped control volume, and mass, energy, and exergy balance equations are written as (Fakhari et al. 2020; Sadi et al. 2021):

$$\sum \dot{m}_{in} = \sum \dot{m}_{out} \tag{9.1}$$

$$\dot{Q} - \dot{W} = \sum \dot{m}_{out} h_{out} - \sum \dot{m}_{in} h_{in} \tag{9.2}$$

$$\dot{E}_Q - \dot{E}_W = \sum \dot{m}_{out} e_{out} - \sum \dot{m}_{in} e_{in} + \dot{E}_D \tag{9.3}$$

Here $\dot{m}$, $\dot{W}$, and $\dot{Q}$ are the mass flow rate, work, and heat, respectively. Also, $\dot{E}_Q$, $\dot{E}_W$, and $\dot{E}_D$ are the exergy of heat, the exergy of work, and exergy destruction. Besides, e is the exergy of each stream as the sum physical (e^{ph}) and chemical (e^{ch}) exergy as follows (Behzadi, Arabkoohsar, & Yang 2020):

$$e = e^{ph} + e^{ch} \tag{9.4}$$

$$e_i^{ph} = (h_i - h_0) - T_0 (s_i - s_0) \tag{9.5}$$

$$e_i^{ch} = \sum_{i=1}^{n} x_i e_{0.i}^{ch} + RT_0 \sum_{i=1}^{n} x_i \; ln(x_i) \tag{9.6}$$

Where $e_{0.i}^{ch}$ is the standard chemical exergy, and x_i is the mole fraction. In the following subsections, the thermodynamic modeling of every component is provided in detail.

9.3.1.1 Waste gasifier

The composition of the produced gas is highly affected by the gasifier type. A moving gate gasifier as efficient equipment generating high enthalpy syngas is implemented in the present work. The global reaction for the waste gasifier, assuming that all the reactions occurred in thermodynamic equilibrium, is written as follows (Basu 2010):

$$\begin{aligned} CH_aO_bN_c + wH_2O + n_1(O_2 + 3.76N_2) + Ash \rightarrow n_2H_2 + n_3CO + n_4CO_2 \\ + n_5H_2O + n_6CH_4 + n_7N_2 + Ash \end{aligned} \tag{9.7}$$

In addition to mass balance equations for carbon, hydrogen, oxygen, and nitrogen, the equilibrium constant equation of methane formation and water gas shift reactions are needed to determine n_2–n_6 (Moran et al. 2010).

$$K_1 = \frac{n_6}{n_2^2} \left(\frac{\frac{P}{P_0}}{n_{tot}} \right)^{-1} \tag{9.8}$$

$$K_2 = \frac{n_2 n_4}{n_3 n_5} \tag{9.9}$$

Table 9.2 The characteristics of municipal solid waste

Carbone	*Hydrogen*	*Oxygen*	*Nitrogen*	*Ash*	*Lower heating value (MJ/kg)*
41.7	3.2	39.9	0.7	14.5	12.81

$$\frac{-\Delta G_1^0}{\underline{R}T_g} = LnK_1 \tag{9.10}$$

$$\frac{-\Delta G_2^0}{\underline{R}T_g} = LnK_2 \tag{9.11}$$

The ultimate analysis of municipal solid waste as a prerequisite data determining the values of a, b, and c (look at Equation (9.1)) is tabulated in Table 9.2.

Furthermore, w as the mole of moisture per mole of inlet municipal solid waste is written as:

$$w = \frac{M_{MSW}MC}{18(1-MC)} \tag{9.12}$$

Finally, the gasifier's energy balance equation is written as Equation (9.13) to determine the air-fuel ratio assuming that the gasification temperature (T_g) is known.

$$\begin{aligned}
&\underline{h}_{f-MSW.daf}^0 + n_{tot-ash@inlet} \times h_{tot-ash@inlet} + w \times \underline{h}_{f-H_2O}^0 + \\
&n_1\left(\underline{h}_{f-air}^0 + \Delta\underline{h}_{air}\right) = n_2\left(\underline{h}_{f-H2}^0 + \Delta\underline{h}_{H_2}\right) + n_3\left(\underline{h}_{f-CO}^0 + \Delta\underline{h}_{CO}\right) + \\
&n_4\left(\underline{h}_{f-CO2}^0 + \Delta\underline{h}_{CO2}\right) + n_5\left(\underline{h}_{f-H2O}^0 + \Delta\underline{h}_{H2O}\right) + n_6\left(\underline{h}_{f-Ch4}^0 + \Delta\underline{h}_{Ch4}\right) + \\
&n_7\left(\underline{h}_{f-N2}^0 + \Delta\underline{h}_{N2}\right) + n_{tot-ash@outlet} \times h_{tot-ash@outlet}
\end{aligned} \tag{9.13}$$

9.3.1.2 Combustion chamber

In the combustion chamber, the syngas and secondary air react with each other and generate combustion products as follows:

$$\begin{aligned}
&n_2H_2 + n_3CO + n_4CO_2 + n_5H_2O + n_6CH_4 + n_7N_2 + n_8\left(O_2 + 3.76N_2\right) \rightarrow \\
&\quad n_9CO_2 + n_{10}H_2O + n_{11}O_2 + \left(n_7 + 3.76n_8\right)N_2
\end{aligned} \tag{9.14}$$

In this equation, it is assumed that complete oxidation takes place. In addition to mass balance equations for carbon, hydrogen, oxygen, and nitrogen, the energy balance for combustion reaction is calculated to determine the unknown coefficients:

$$\sum_j N_j \left(h^0_{f_j} + \Delta h\right)_{produced\ gas} + \sum_j N_j \left(h^0_{f_j} + \Delta h\right)_{Air} = \sum_j N_j \left(h^0_{f_j} + \Delta h\right)_{products} \tag{9.15}$$

9.3.1.3 Thermoelectric generator

TEG is a promising innovative option for direct conversion of heat into electricity due to the reliability, silently operated, and fuel flexibility. The TEG output power as a function of TEG efficiency (η_{TEG}) and heat ($\dot{Q}_{TEG}$) is calculated as follows (Siddique, Mahmud, & Heyst 2017):

$$\dot{W}_{TEG} = \eta_{TEG}\dot{Q}_{TEG} \tag{9.16}$$

$$\dot{Q}_{TEG} = \dot{m}_{20}\left(h_{21} - h_{20}\right) \tag{9.17}$$

$$\eta_{TEG} = \eta_{Carnot}\frac{\sqrt{1+ZT_M} - 1}{\sqrt{1+ZT_M} + \frac{T_c}{T_h}} \tag{9.18}$$

$$\eta_{Carnot} = 1 - \frac{T_c}{T_h} \tag{9.19}$$

Here T_c is the cold side temperature, and T_h denotes the hot side temperature. Moreover, ZT_M, which is a significant parameter evaluating the inner performance efficiency of the TEG, is defined as (Siddique, Mahmud, & Heyst 2017):

$$ZT_M = \frac{\psi^2 T_m}{KR} \tag{9.20}$$

$$T_M = \frac{1}{2}\left(T_c + T_h\right) \tag{9.21}$$

$$\psi = \frac{-\Delta V}{\Delta T} \tag{9.22}$$

Here K is the thermal conductivity, and R is the thermal resistance inside the TEG.

9.3.1.4 Exergoeconomic evaluation

Exergoeconomic assessment is a significant criterion to determine every components' cost-effectiveness by comparing the investment and exergy destruction costs. For each component, the cost balance equation contemplating the component ($\dot{Z}_k^{PY}$) and the exergy stream ($\dot{C}$) costs can be written as (Mojaver, Khalilarya, & Chitsaz 2020):

$$\sum \dot{C}_{out.k} + \dot{C}_{w.k} = \sum \dot{C}_{in.k} + \dot{C}_{q.k} + \dot{Z}_k^{PY} \tag{9.23}$$

$$\dot{Z}_k^{PY} = \dot{Z}_k^{CI} + \dot{Z}_k^{OM} \tag{9.24}$$

$$\dot{Z}_k^{PY} = \frac{cost\ index^{PY}}{cost\ index^{RY}} \times \dot{Z}_k^{RY} \tag{9.25}$$

$$\dot{C}_{in} = c_{in} \cdot \dot{E}_{in} \tag{9.26}$$

$$\dot{C}_{out} = c_{out} \cdot \dot{E}_{out} \tag{9.27}$$

$$\dot{C}_q = c_q \cdot \dot{E}_q \tag{9.28}$$

$$\dot{C}_w = c_w \cdot \dot{W} \tag{9.29}$$

A Marshall and Swift cost equation, Equation (9.25), is applied to convert the component cost from the reference to the present year (2021, with the cost index of 681.47). Moreover, the component cost is the sum of capital investment ($\dot{Z}_k^{CI}$) and operating and maintenance ($\dot{Z}_k^{OM}$) costs as follows (Vojdani, Fakhari, & Ahmadi 2021):

$$\dot{Z}_k^{CI} = \left(\frac{CRF}{\tau}\right) Z_k \tag{9.30}$$

$$\dot{Z}_k^{OM} = \left(\frac{\gamma_k}{\tau}\right) Z_k + \omega_k \dot{E}_{P.k} + \dot{R}_k \tag{9.31}$$

$$CRF = \frac{i_r\left(1+i_r\right)^n}{\left(1+i_r\right)^n - 1} \tag{9.32}$$

Where τ, γ, and ω are the yearly operation hours, fixed and variable operating cost coefficients. Also, $\dot{E}_P$ denotes the produced exergy and $\dot{R}_k$ is the other operating and maintenance costs, which is negligible. Besides, i_r is the interest rate, and n is the plant working hours. Eventually, Z is the component cost as listed in Table 9.3.

After solving the cost equation and finding the investment costs for each component, the exergoeconomic assessment is accomplished by calculating the exergoeconomic factor (f_k) as follows (Mojaver et al. 2020):

$$f_k = \frac{\dot{Z}_k}{\dot{Z}_k + \dot{C}_{D.k+}\dot{C}_{L.k}} \tag{9.33}$$

$$c_{F.k} = \frac{\dot{C}_{F.k}}{\dot{E}_{F.k}} \tag{9.34}$$

$$c_{P.k} = \frac{\dot{C}_{P.k}}{\dot{E}_{P.k}} \tag{9.35}$$

Table 9.3 The purchased cost and reference year for each component

Component	*Purchased cost*	*Reference year*	*Cost index*
Gasifier	$Z_G = 1600\left(\dot{m}_{MSW}\right)^{0.67}$	2013	530.6
Combustion chamber	$Z_{cc} = 48.64\dot{m}_{air}\left(1 + exp\ exp\left(0.018T_{out} - 26.4\right)\right)\frac{1}{0.995 - \frac{P_{out}}{P_{in}}}$	2013	530.6
Steam generator	$\dot{Z}_{SG} = 6570\left(\left(\frac{Q_{ec}}{\Delta T_{lm,ec}}\right)^{0.8} + \left(\frac{Q_{ev}}{\Delta T_{lm,ev}}\right)^{0.8}\right) + 21276\dot{m}_{water} + 1184.4\dot{m}_{syngas}{}^{1.2}$	1997	390.1
Condenser	$Z_{COND} = 1773\dot{m}_{water}$	2010	522.8
Turbine	$Z_{ST} = 6000\dot{W}_{ST}{}^{0.7}$	2010	522.8
Pump	$Z_{Pm} = 3540\dot{W}_{Pm}{}^{0.71}$	2010	522.8
TEG	$Z_{TEG}=1500\dot{W}_{TEG}$	2013	530.6
ORC pump	$Z_{ORC,Pm} = 200\,W_{ORC,Pm}{}^{0.65}$	2010	522.8
ORC turbine	$Z_{ORC,T} = 4750\dot{W}_{ORC,T}{}^{0.75}$	2010	522.8
ORC evaporator	$Z_{ORC,Ev} = 309.14A_{HROG}{}^{0.85}$	2003	402.3
ORC condenser	$Z_{ORC,Cond} = 516.62A_{ORC,Cond}{}^{0.6}$	2005	468.2

Sources: Fakhari et al. (2021); Gholamian, Mahmoudi, and Zare (2016).

$$\dot{C}_{D.k} = c_{F.k}\dot{E}_{D.k} \tag{9.36}$$

$$\dot{C}_{L.k} = c_{F.k}\dot{E}_{L.k} \tag{9.37}$$

Here $c_{F,k,}$ and $c_{P,k}$ are the fuel and product unit costs for each component. Also, $\dot{C}_{D,k}$ and $\dot{C}_{L,k}$ are the exergy destruction and loss cost rates, respectively.

9.3.1.5 *Performance criteria*

Crucial objectives, comprising net produced power (NPP), energy and exergy efficiencies, total cost rate (TCR), levelized cost of power (LCOP), payback period (PP), sustainability index (SI), and levelized carbon dioxide emission (LCE), are calculated to evaluate and compare each system's performance from performance, economic, and environmental standpoints. Energy and exergy efficiency is defined as follows:

$$\eta_I = \frac{NPP}{2\times \dot{m}_{MSW} LHV_{MSW}} \tag{9.38}$$

$$\eta_{II} = \frac{NPP}{2\times \dot{m}_{MSW}\dot{E}_{in}} \tag{9.39}$$

$$\text{Conventional plant}: NPP = 2\times\left(\dot{W}_{ST} - \dot{W}_{Pu}\right) + \dot{W}_{ORC,T} - \dot{W}_{ORC,Pu} \tag{9.40}$$

$$\text{Proposed plant}: NPP = 2\times\left(\dot{W}_{ST} + \dot{W}_{TEG} - \dot{W}_{Pu}\right) + \dot{W}_{ORC,T} + \dot{W}_{ORC,TEG} - \dot{W}_{ORC,Pu} \tag{9.41}$$

$$\dot{E}_{in} = e^{ch}_{MSW} + we^{ch}_{water} \tag{9.42}$$

$$e^{ch}_{MSW} = LHV_{MSW}\frac{1.044 + 0.016\frac{z_H}{z_C} - 0.3493\frac{z_O}{z_C}\left(1 + 0.0531\frac{z_H}{z_C}\right)}{1 - 0.4124\frac{z_O}{z_C}} \tag{9.43}$$

Moreover, the economic objectives including, total cost rate, levelized cost of power, and the payback period can be calculated as follows, respectively:

$$TCR = \sum_{k=1}^{nk}\dot{Z}_k + \sum_{i=1}^{nF}\dot{C}_{F_i} \tag{9.44}$$

$$LCOP = \frac{TCR}{NPP} \tag{9.45}$$

$$PP = \frac{\sum_{k=1}^{n_k} Z_k}{NPP \times \tau \times c_{ele}} \tag{9.46}$$

Additionally, the sustainability index as criteria assessing the influence of exergy destruction reduction on the improvement of the system's environmental friendliness is defined as:

$$SI = \frac{1}{D_P} \tag{9.47}$$

In this equation, D_P denotes the depletion factor, which is equal to the ratio of destructed exergy to the input exergy as follows:

$$D_P = \frac{\sum_{k=1}^{n_k}\dot{E}_{D,k}}{2\times \dot{m}_{MSW}\dot{E}_{in}} \tag{9.48}$$

Finally, the environmental comparison of each system is compared by assessing the levelized carbon dioxide emission in t/MWh as:

$$LCE = \frac{2\times \dot{m}_{CO_2}}{NPP} \tag{9.49}$$

In this equation, $\dot{m}_{CO_2}$ is the produced carbon dioxide emitted to the atmosphere.

Table 9.4 The range of significant decision parameters for optimization

Variable	*Lower bound*	*Higher bound*
MC	0.3	0.4
$T_{Combustion}$ (°C)	1,050	1,250
TIP (kPa)	1,200	3,200
ΔT_{PP} (°C)	170	230
ZT_m	0.4	1.2
ORC TIP (kPa)	1,200	2,800

Table 9.5 Tuning parameters used for the genetic algorithm optimization approach

Parameter	*Value*
Population size	150
Maximum number of generation	200
Crossover probability	85%
Mutation probability	1%
Selection process	Tournament
Tournament size	2

9.3.1.6 *Multi-criteria genetic optimization*

A multi-criteria optimization is a robust tool facing more than one objective function required to be optimized simultaneously. It is implemented in thermal energy systems to obtain a better performance efficiency while decreasing the cost and environmental contamination. Among different optimization approaches, many researchers have widely used genetic algorithms due to the capability to solve problems that cannot be solved with other techniques. Furthermore, a genetic algorithm is superior to other optimization methods from calculation speed and fitness value points of view. The genetic algorithm method using developed MATLAB code is applied to ascertain the most favorable techno-economic-environmental condition in the present study. The optimum value of design parameters is determined to maximize the exergy efficiency and minimize the levelized carbon dioxide emission and total cost rate. The logical domain of significant design parameters is tabulated in Table 9.4. Moreover, the input parameters used for executing the genetic algorithm are listed in Table 9.5.

9.4 RESULTS AND DISCUSSION

Thermodynamic, exergoeconomic, sustainability, and environmental comparison of each system is investigated by evaluating the effect of significant operational parameters on net produced power, total cost rate, energy and exergy efficiencies, levelized cost of power, payback period, sustainability index, and levelized carbon dioxide emission through a parametric study. The exergy destruction rate is then evaluated to compare each component's performance from the second law of thermodynamics viewpoint. Finally, the proposed innovative waste-fired power plant is optimized, contemplating levelized cost of power, exergy efficiency, and levelized carbon dioxide emission as objectives.

Because the performance of the gasifier is highly affected by the specification of the municipal solid waste, the influence of moisture content on the techno-economic, sustainability, and environmental standpoints of each system are presented and compared in Figure 9.2. By increasing the moisture content, the enthalpy of generated syngas and the heat transfer rate to the Rankine cycle and ORC unit decrease; therefore, the mass flow rate of water and R123 will decrease. Accordingly, when the moisture content increases from 0.3 to 0.4, the net produced power and energy and exergy efficiencies for both systems decrease, as illustrated in Figure 9.2. As depicted, when the moisture content increases, conventional and proposed power plants' total cost rate decreases to about 55.3 \$/h and 59 \$/h, respectively. This is reasonable since the purchased cost of the turbine is a direct function of produced power. The figure further shows that moisture content growth leads to a higher levelized cost of power and a payback period that is not suitable. Higher moisture content leads to more irreversibility of the gasifier, so the total exergy destruction increases, and the sustainability index decreases, as shown in Figure 9.2(d). Finally, the figure indicates that an unfavorable environmental condition is obtained when the moisture content rises in addition to the negative effect on techno-economic facets. Furthermore, as the moisture content increases from 0.3 to 0.4, the levelized carbon dioxide emission increases from 2.47 t/MWh and 2.34 t/MWh to 2.59 t/MWh and 2.46 t/MWh for the conventional and proposed system, respectively.

Figure 9.3 shows the net produced power variation, total cost rate, energy and exergy efficiencies, levelized cost of power, payback period, sustainability index, and levelized carbon dioxide emission with the combustion temperature. According to the figure, when the combustion temperature increases, the performance indicators, i.e., the net present value and

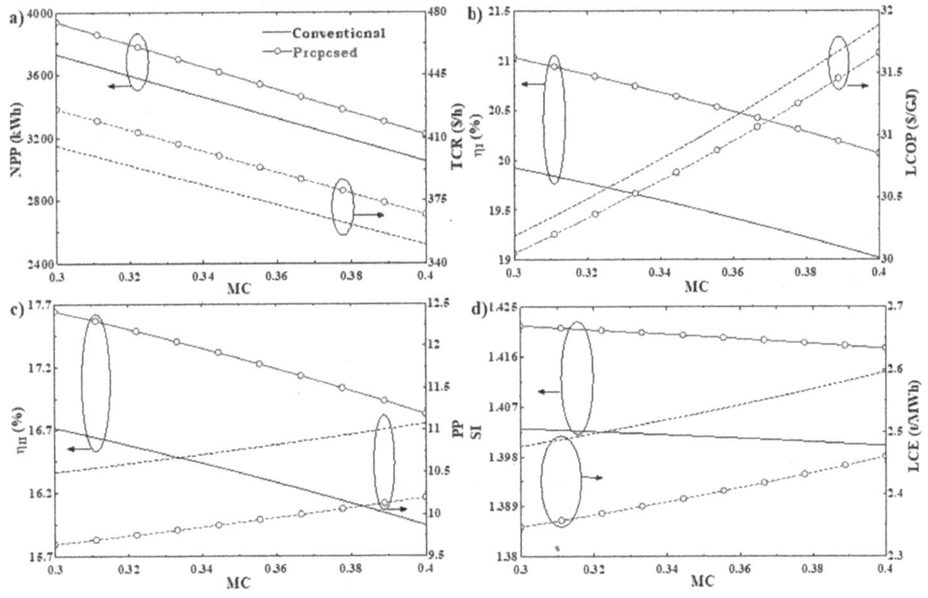

Figure 9.2 The variation of techno-economic-environmental metrics with the municipal solid waste moisture content.

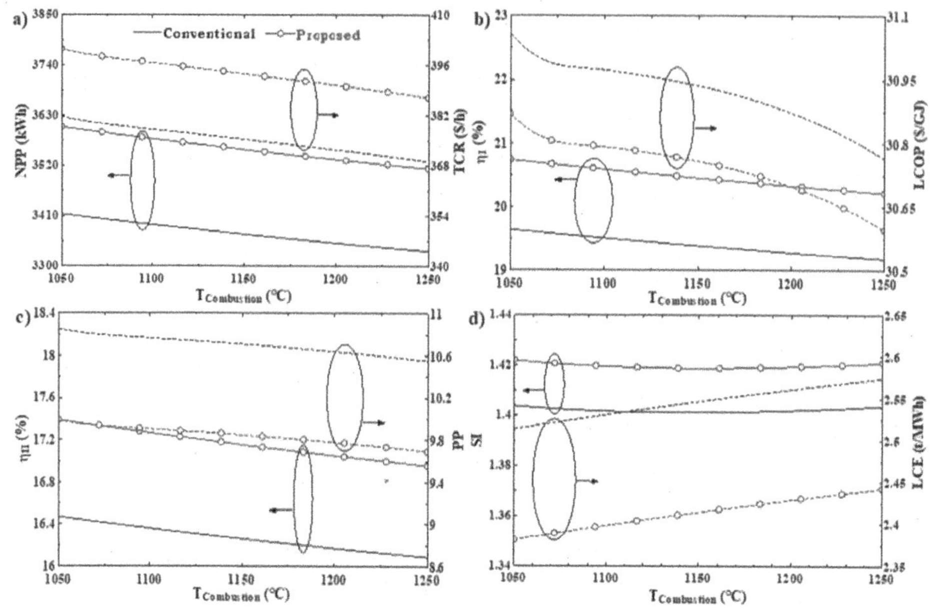

Figure 9.3 The variation of techno-economic-environmental metrics with the combustion temperature.

energy and exergy efficiencies, will decrease. This is justified because by increasing the combustion temperature, while the steam turbine generated electricity increases, the ORC turbine power decreases due to the enthalpy reduction of exhaust gases going into the ORC evaporator (at a constant pinch point temperature). The figure further illustrates that the increase in combustion temperature is environmentally favorable since the total cost rate, levelized cost of power, and payback period for each system decrease. In contrast, Figure 9.3(d) shows that undesirable environmental conditions are achieved as the combustion temperature increases from 1,050°C to 1,250°C. What stands out from the figure is that the variation of combustion temperature has a neutral effect on each system's sustainability index. According to Figure 9.3, in the whole domain of combustion temperature, the proposed novel power plant is superior to the conventional system from techno-economic-environmental aspects due to the higher net produced power, performance efficiencies, and sustainability, lower levelized cost of power, payback period, and levelized carbon dioxide emission.

More turbine inlet pressure results in a higher steam enthalpy; therefore, the net produced power and performance efficiencies will increase, as presented in Figure 9.4. In contrast, the figure indicates that the increase in TIP negatively affects the plants' initial costs because of the increase in the total cost rate. As indicated, when the TIP increases from 1,200 kPa to 3,200 kPa, the levelized cost of power and payback period for each plant decrease, which is economically favorable. According to Figure 9.4(d), each plant's sustainability index increases as the TIP increases. This is justified since, for a constant input exergy value, the rate of total exergy destruction declines. The figure further presents that the increase in TIP is environmentally suitable since the levelized carbon dioxide emission of the conventional

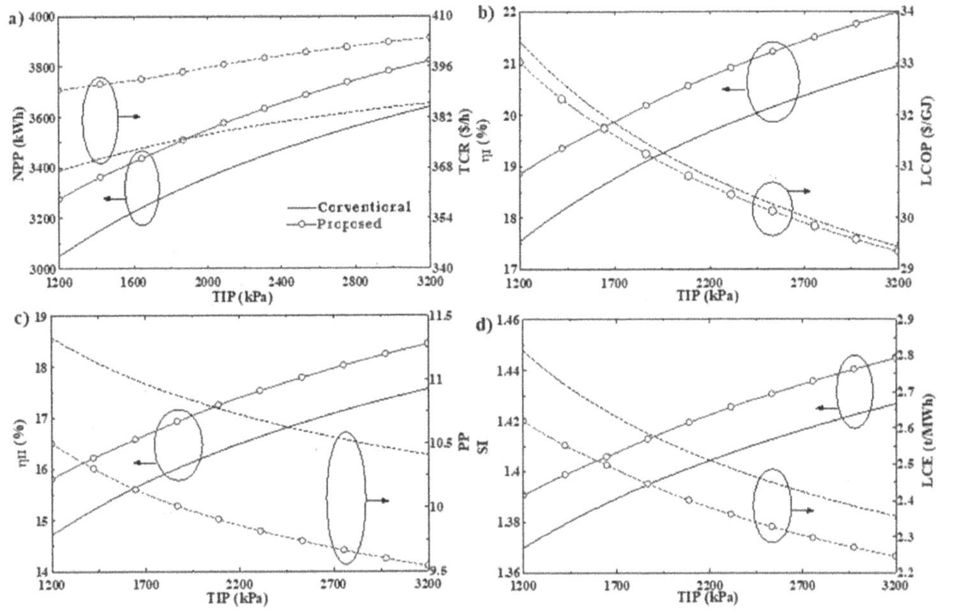

Figure 9.4 The variation of techno-economic-environmental metrics with the turbine inlet pressure.

and proposed novel plants decreases about 0.49 t/MWh and 0.38 t/MWh. Eventually, it can be concluded that, as the previous figure, in all TIP domains, the proposed innovative plant integrated with TEG is an excellent option from all aspects.

The influence of evaporator pinch point temperature difference on plants is illustrated in Figure 9.5. According to the figure, as the temperature difference increases, the net produced power and performance efficiencies for both systems increase. The figure further depicts that when the temperature difference rises from 170°C to 230°C, the conventional and proposed plants' total cost rate increase from 365.8 \$/h and 383.1 \$/h to 379.6 \$/h to 398.3 \$/h, respectively. According to Figure 9.5(b), for the conventional and proposed plants, when the temperature difference increases, the levelized cost of power increases up to 31.3 \$/GJ and 31.1 \$/GJ, respectively, and then decreases considerably, which is favorable. It can also be concluded that the temperature difference has a relatively neutral effect on the payback period. This is rational since the total cost rate and net produced power will increase simultaneously, so their ratio remains unchanged. The more pinch point temperature difference results in an upper-temperature difference between the cold and hot sides; ergo, higher irreversibility is obtained, and each system's sustainability index will decrease, as illustrated in Figure 9.5(d). The simultaneous increase in the total cost rate, which is unfavorable, and net produced power, which is suitable, illustrates the significance of multi-criteria optimization trying to find a trade-off between conflictive objectives.

In Figure 9.6, the influence of the figure of merit, which is a significant parameter affecting the TEG performance, is investigated. According to the figure, as the figure of merit rises, the TEG produced power increases too. Therefore, the net produced power and thermodynamic

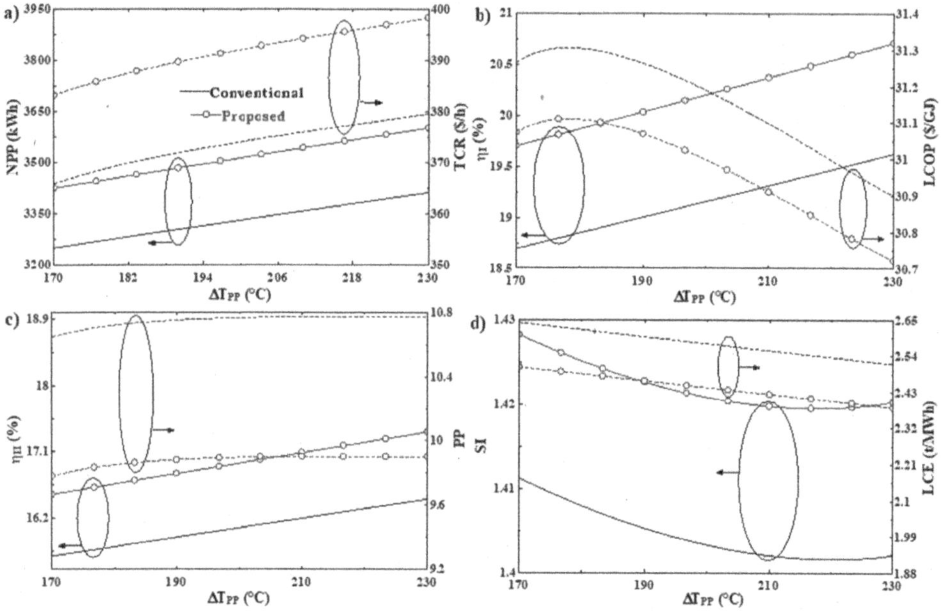

Figure 9.5 The variation of techno-economic-environmental metrics with the pinch point temperature difference.

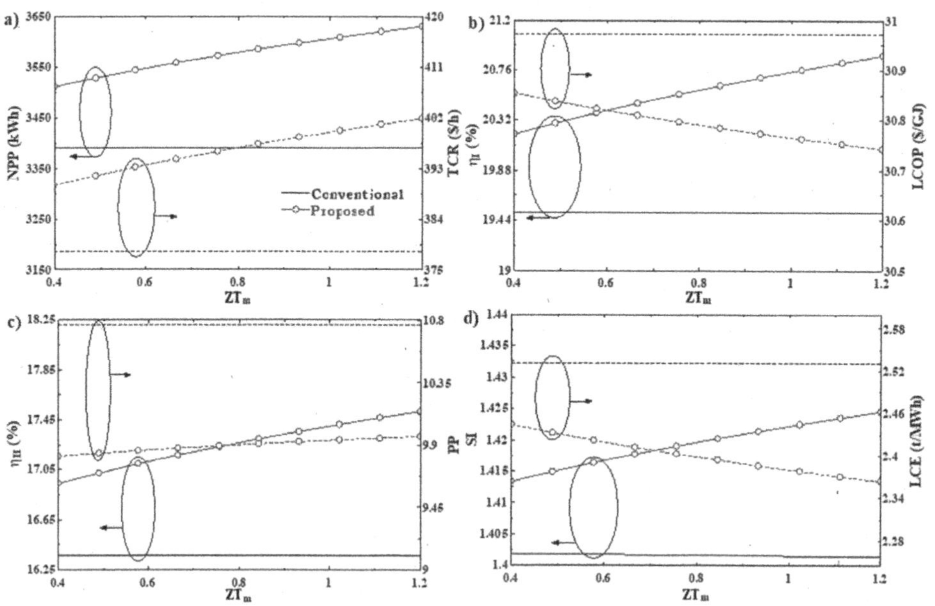

Figure 9.6 The variation of techno-economic-environmental metrics with the figure of merit.

efficiencies of the proposed novel system will increase. A higher TEG produced power leads to a higher TEG purchased cost; hence, the proposed plant's total cost rate increases as shown in the figure. It can also be obtained that the levelized cost of power and carbon dioxide emission will decrease by increasing the figure of merit. Moreover, when the figure of merit increases from 0.4 to 1.2, the proposed novel system's sustainability index increases from 1.413 to 1.435. The figure further shows that the proposed innovative plant's payback period is not highly affected by the figure of merit.

Figure 9.7 illustrates the exergy analysis results comprising the destructions and losses for both systems. According to the figure, the waste gasifier has the highest destruction rate because of the mixing and chemical reaction of inlet waste and air and the high temperature difference between the reactants and products. The figure further indicates that the steam generator is another deficient component from the quality of energy component with the second-highest destruction rate (2035 kW). The comparison of the exergy destruction rate of the condenser (Figure 9.7(a)) and TEG (Figure 9.7(b)) illustrates the excellence of TEG from the irreversibility standpoint due to a lower destruction rate of 80.5 kW. According to the figure, the ORC unit integrated with TEG (the proposed innovative system) operates better than the conventional ORC equipped with a condenser from the thermodynamic aspect due to a lower destruction rate of 26.26 kW as depicted in the figure. It can also be concluded that the flue gas condensation process is of great significance to recover the waste energy since about 11% of the total exergy destruction corresponds to the exergy loss, which is considerable.

The comparative parametric study results reveal that the proposed innovative waste-driven power plant integrated with ORC and TEG unit is the superior option from all aspects. Therefore, the proposed system is optimized to simultaneously find the most favorable conditions from techno-economic and environmental standpoints. The Pareto frontier diagram of exergy efficiency, levelized carbon dioxide emission, and total cost rate for the proposed innovative plant are illustrated in Figure 9.8. According to the figure, the ideal

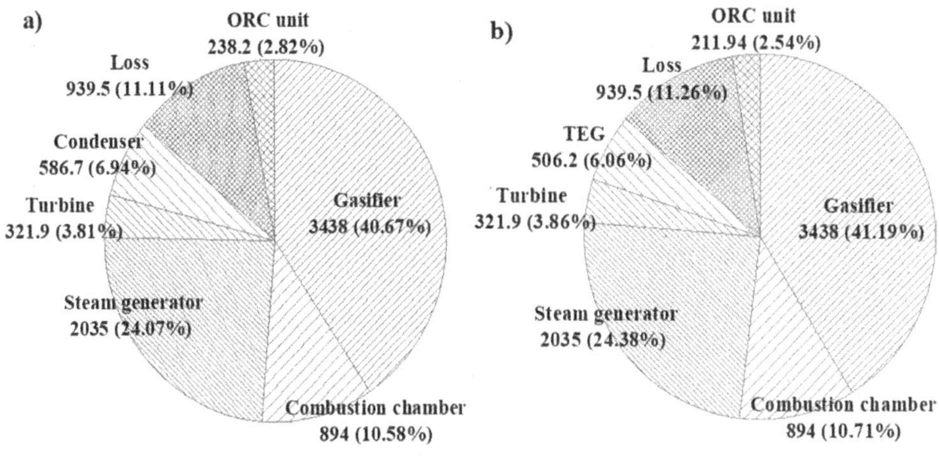

Figure 9.7 Results of exergy analysis (in kW) for (a) conventional plant and (b) proposed innovative plant.

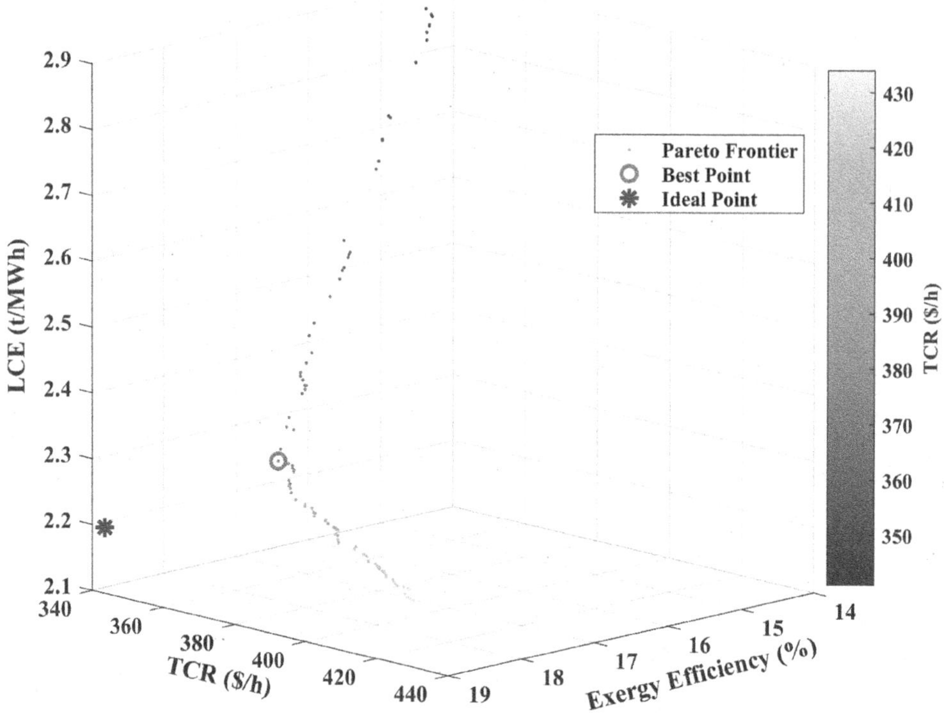

Figure 9.8 Three-dimensional Pareto frontier for the proposed system.

point with the maximum exergy efficiency and minimum total cost rate, and levelized carbon dioxide emission is not on the Pareto curve. Therefore, the nearest point to the ideal point is designated as the best optimization answer.

Furthermore, a two-dimensional Pareto projection of objective functions is presented in Figure 9.9 to better understand the Pareto frontier diagram.

Detailed Pareto frontier diagram information, including the optimum value of objective functions and significant decision variables at the best point, is tabulated in Table 9.6. As indicated, the MSW moisture content, turbine inlet pressure, pinch point temperature difference, the figure of merit, and ORC turbine inlet pressure have optimum values of 0.39, 1,209.35°C, 3,175.85 kPa, 206.42°C, 0.74, and 1,404.96 kPa. The table further indicates that the optimum values of exergy efficiency, total cost rate, and levelized carbon dioxide emission is 17.93%, 370.81, and 2.31 t/MWh.

Figure 9.10 illustrates the scatter distribution of significant decision variables for better comprehension of the optimum domains. According to the figure, the optimum moisture content points are distributed between 0.32 and 0.4, where more points are near the upper bound. The figure further shows that a high value of turbine inlet pressure should be selected to obtain the optimum condition, illustrating a sensitive parameter. It can also be concluded that most combustion chamber temperature optimum points are dispensed from 1,000°C to 1,170°C. As depicted, the figure of merit and pinch point temperature difference are not

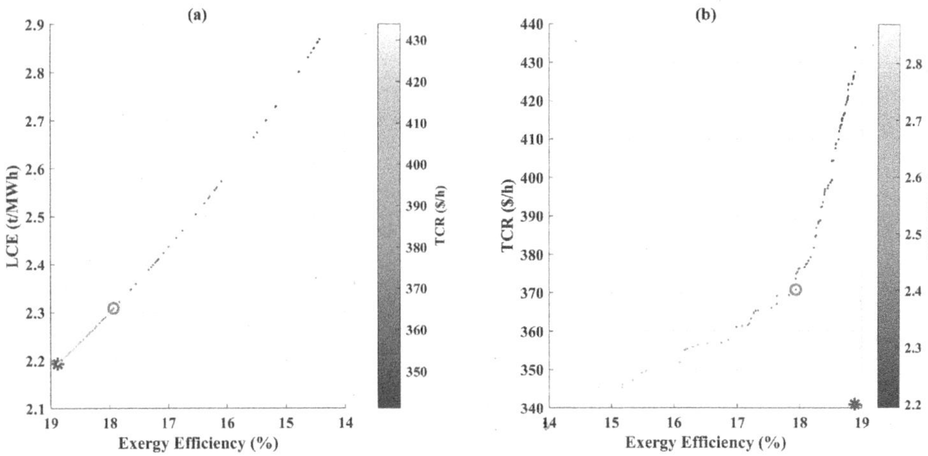

Figure 9.9 Two-dimensional Pareto projection for the proposed system.

Table 9.6 The value of objective functions and significance decision variables at the best point

		Significant parameters			
MC	*$T_{Combustion}$ (°C)*	*TIP (kPa)*	*ΔT_{PP} (°C)*	*ZT_m*	*ORC TIP (kPa)*
0.39	1,209.35	3,175.85	206.42	0.74	1,404.96

	Objective functions	
Exergy efficiency (%)	*Total cost rate ($/h)*	*Levelized carbon dioxide emission (t/MWh)*
17.93	370.81	2.31

sensitive parameters while more points are adjacent to their higher bounds. What stands out from the figure is that the ORC turbine inlet pressure should be kept between 1,400 kPa and 2,400 kPa to reach the best optimal condition.

9.5 CONCLUSION

This chapter introduces a novel waste-driven plant integrated with a gasifier, organic Rankine cycle, and thermoelectric generator for lower emission and higher power production and sustainability. The proposed system's performance is examined and compared against the conventional plant equipped with condenser from exergoeconomic, sustainability, and environmental viewpoints. For this, a detailed comparative parametric investigation is accomplished to assess the influence of significant variables on produced power, performance efficiencies, total cost rate, levelized cost of power, payback period, sustainability index, and levelized carbon dioxide emission of each system. The proposed plant is then optimized, applying the genetic algorithm method to obtain higher exergy efficiency, lower

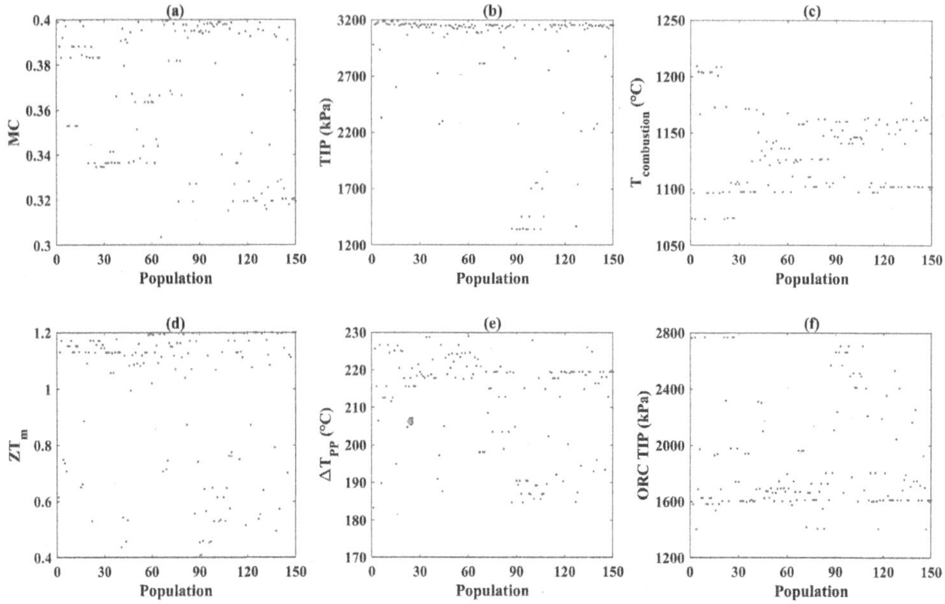

Figure 9.10 Scatter distribution of (a) moisture content, (b) combustion chamber temperature, (c) turbine inlet pressure, (d) figure of merit, (e) pinch point temperature difference, (f) ORC turbine inlet pressure.

total cost rate, and levelized carbon dioxide emissions. The main performance comparison of the proposed system and the conventional waste-driven plant is summarized as:

- The recommended system is techno-economically superior to the conventional system due to the higher produced power and energy and exergy efficiencies of 187 kW, 1.08%, and 0.91% and lower payback period and levelized cost of power of 0.87 year and 0.18 $/GJ, respectively.
- Higher sustainability index and lower levelized carbon dioxide emission indicate that in addition to techno-economic excellence, the integration of the conventional power plant with organic Rankine cycle and thermoelectric generator is a promising option from an environmental standpoint.
- The exergy assessment results reveal that the thermoelectric generator operates better than the condenser from the irreversibility aspect because of the lower exergy destruction rate.

The significant parametric investigation findings are written as below:

- Lower moisture content and combustion temperature result in considerably higher performance efficiencies.
- By increasing the pinch point temperature difference and turbine inlet pressure while performance indicators improve, unfavorable economic conditions are obtained revealing the importance of multi-criteria optimization.

- The sustainability index and emission indicators are susceptible to the moisture content and turbine inlet pressure.

Eventually, the main optimization results could be outlined as:

- Multi-criteria optimization of the proposed system leads to 0.66% more exergy efficiency and 25.89 \$/h lower total cost rate, and 0.11 t/MWh lower levelized carbon dioxide emission than the operating condition.
- Turbine inlet pressure and moisture content should be kept at their highest domain to achieve the best optimum condition.
- The figure of merit is an insensitive parameter because its optimum point is dispensed in the whole domain of the scatter distribution.

NOMENCLATURE AND ABBREVIATIONS

c	Exergy cost (\$/GJ)
$\dot{C}$	Cost rate (\$/hr)
CIP	Condenser inlet pressure (kPa)
$\dot{E}$	Exergy rate (kW)
f	Exergoeconomic factor
$\underline{\Delta G}^{0}$	Change in Gibbs function (kJ/kmole)
h	Specific enthalpy (kJ/kg)
$\underline{h}_f^c$	Enthalpy of formation (kJ/kmole)
i_r	Interest rate
K	Equilibrium constant
LCE	Levelized carbon dioxide emission (t/MWh)
$LCOP$	Levelized cost of power (\$/GJ)
$\dot{m}$	Mass flowrate (kg/s)
MC	Moisture content (%)
NPV	Net produced power (kW)
$\dot{Q}$	Heat (kW)
SI	Sustainability index
TCR	Total cost rate (\$/h)
$\Delta T_{\text{Cooling air}}$	Condenser/TEG cooling air temperature difference (°C)
$\Delta T_{\text{Cooling water}}$	Condenser/TEG cooling water temperature difference (°C)
ΔT_{PP}	Pinch point temperature difference (°C)
ΔT_{SH}	Superheat temperature difference (°C)
TIP	Turbine inlet pressure (kPa)
w	a mole of moisture per mole of MSW
X_i	The ith component mole fraction
$\dot{Z}$	Investment cost rate of component (\$/hr)

Subscript and abbreviations

0	Dead state
1,2, 3...	State points
c	The cold side
CC	Combustion chamber
Ch	Chemical
CI	Capital investment
CRF	Capital recovery factor
D	Destruction
EES	Engineering equation solver
h	The hot side
i	Inlet
L	Loss
lm	Log mean temperature difference
ORC	Organic Rankine cycle
P	Product
Pu	Pump
ST	Steam turbine
SG	Steam generator
TEG	Thermoelectric generator
tot	Total

Greek symbols

η_{II}	Exergy efficiency
η_{I}	Energy efficiency
η_{is}	Isentropic efficiency
τ	Annual plant operation hours (hr)

REFERENCES

Aliahmadi, Mohammad, Ali Moosavi, & Hani Sadrhosseini. 2021. "Multi-objective optimization of regenerative ORC system integrated with thermoelectric generators for low-temperature waste heat recovery." *Energy Reports* 7: 300–313. https://doi.org/10.1016/j.egyr.2020.12.035.

Arabkoohsar, Ahmad, Amirmohammad Behzadi, & Ali Sulaiman Alsagri. 2021. "Techno-economic analysis and multi-objective optimization of a novel solar-based building energy system: An effort to reach the true meaning of zero-energy buildings." *Energy Conversion and Management* 232: 113858. https://doi.org/10.1016/j.enconman.2021.113858.

Arabkoohsar, Ahmad, & Hossein Nami. 2019. "Thermodynamic and economic analyses of a hybrid waste-driven CHP–ORC plant with exhaust heat recovery." *Energy Conversion and Management* 187: 512–522. https://doi.org/10.1016/j.enconman.2019.03.027.

Arora, Ranjana, & Rajesh Arora. 2018. "Multiobjective optimization and analytical comparison of single- and 2-stage (series/parallel) thermoelectric heat pumps." *International Journal of Energy Research* 42(4): 1760–1778. https://doi.org/10.1002/er.3988.

Basu, Prabir. 2010. *Biomass Gasification and Pyrolysis: Practical Design and Theory*. Cambridge, MA: Academic Press.

Behzadi, Amirmohammad, Ahmad Arabkoohsar, & Ehsan Gholamian. 2020. "Multi-criteria optimization of a biomass-fired proton exchange membrane fuel cell integrated with organic Rankine cycle/thermoelectric generator using different gasification agents." *Energy* 201: 117640. https://doi.org/10.1016/j.energy.2020.117640.

Behzadi, Amirmohammad, Ahmad Arabkoohsar, & Vedran S. Perić. 2021. "Innovative hybrid solar-waste designs for cogeneration of heat and power, an effort for achieving maximum efficiency and renewable integration." *Applied Thermal Engineering* 190. https://doi.org/10.1016/j.applthermaleng.2021.116824.

Behzadi, Amirmohammad, Ahmad Arabkoohsar, & Yongheng Yang. 2020. "Optimization and dynamic techno-economic analysis of a novel PVT-based smart building energy system." *Applied Thermal Engineering* 181: 115926. https://doi.org/10.1016/j.applthermaleng.2020.115926.

Behzadi, Amirmohammad, Ehsan Gholamian, Ehsan Houshfar, & Ali Habibollahzade. 2018. "Multi-objective optimization and exergoeconomic analysis of waste heat recovery from Tehran's waste-to-energy plant integrated with an ORC unit." *Energy* 160: 1055–1068. https://doi.org/10.1016/j.energy.2018.07.074.

Behzadi, Amirmohammad, Ali Habibollahzade, Pouria Ahmadi, Ehsan Gholamian, & Ehsan Houshfar. 2019. "Multi-objective design optimization of a solar based system for electricity, cooling, and hydrogen production." *Energy* 169. https://doi.org/10.1016/j.energy.2018.12.047.

Biswas, Souvik, Ayan Roynaskar, Chetan Kumar Hirwani, & Subrata Kumar Panda. 2018. "Design and fabrication of thermoelectric waste heat reutilization system: Possible industrial application." *International Journal of Energy Research* 42(12): 3977–3986. https://doi.org/10.1002/er.4157.

Eboh, Francis Chinweuba, Bengt-Åke Andersson, & Tobias Richards. 2019. "Economic evaluation of improvements in a waste-to-energy combined heat and power plant." *Waste Management* 100: 75–83. https://doi.org/10.1016/j.wasman.2019.09.008.

Fakhari, Iman, Amirmohammad Behzadi, Ehsan Gholamian, Pouria Ahmadi, & Ahmad Arabkoohsar. 2020. "Design and tri-objective optimization of a hybrid efficient energy system for tri-generation of power, heat, and potable water." *Journal of Cleaner Production*, 290: 125205. https://doi.org/10.1016/j.jclepro.2020.125205.

Fakhari, Iman, Amirmohammad Behzadi, Ehsan Gholamian, Pouria Ahmadi, & Ahmad Arabkoohsar. 2021. "Comparative double and integer optimization of low-grade heat recovery from PEM fuel cells employing an organic Rankine cycle with zeotropic mixtures." *Energy Conversion and Management* 228: 113695. https://doi.org/10.1016/j.enconman.2020.113695.

Fang, Yuwen, Fubin Yang, & Hongguang Zhang. 2019. "Comparative analysis and multi-objective optimization of organic Rankine cycle (ORC) using pure working fluids and their zeotropic mixtures for diesel engine waste heat recovery." *Applied Thermal Engineering* 157: 113704. https://doi.org/10.1016/j.applthermaleng.2019.04.114.

Georgousopoulos, Serafim, Konstantinos Braimakis, Dimitrios Grimekis, & Sotirios Karellas. 2021. "Thermodynamic and techno-economic assessment of pure and zeotropic fluid orcs for waste heat recovery in a biomass IGCC plant." *Applied Thermal Engineering* 183(P2): 116202. https://doi.org/10.1016/j.applthermaleng.2020.116202.

Gholamian, Ehsan, Ali Habibollahzade, & Vahid Zare. 2018. "Development and multi-objective optimization of geothermal-based organic Rankine cycle integrated with thermoelectric generator and proton exchange membrane electrolyzer for power and hydrogen production." *Energy Conversion and Management* 174: 112–125. https://doi.org/10.1016/J.ENCONMAN.2018.08.027.

Gholamian, Ehsan, S. M. Seyed Mahmoudi, & Vahid Zare. 2016. "Proposal, exergy analysis and optimization of a new biomass-based cogeneration system." *Applied Thermal Engineering* 93: 223–235. https://doi.org/10.1016/j.applthermaleng.2015.09.095.

Holik, Mario, Marija Zivi, Zdravko Virag, Antun Barac, & Milan Vujanovi. 2021. "Thermo-economic optimization of a Rankine cycle used for waste-heat recovery in biogas cogeneration plants." *Energy Conversion and Management* 232: 113897. https://doi.org/10.1016/j.enconman.2021.113897.

Hussain, Danish, Meeta Sharma, & Anoop Kumar Shukla. 2021. "Investigative analysis of light duty diesel engine through dual loop organic Rankine cycle." *Materials Today: Proceedings* 38: 146–52. https://doi.org/10.1016/j.matpr.2020.06.166.

Khan, Razzak, Anoop Kumar Shukla, Meeta Sharma, Rakesh Kumar Phanden, & Shivam Mishra. 2021. "Thermodynamic investigation of intercooled reheat gas turbine combined cycle with carbon capture and methanation." *Materials Today: Proceedings* 38: 449–455. https://doi.org/10.1016/j.matpr.2020.07.680.

Khanmohammadi, Shoaib, Morteza Saadat-Targhi, Abdullah A. A. A. Al-Rashed, & Masoud Afrand. 2019. "Thermodynamic and economic analyses and multi-objective optimization of harvesting waste heat from a biomass gasifier integrated system by thermoelectric generator." *Energy Conversion and Management* 195: 1022–1034. https://doi.org/10.1016/j.enconman.2019.05.075.

Lu, Fulu, Yan Zhu, Mingzhang Pan, Chao Li, Jiwen Yin, & Fuchuan Huang. 2020. "Thermodynamic, economic, and environmental analysis of new combined power and space cooling system for waste heat recovery in waste-to-energy plant." *Energy Conversion and Management* 226: 113511. https://doi.org/10.1016/j.enconman.2020.113511.

Maria, Francesco Di, Stefano Contini, Gianni Bidini, Antonio Boncompagni, Marzio Lasagni, & Federico Sisani. 2016. "Energetic efficiency of an existing waste to energy power plant." *Energy Procedia* 101: 1175–1182. https://doi.org/10.1016/j.egypro.2016.11.159.

Mojaver, Parisa, Ata Chitsaz, Mohsen Sadeghi, & Shahram Khalilarya. 2020. "Comprehensive comparison of SOFCs with proton-conducting electrolyte and oxygen ion-conducting electrolyte: thermoeconomic analysis and multi-objective optimization." *Energy Conversion and Management* 205: 112455. https://doi.org/10.1016/j.enconman.2019.112455.

Mojaver, Parisa, Shahram Khalilarya, & Ata Chitsaz. 2020. "Multi-objective optimization and decision analysis of a system based on biomass fueled SOFC using couple method of entropy/VIKOR." *Energy Conversion and Management* 203: 112260. https://doi.org/10.1016/j.enconman.2019.112260.

Moran, Michael J., Howard N. Shapiro, Daisie D. Boettner, & Margaret B. Bailey. 2010. *Fundamentals of Engineering Thermodynamics*. New York: John Wiley & Sons.

Musharavati, Farayi, Shoaib Khanmohammadi, Amir Hossein Pakseresht, & Saber Khanmohammadi. 2021. "Enhancing the performance of an integrated CCHP system including ORC, Kalina, and refrigeration cycles through employing TEG: 3E analysis and multi-criteria optimization." *Geothermics* 89: 101973. https://doi.org/10.1016/j.geothermics.2020.101973.

Özahi, Emrah, Alperen Tozlu, & Ayşegül Abuşoğlu. 2018. "Thermoeconomic Multi-objective optimization of an organic Rankine cycle (ORC) adapted to an existing solid waste power plant." *Energy Conversion and Management* 168: 308–319. https://doi.org/10.1016/j.enconman.2018.04.103.

Pan, Mingzhang, Fulu Lu, Yan Zhu, Guicong Huang, Jiwen Yin, Fuchuan Huang, Guisheng Chen, & Zhaohui Chen. 2020. "Thermodynamic, exergoeconomic and multi-objective optimization analysis of new ORC and heat pump system for waste heat recovery in waste-to-energy combined heat and power plant." *Energy Conversion and Management* 222: 113200. https://doi.org/10.1016/j.enconman.2020.113200.

Sadi, Meisam, Krishna Hara Chakravarty, Amirmohammad Behzadi, & Ahmad Arabkoohsar. 2021. "Techno-economic-environmental investigation of various biomass types and innovative biomass-firing technologies for cost-effective cooling in India." *Energy* 219: 119561. https://doi.org/10.1016/j.energy.2020.119561.

Sharma, Sudhanshu, V. K. Dwivedi, & S. N. Pandit. 2014. "Exergy analysis of single-stage and multi stage thermoelectric cooler." *International Journal of Energy Research* 38(2): 213–222. https://doi.org/10.1002/er.3043.

Siddique, Abu Raihan Mohammad, Shohel Mahmud, & Bill Van Heyst. 2017. "A review of the state of the science on wearable thermoelectric power generators (TEGs) and their existing challenges." *Renewable and Sustainable Energy Reviews* 73: 730–744. https://doi.org/10.1016/j.rser.2017.01.177.

Vojdani, Mehrdad, Iman Fakhari, & Pouria Ahmadi. 2021. "A Novel triple pressure HRSG integrated with MED/SOFC/GT for cogeneration of electricity and freshwater: techno-economic-environmental assessment, and multi-objective optimization." *Energy Conversion and Management* 233: 113876. https://doi.org/10.1016/j.enconman.2021.113876.

Wang, Lingbao, Xianbiao Bu, & Huashan Li. 2020. "Multi-objective optimization and off-design evaluation of organic Rankine cycle (ORC) for low-grade waste heat recovery." *Energy* 203: 117809. https://doi.org/10.1016/j.energy.2020.117809.

Yang, Fubin, Heejin Cho, Hongguang Zhang, & Jian Zhang. 2017. "Thermoeconomic multi-objective optimization of a dual loop organic Rankine cycle (ORC) for CNG engine waste heat recovery." *Applied Energy* 205(100): 1100–1118. https://doi.org/10.1016/j.apenergy.2017.08.127.

Yatsunthea, Theppasit, & Nattaporn Chaiyat. 2020. "A very small power plant – municipal waste of the organic Rankine cycle and incinerator from medical and municipal wastes." *Thermal Science and Engineering Progress* 18: 100555. https://doi.org/10.1016/j.tsep.2020.100555.

Zare, Vahid, & Vahid Palideh. 2018. "Employing thermoelectric generator for power generation enhancement in a Kalina cycle driven by low-grade geothermal energy." *Applied Thermal Engineering* 130: 418–428. https://doi.org/10.1016/j.applthermaleng.2017.10.160.

Ziapour, Behrooz M., Mohammad Saadat, Vahid Palideh, & Sadegh Afzal. 2017. "Power generation enhancement in a salinity-gradient solar pond power plant using thermoelectric generator." *Energy Conversion and Management* 136: 283–293. https://doi.org/10.1016/j.enconman.2017.01.031.

Chapter 10

4E-analysis of sustainable hybrid tri-generation system

Meisam Sadi
Department of Engineering, Shahrood Branch, Islamic Azad University, Shahrood, Iran

Ahmad Arabkoohsar
Department of Energy Technology, Aalborg University, Denmark

10.1 INTRODUCTION

According to recent studies, the residential sector has a high potential of savings, nearly 42 TWh per year (Hieminga 2013). In Europe, nearly 40% of the overall energy consumption is for residential communities. The high potential of saving with considering the consequences of climate change led to the establishment of a goal of 60% reduction in residential energy consumption by 2050 (Janda 2009). One approach in the reduction of energy demand for this high-energy consumer sector is the implementation of district heating. The widespread use of district heating and integration with heating and power production units is one of the most effective ways to increase the energy efficiency of the residential sector. To have a sustainable energy system in the future, district heating is an important component; thus, this issue requires more concentration and consideration.

The increase in population and the change in social lifestyle have created significant waste generation challenges for the world as waste production increases even faster than the rate of urbanization (Hoornweg & Bhada-Tata 2012). It seems that by 2025, waste generation will grow to 2.2 billion tonnes annually. This waste has a negative impact on the local environment. As a solution, the combustion of municipal solid waste (MSW) and energy extraction has been introduced for many years in the world to address this challenge. This approach brings many benefits for society including a decrease in the quantity of waste, efficient waste management, production of heat and power, reduction of pollution with the aid of filters, saves on transportation of waste, and is a prevention solution for CH_4 production. Furthermore, da Silva et al. (2020) estimated the energy potential of MSW incineration in the function of the population and stated that the community can achieve 15% of the required energy from the waste they generate. In addition to the benefits, waste incineration includes a few disadvantages including being very expensive to build and maintain (i.e., high capital cost), the need for skilled operators, and the production of hazard ash.

The growth of world population, MSW generation, and energy have pushed a growing demand for the establishment of waste incineration combined heat and power (WI CHP) plants in the coming years. Consequently, WI CHP has become the popular technical knowledge of MSW dumping in several countries in the world. Kythavone and Chaiyat (2020) investigated the environmental impact of a very small organic Rankine cycle integrated with a municipal solid waste incinerator. Liu et al. (2020) studied the effect of incinerator loads on the combustion efficiency and heat loss of the process. Mohammadi and Harjunkoski (2020) proposed a mixed-integer optimization method for a waste-to-energy supply chain problem, to improve the fuel and energy production from MSW. Chen et al. (2020) studied a CO_2

DOI: 10.1201/9781003213741-10

power cycle integrating with a WI unit. The waste-to-electricity efficiency is surprisingly increased with a much higher net present value and a much shorter payback period. Liuzzo, Verdone, and Bravi (2007) investigated the important operative and environmental gains of flue gas recirculation in a WI unit and clarified that the mercury and pollutant emissions could be considerably decreased by flue gas recirculation. The impact of combustion with higher amounts of oxygen on the flue gas emission and combustion performance of MSW was investigated in a full-scale WI unit by Ma et al. (2019). The results showed that the average temperature in three grates of an incinerator increased as oxygen concentration increased, while the unburnt rate reduced little by little. Miao et al. (2021) investigated the integration of a WI power plant and coal-fired unit. This integration increased the waste to energy efficiency from 20% to 31.6%, and the MSW incinerator's operation investment decreased by 40% with important economic profits. Mondal, Barman, and Samanta (2021) developed a thermodynamic model of an integrated MSW-fueled externally fired air turbine plant and hot water generation plant for Indian cities. Two types of MSW are utilized in the combustion process: dry and segregated MSW. Also, to address the hot water demand of the city, exhaust heat from the dual combustor-air heating unit is utilized.

Solar energy utilization has been considered a promising way to solve the problems of global warming, climate change, and fossil fuel depletion (Lewis 2016). Solar district heating system is a bright approach to ease the adoption of large-scale, solar energy-based technologies. Denmark, as one of the world-leading countries within the field of integrating large-scale solar heating into district heating, has significantly increased the number of large solar heating plants during the last couple of years (Tian et al. 2019). This is due to solar heating technology becoming cheaper and more readily available and therefore a more viable alternative to fossil fuel, especially during the summer. Also, solar thermal heating is exempt from fuel tax (Burnett, Barbour, & Harrison 2014). In this regard, Sadi and Arabkoohsar (2020a; Arabkoohsar & Sadi 2020b) proposed a hybrid concentrating solar power-plant and investigated the economic and environmental aspects of the proposed system. They investigated the parabolic trough collector (PTC)-driven power plant to increase the share of solar energy into the existing energy matrices. Arabkoohsar and Sadi (2020a) investigated a PTC-powered absorption cooling unit for co-supply of district heating/cooling networks. This approach illustrated good results when the district heating system did not have enough load. Huang, Fan, and Furbo (2020) studied experimentally and theoretically the first large-scale solar-assisted ground source heat pump for heating and cooling of a village in China. Chen, Wang, and Lund (2020) investigated a novel district heating system with photovoltaic/thermal collector and geothermal heat pump. Abokersh et al. (2020) utilized the optimization approaches to determine the optimal structure, sizing, and operation of the solar district heating systems. In a few pieces of research, the seasonal solar energy storage for district heating applications has gained much attention (Abokersh et al. 2021; Saloux and Candanedo 2020; Bai et al. 2021). A district heating unit driven by solar PTC with a seasonal thermochemical energy storage system and a chemical reactor has been erected (Li et al. 2020). The integration of seasonal energy storage would significantly help the performance of the system.

Although several pieces of research could be found in the literature working on WI CHP, the synergy between waste incineration plants with solar fields and absorption cooling unit has not gained much attention. Very few articles have comprehensively considered the hot season challenges of district heating as they will be considered in this research. Then the proposed system will be examined to find out how it can solve or remove the problem or improve the overall performance of the system. In this regard, the best size of the system will be estimated by the implementation of the optimization genetic algorithm and a 4E analysis is presented to investigate the energy, exergy, economic aspects, and environmental impacts of the proposed cycle.

10.2 DESCRIPTION OF THE PROPOSED SYSTEM

The schematic of the proposed system is presented in Figure 10.1. This system is made up of a waste incineration unit, Rankine cycle, solar thermal field, absorption cooling unit, and two heat exchangers. In the Rankine cycle, there are steam turbines, condenser, pumps, a boiler and a heat exchanger. In the incineration unit and after preparation treatment on the MSW, the fuel is sent to the incineration chamber. The released heat of fuel in this unit provides the required heating energy of the Rankine cycle. A part of this energy should be removed from the condenser which can be used as a heat source for cold production or district heating units. Due to the high demand for district heating units in winter, if the plant is responsible for heat production, a high amount of MSW is required to be burnt in the winter. While in summer the heating requirement of the DHS decreases to its minimum value during the year. Along with the reduction in heating requirement of district heating, the demand for cooling

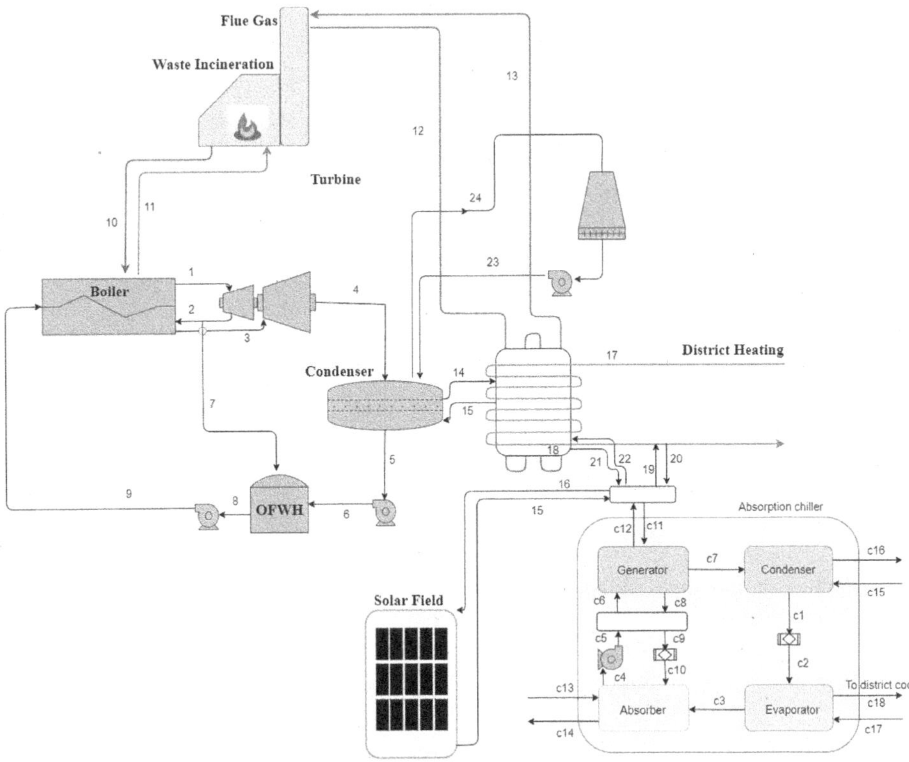

Figure 10.1 Combined cooling, heating, and power unit proposal in the response of district unit loads.

will increase. With the potential of extra unnecessary heat from the condenser and gas condensation unit, the absorption cooling unit would be able to provide the required cooling of the system during the summer.

10.3 METHODOLOGY: THERMODYNAMIC MODELING

10.3.1 WI power plant

In the power plant unit, the power is produced in the turbine, then it is converted into electricity. By considering the first law of thermodynamics for the whole cycle, the net produced work of the Rankine cycle can be calculated as:

$$\dot{W}_{net} = \dot{W}_{hpt} + \dot{W}_{lpt} - \sum \dot{W}_{p} \tag{10.1}$$

Where, $\dot{W}_{net}$ is the net produced work and hpt and lpt denote the high- and low-pressure. $\sum \dot{W}_{p}$ is the consuming work of pumps. The extra heating of the Rankine cycle is released in the condenser as:

$$q'_{cond} = (h_4 - h_5);\ \dot{Q}_{cond} = \dot{m}_4 q'_{cond} \tag{10.2}$$

q'_{cond} is the amount of heat released from the condenser. By applying the first law of thermodynamics for the open feed water tank and the mixing chamber:

$$\dot{m}_7 h_7 + \dot{m}_8 h_8 = \dot{m}_9 h_9 + \dot{m}_{10} h_{10} \tag{10.3}$$

$$\dot{m}_{12} h_{12} = \dot{m}_{11} h_{11} + \dot{m}_9 h_9 \tag{10.4}$$

In the boiler, the provided heat inside the incineration unit is implemented to increase the energy level of the working fluid of the Rankine cycle as:

$$\dot{m}_{WI} . \Delta h_{WI} = \dot{m}_1 \left[(h_1 - h_{12}) + (h_3 - h_2) \right] \tag{10.5}$$

10.3.2 Absorption chiller

The absorption unit is implemented to provide cooling by getting enough heating through its generator. The absorption unit includes a generator, an absorber, a condenser, an evaporator, a heat exchanger, and some valves. For the mentioned components, the mass and energy balance and first law of thermodynamics should be written as follows:

$$\dot{m}_{c9} = \dot{m}_{c10} \tag{10.6}$$

$$\dot{Q}_{c-eva} = \dot{m}_{c10}h_{c10} - \dot{m}_{c9}h_{c9} \tag{10.7}$$

$$\dot{m}_{c1} = \dot{m}_{c10} + \dot{m}_{c6} \tag{10.8}$$

$$\dot{m}_{c1}x_{c1} = \dot{m}_{c6}x_{c6} \tag{10.9}$$

$$\dot{Q}_{c-abs} = \dot{m}_{c10}h_{c10} + \dot{m}_{c6}h_{c6} - \dot{m}_{c1}h_{c1} \tag{10.10}$$

$$\dot{m}_{c3} = \dot{m}_{c4} + \dot{m}_{c7} \tag{10.11}$$

$$\dot{Q}_{c-des} = \dot{m}_{c4}h_{c4} + \dot{m}_{c7}h_{c7} - \dot{m}_{c3}h_{c3} \tag{10.12}$$

$$\dot{m}_{c3} = \dot{m}_{c4} + \dot{m}_{c7} \tag{10.13}$$

$$\dot{Q}_{c-cond} = \dot{m}_{c7}\left(h_{c7} - h_{c8}\right) \tag{10.14}$$

$$\dot{m}_{c4}c_{c4}\left(T_{c4} - T_{c5}\right) = \dot{m}_{c2}c_{c2}\left(T_{c3} - T_{c2}\right) \tag{10.15}$$

10.3.3 Solar evacuated thermal collector

Here, a solar field is considered to increase the renewable energy penetration into the energy generation network. In the previous works, many types of solar collectors have been employed in several solar power plants depending on the design of the temperature field. As mentioned in the later section, the single-effect absorption chiller is implemented to provide the cooling demand of the system. In this regard, the required temperature of the working fluid at the inlet of the absorption chiller should be considered at about 90°C. Thus, an evacuated heat pipe collector which is composed of two glasses with a vacuum in between, absorber coating, copper heat pipe, heat transfer pin, and spring clip is implemented to be integrated with the hybrid system and depicted in Figure 10.2(a). The thermal resistances of the different sections of this collector could be seen in Figure 10.2(b).

For this collector, the amount of solar energy which passes through the glasses and irradiates on the absorber plate can be presented as (Chopra et al. 2020):

$$Q_{ap} = I_t \alpha_{ap} \tau_{ig} \tau_{og} A_{ap} \tag{10.16}$$

Where inner and outer, plate and glass are denoted by *i*, *o*, *ap* and *g*. α and τ can be described as absorptivity and transmissivity.

As the absorber plate receives solar radiation, the temperature increases, and an amount of this generated heat is wasted in surrounding (Chopra et al. 2020):

$$Q_{Loss} = \left(T_{ap} - T_a\right) / R_{eq} \tag{10.17}$$

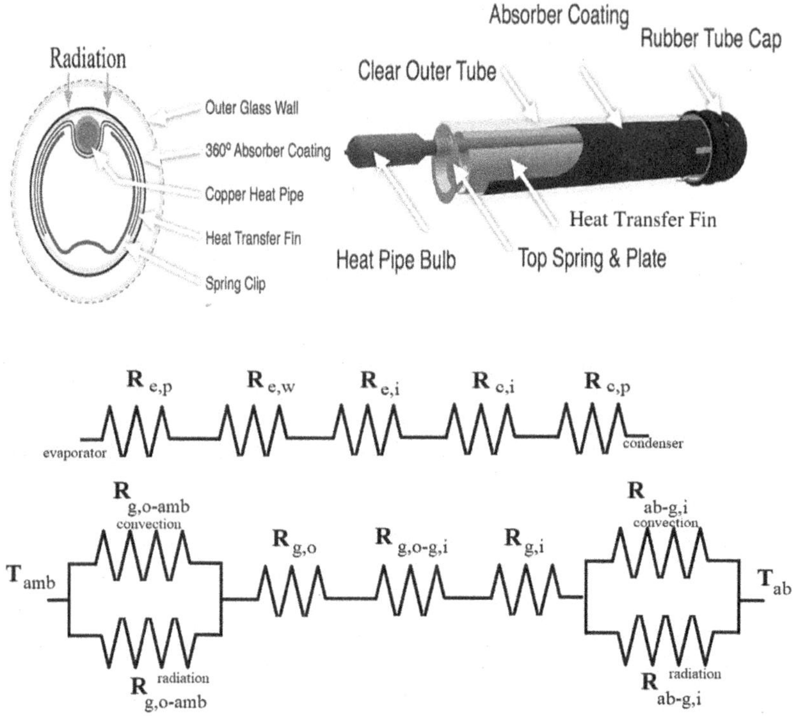

Figure 10.2 Schematic of the evacuated heat pipe collector. (Sources: Shafieian & Khiadani 2020; Apricus 2006.)

The equivalent radiative resistance can be evaluated as (Chopra et al. 2020):

$$R_{eq} = \frac{R_{r,og-a}.R_{c,og-a}}{R_{r,og-a} + R_{c,og-a}} + \frac{R_{r,ig-og}.R_{c,ig-og}}{R_{r,ig-og} + R_{c,ig-og}} + R_{ap-ig} \tag{10.18}$$

$$R_{ap-ig} = \frac{T_{ap} - T_{ig}}{\varepsilon_{ap}\sigma A_{ap}\left(T_{ap}^4 - T_{ig}^4\right)} \tag{10.19}$$

$$R_{r,ig-og} = \frac{T_{ig} - T_{og}}{\varepsilon_{g}\sigma A_{ig}\left(T_{ig}^4 - T_{og}^4\right)} \tag{10.20}$$

$$R_{r,og-a} = \frac{T_{og} - T_{a}}{\varepsilon_{g}\sigma A_{og}\left(T_{og}^4 - T_{a}^4\right)} \tag{10.21}$$

The heat pipe evaporator/condenser wall conductive represented as (Chopra et al. 2020):

$$R_{cond,eva} = \frac{ln\left(\frac{r_{eva,o}}{r_{eva,i}}\right)}{2\pi L_{eva}k_{cu}}, \quad R_{cond,con} = \frac{ln\left(\frac{r_{con,o}}{r_{con,i}}\right)}{2\pi L_{con}k_{cu}} \tag{10.22}$$

To include the heat pipe heat transfer, the total thermal resistance of a heat pipe, R_{hp}, is defined as the summation of several resistances:

$$R_{hp} = R_{e,p} + R_{e,w} + R_{e,i} + R_v + R_{c,i} + R_{c,p} \tag{10.23}$$

$R_{e,p}$, $R_{e,w}$, $R_{e,i}$, $R_{c,i}$, and $R_{c,p}$ refer to the evaporator wall resistance, the wick resistance, the internal resistance, the condenser resistance, and the condenser wall resistance and are defined as:

$$R_{e,p} = \frac{ln\left(\frac{d_o}{d_i}\right)}{2\pi k_p L_e} \tag{10.24}$$

$$R_{e,w} = \frac{ln\left(\frac{d_{o,w}}{d_{i,w}}\right)}{2\pi k_w L_e} \tag{10.25}$$

$$R_{e,w} = \frac{2}{h_e \pi d_i L_e} \tag{10.26}$$

$$R_{c,p} = \frac{ln\left(\frac{d_o}{d_i}\right)}{2\pi k_p L_c} \tag{10.27}$$

$$R_{c,p} = \frac{ln\left(\frac{d_o}{d_i}\right)}{\pi h_{c,i} d_i L_c} \tag{10.28}$$

In these equations, the effective thermal conductivity of the wick structure and the film coefficient of the internal resistance are symbolized by k_w (W/mK) and h_e (W/m^2K).

$$k_w = \frac{k_l\left[k_l + k_s - (1-\varepsilon_w)(k_l - k_s)\right]}{k_l + k_s + (1-\varepsilon_w)(k_l - k_s)} \tag{10.29}$$

$$h_e = \frac{k_l}{t_w} \tag{10.30}$$

and $h_{c,i}$ (W/m^2K) is the heat transfer coefficient of the condensation process and can be expressed as:

$$h_{c,i} = 0.728\left[\frac{g\ sin\theta\rho_l\left(\rho_l - \rho_v\right)k^3 h_{fg}}{D\mu_i\Delta T_i}\right] \tag{10.31}$$

g (m/s^2), θ (°), ρ (kg/m^3), hfg (J/kg), and μ ($Pa\ s$) are the gravitational acceleration, the inclination angle, the density, the latent heat of evaporation, and the dynamic viscosity, respectively.

10.3.4 Economic analysis

The economic analysis is implemented to document a reasonable expected return on investment. The economic analysis is used to document that the project is a net benefit to the system. Levelized cost of energy (LCOE) represents the unit cost of energy production as:

$$LCOE = \frac{\sum_{i=1}^{N}\frac{I_i + M_i + F_i}{\left(1+r\right)^i}}{\sum_{i=1}^{N}\frac{E}{\left(1+r\right)^i}} \tag{10.32}$$

In this equation, r presents the interest rate. Investment cost, maintenance cost, and fuel cost are symbolized with I, M and F. E refers to the amount of energy produced by the hybrid cycle. For the economic evaluation, the latest investment cost of components is listed in Table 10.1.

10.3.5 Environmental analysis

This section aims to establish how environmental analysis is used as a determinant of different circumstances. The ratio of generated carbon dioxide to the produced energy as the

Table 10.1 Investment cost of components

Component	*Cost*
Evacuated plate collector with piping	500 $/m^2
Heat storage tank and piping	250 $/m^3
Single-effect absorption chiller	800–1,100 $/TR
Tax	35 $/MWhr
District heating pipe, D=50 cm	160 $/m
WI power plant	1,531 $/kW

Sources: LOGSTOR (n.d.); PE Continuing Education (n.d.); EIA (2017); Sadi and Arabkoohsar (2020b); Behzadi et al. (2018).

CO_2 emission index is considered as a determinant to investigate the environmental impact of a specified system from an environmental aspect and can be defined as:

$$\varepsilon = \frac{mass\ of\ generated\ carbon\ dioxide}{produced\ cooling\ load} \tag{10.33}$$

10.3.6 Exergy analysis

The steady-state exergy balance of the exergy is expressed as:

$$\dot{E}x_D = \sum \dot{E}x_{in} + \dot{Q}\left(1 - \frac{T_0}{T}\right) - \sum \dot{E}x_{out} - \dot{W} \tag{10.34}$$

This equation means that the irreversibilities cause the exergy destruction and it is equal to the difference of inlet and outlet exergy flows and the variation of exergy due to the heat transfer and the work of the system. Exergy includes two parts; physical and chemical exergy as follows:

$$\dot{E}x = \dot{E}x_{ch} + \dot{E}x_{ph} \tag{10.35}$$

$$\dot{E}x_{ph} = h - h_0 - T\left(s - s_0\right) \tag{10.36}$$

h is the specific enthalpy and s is the specific entropy. Chemical exergy can be disregarded. According to the Equation (10.34), exergy destruction of components of the power plant is listed in Table 10.2.

For MSW, the exergy is calculated as (Song, Shen, and Xiao 2011):

$$\dot{E}X_{MSW} = \dot{m}_{MSW}\begin{pmatrix}1812.5 + 295.606C + 587.354H + 17.506O + 17.735N + \\ 95.615S - 31.8Ash\end{pmatrix} \tag{10.37}$$

where, the parameters C, H, O, N, S, and Ash are the mass of compositions Carbon, hydrogen, oxygen, nitrogen, sulfur, and ash, respectively. For ash, the exergy is calculated as:

$$\dot{E}X_A = \dot{m}_A\left(0.0004056T_{Ash}^{\ 2} + 0.01057T_{Ash} - 54.44\right) \tag{10.38}$$

Table 10.2 Exergy destruction relations for the power plant component

Component	*Exergy destruction*
Boiler	$\dot{E}X_{10} - \dot{E}X_{11} - (\dot{E}X_1 - \dot{E}X_9 + \dot{E}X_3 - \dot{E}X_{2a})$
Condenser	$\dot{E}X_4 - \dot{E}X_5 - (\dot{E}X_{14} - \dot{E}X_{15})$
Turbine	$\dot{E}X_1 - \dot{E}X_2 + \dot{E}X_3 - \dot{E}X_4$
Pump	$\dot{E}X_5 + \dot{E}X_6$
Open feedwater heater	$\dot{E}X_7 + \dot{E}X_6 - \dot{E}X_8$

T_{Ash} is the temperature of the ash in K. For flue gas, the exergy is calculated by the following equation:

$$\dot{E}X_{FG} = \dot{m}_{FG}\left(\dot{E}X_{FG}^{ch} + c_{p,FG}\left(T_{FG} - T_o\right) - T_o\left(c_{p,FG}ln\left(\frac{T_{FG}}{T_o}\right) - Rln\left(\frac{P_{FG}}{P_o}\right)\right)\right) \quad (10.39)$$

$\dot{m}_{FG}, \dot{E}X_{FG}^{ch}$, T_{FG} and P_{FG} denote the mass flow rate, chemical exergy, temperature, and pressure of the flue gas exiting from the stack. Since air enters the chamber at ambient temperature, the exergy for air is considered zero.

For solar field, the input solar exergy would be represented as (Achenbach and Riensche 1994):

$$\dot{E}X_{in,sol} = I_b A_a\left(1 - \frac{4T_o}{3T_s} + \frac{1}{3}\left(\frac{T_o}{T_s}\right)^4\right) \quad (10.40)$$

in which, T_o and T_s are the dead state temperature and the sun's surface temperature, respectively. The increased exergy from the solar would be described as:

$$\dot{E}X_{gain,sol} = I_b A_a\left(\dot{E}X_{in,sol}\right) - \dot{E}X_{sol}^{De} - \dot{E}X_{sol}^{Loss} \quad (10.41)$$

$\dot{E}X_{PTC}^{Loss}$ refers to exergy losses which is due to optical error and heat transfer losses from the solar receiver and $\dot{E}X_{PTC}^{De}$ refers to the exergy destruction due to heat transfer between the absorber and the solar working fluid.

For the absorption chiller, the exergy equations are tabulated in Table 10.3.

In Table 10.4, data about the type of MSW and the specification of the waste incineration unit is presented.

10.4 MULTI-OBJECTIVE OPTIMIZATION

In this chapter, an integrated hybrid system is working in an integrated manner. Solar field, Rankine power plant, and absorption cooling unit are integrated with the district heating/

Table 10.3 Exergy destruction for the absorption chiller components

Component	*Exergy destruction*
Generator	$\dot{E}X_{15} - \dot{E}X_{16} + \dot{E}X_{21} - \dot{E}X_{22} + \dot{E}X_{23} - \dot{E}X_{20} - (\dot{E}X_8 + \dot{E}X_7 - \dot{E}X_6)$
Condenser	$\dot{E}X_7 - \dot{E}X_1 + (\dot{E}X_{16} - \dot{E}X_{15})$
Evaporator	$\dot{E}X_2 - \dot{E}X_3 - (\dot{E}X_{24} - \dot{E}X_{25})$
Absorber	$\dot{E}X_{3s} + \dot{E}X_{10} - \dot{E}X_4 + (\dot{E}X_{14} - \dot{E}X_{13})$

Table 10.4 MSW specifications and incineration unit

	Type of waste	*Municipal solid waste*
MSW incineration unit	Waste compositions (weight percent)	5.91% Ash 47.18% Carbon 6.25% Hydrogen 39.57% Oxygen 0.91% Nitrogen 0.18% Sulfur
	LHV of the waste (kJ/kg)	12,500
	Excess air in the incineration process	80%
	Combustion production temperature (K)	1,100

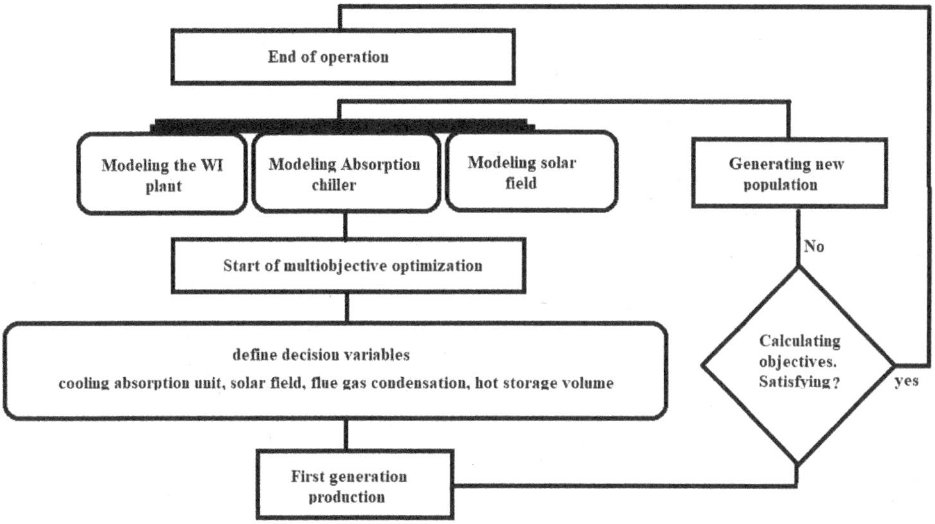

Figure 10.3 Flow chart of the solution process of the problem and the optimization method.

cooling unit to provide the required cooling, heating, and power of the system. So, the determination of the right values of the component size requires a multiobjective optimization approach. The multiobjective genetic algorithm is implemented to extract the best size of the components. Figure 10.3 shows the flow chart of the steps that are used as well as the optimization approach.

10.5 CASE STUDY AND THE CHALLENGES

A district in Esbjerg, Denmark, is considered to include many buildings with a WI CHP power plant of Energnist. This plant supplies heat and electricity to a network of about

25,000 homes (Fabricius et al. 2020). This plant produces 20 MW electricity, 60 MW heat in the condenser, and 13 MW heat in the flue gas condensation unit. In the summer, due to a decrease in the heat demand of district heating, a lower amount of MSW is required to be burnt, thus a higher amount of MSW is sent to the high-capacity MSW incinerator which is 300 km away from the city and this causes more cost for the system which is about €135 per tonne of MSW (Energist n.d.).

To assess the performance of the proposed system, it would be essential that one knows how temperature and solar radiation change in this location. For this reason, Figure 10.4 shows the temperature and solar radiation over a year. This figure shows that the maximum temperature occurs in July at about 30°C and the minimum temperature will occur in December and January. The same trend will be seen for solar radiation. The average solar radiation occurs in June and July and the lowest values are for December and January.

Another action before evaluating the performance of the system is to estimate the amount of cooling and heating for the system during the year. It would be necessary that one knows how much heating/cooling will be required during each hour. For this reason, Figure 10.5 shows the amount of required hourly heating and cooling load. It could be seen that in winter and fall, the highest amount of heat is required for the network. In summer and when the ambient temperature increases, the cooling demand appears while it fluctuates based on the ambient temperature. This figure also shows that a minimum cooling and heating load can be seen during the year, for example, the heating load is always greater than 17 MW for the network during the year due to the residential hot water consumption. For heating, three sources of consumption are hot water demand, heating demand on cold days, and heating demand for cold production on hot days of the year. The heating demand varies from 17 to 70 MW for this network. For cooling requirement of the network, a variation in the range between 3.5 up to 51 MW is observed.

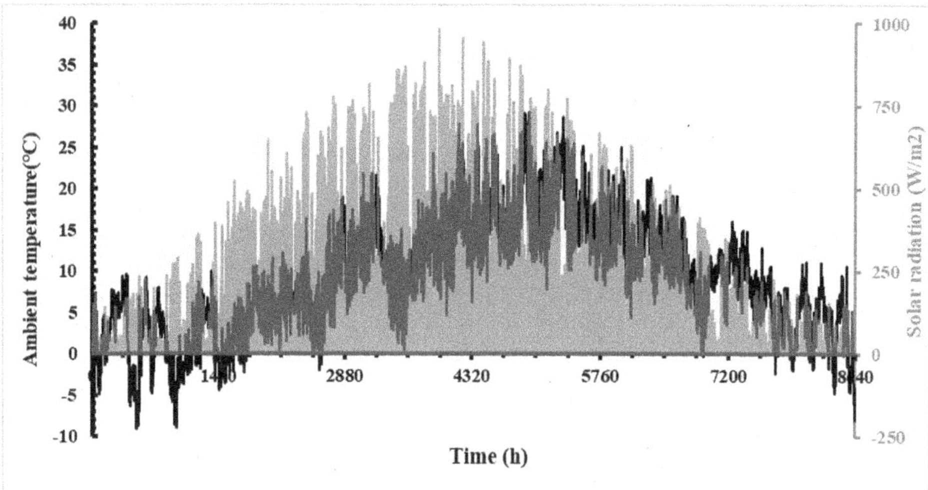

Figure 10.4 Solar irradiation and temperature of the ambient for the case study.

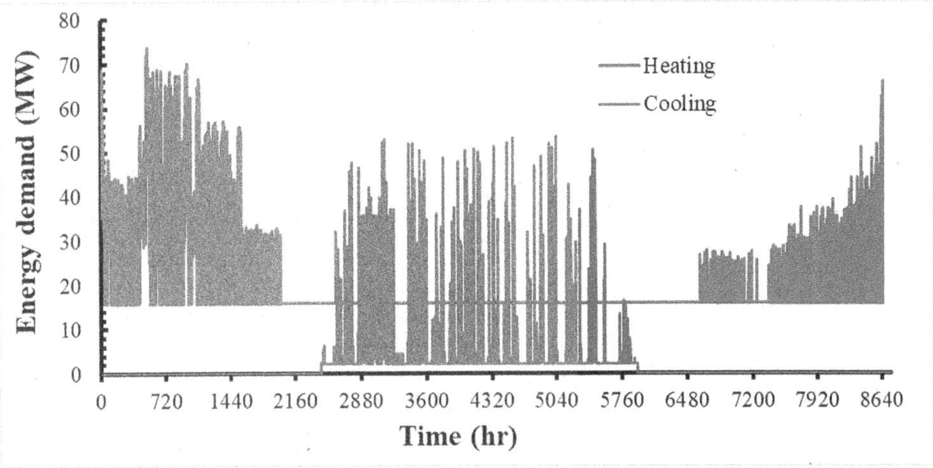

Figure 10.5 Hourly heating and cooling demand of one building in Esbjerg.

Then for the specified location, one would know the cost of energy as energy price has great influence on the economic analysis. In the determination of the optimum system from an economic point of view, we must know the price of energy at each hour. It is worth mentioning that the energy price fluctuates a lot due to evolution in the short-term market. Figure 10.6 shows the spot price of electricity and heat in the western Denmark market over an entire year. It could be seen that most of the time the electricity price is between 16 and 32 $/MWhr, while the price will be negative for a few hours. The negative price means that wholesale customers get money from power suppliers for buying electrical energy. The heat price is almost in the range of 32 to 50 $/MWhr.

As we know high velocities of the working fluid through pipelines lead to many operational and maintenance issues like vibration, noise, and material erosion. Also, the low velocities of the working fluid through pipelines lead to sedimentation. The magnitude of flow velocity is one of the major factors governing the rate of sedimentation. Therefore, increasing flow velocity in the pipeline can significantly reduce the occurrence rate of this problem. Figure 10.7 shows the variation of working fluid velocity for two sample days in winter and summer. As could be seen in summer, the magnitude of velocity is constant and is at the lowest value of 0.45 m/s, while the velocity in winter is in an appropriate range. Having constant velocity in summer is for this reason to supply the hot water consumption of the network. The low velocity brings two difficulties; sedimentation which is a consequence of this situation while when the velocity is low, the lower heat should be provided for district heating and it means that a lower amount of MSW should be burned at the WI power plant. According to the amount of MSW collected during summer, manipulation of this extra amount of MSW that should not be combusted becomes a serious problem for the municipalities and responsible authorities.

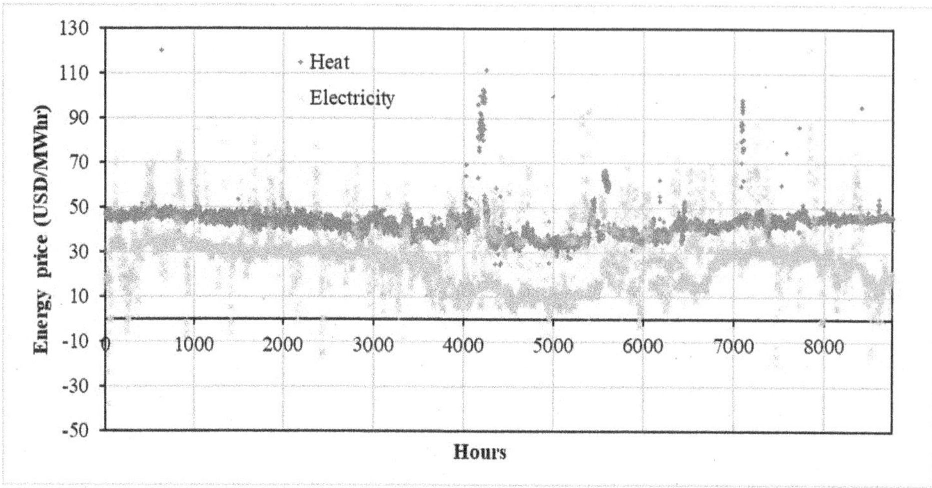

Figure 10.6 The annual price of heat and electricity in the western Denmark market.

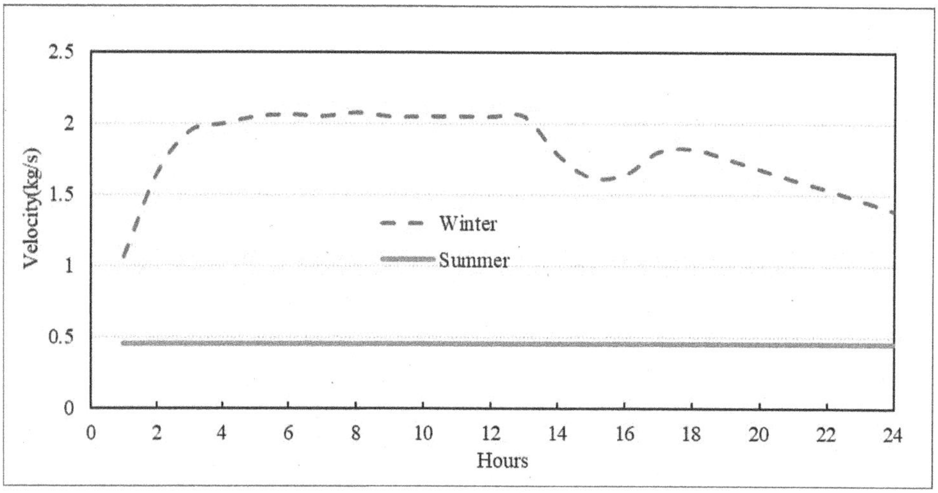

Figure 10.7 The velocity of working fluid inside the pipeline for two sample days in winter and summer.

10.6 METHODOLOGY

For the simulation of the current system, MATLAB 2019 is implemented to provide a user-friendly environment to develop numerical code. This code is used to investigate the performance of the hybrid system which is an integration of waste incineration power plant, district cooling, and heating network, solar thermal system, and absorption cooling machine.

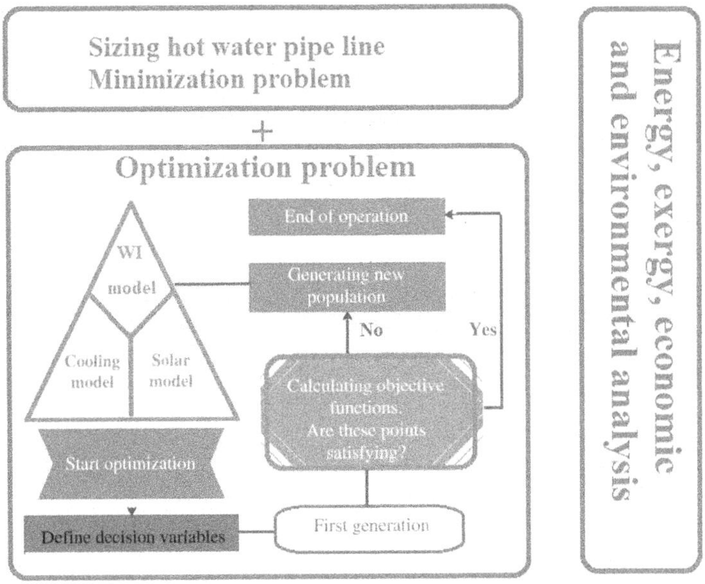

Figure 10.8 The considered strategy of the work.

In this regard, the hot water transfer line should be sized as a cost minimization problem. In this minimization problem, we should consider the pressure loss occurring through the pipelines, the total cost of installation, and the electricity cost of pumping. Then by considering the proposed pipeline, the whole system should be considered to find out in what size of the components, the optimum performance of the whole system occurs according to the variations of decision variables. After selecting the appropriate size of components, the system is considered to be investigated from an energy, exergy, economic, and environmental point of view and then the results will be presented. Figure 10.8 shows the strategy considered in this work according to the mentioned topics of this section.

10.7 RESULTS AND DISCUSSION

The hybrid system shows that the hot water is supposed to be prepared in a heat exchanger connected to the condenser and flue gas condensation of the WI power plant. As we know this equipment is installed once and is used during the period of the project which is considered 20 years for this case. Consequently, there would be three types of expenses during this project; expenses related to the installation at the first year of the project, the expenses related to the heat loss through the pipes, and finally the electricity cost of pumping during the project period. To select the best feeding pipeline size, all these expenses should be investigated during the project time. For this reason, for different sizes of diameter, these expenses are represented in Figure 10.9. When the pipeline diameter changes, the first point is that the cost of the pipeline changes. Then this change would affect the electricity consumption. When the diameter changes, the velocity of the working fluid inside of the pipeline varies. According to

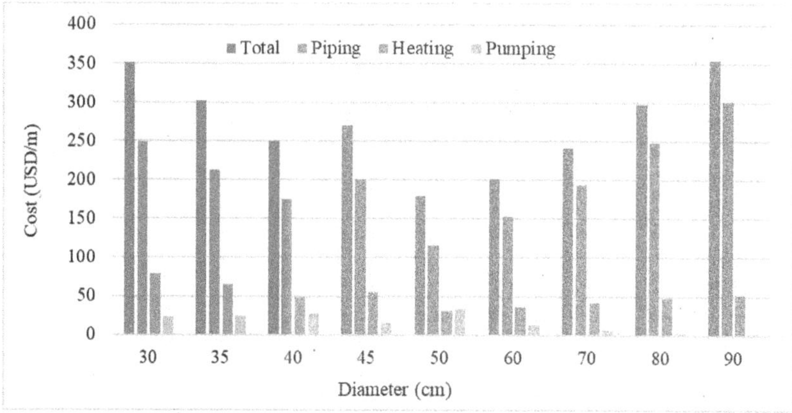

Figure 10.9 Sizing the district heating feeding pipeline.

the velocity limit of the district heating pipeline, a decrease in diameter could not happen as much as possible. Thus, due to these controversial effects, the figure shows that for a diameter of 50 cm, the minimum pipeline cost will occur. It is worth mentioning that when the diameter decreases, the velocity of working fluid inside of the pipe exceeds its limit, thus the number of pipelines should be increased, which means the huge extra expense of pipelines and installation. This figure shows that the cost of heat loss occurring through the pipeline wall minimizes when the diameter is 50 cm. When the diameter is larger, it means a higher surface for heat transfer and consequently higher heat loss. For smaller diameter, it should be mentioned that the number of pipelines causes the surface of heat transfer to increase, and thus the optimum diameter from the heat loss point of view would be considered 50 cm. The third factor affecting the expenses of the project is the amount of electricity used for pumping. This parameter decreases when the diameter increases because the friction factor of the pipeline is a function of diameter and it reduces when diameter increases. However, pumping expenses are smaller in comparison with the two other expenses. By considering all three factors, the total cost of pipeline experiences a minimum value when the diameter is 50 cm.

The sizing of the feeding pipeline made obvious the system that is going to be serviced by the hybrid power plant. The difficulties of the current system are the lower velocity of working fluid, accumulation of MSW at the power plant, and decrease of working fluid temperature due to the low velocity. To address these issues, the power plant is designed to burn more MSW in summer and produce more heat to warm the working fluid of district heating. In this way, more working fluid is injected into the pipeline and the low velocity and low temperature problem of the working fluid is removed. Also, the working fluid is going through the generator of an absorption cooling unit and provides the heating requirement of the unit. This cooling facility is linked to a solar field which provides a portion of the heat requirement of the cooling system. The integration of these subsystems requires a precise modeling and optimization approach because the designers should know in what size the system operates on the best efficiency. For this reason, one would like to know the optimum size of the system. Two objectives are considered in this problem to find out the best solution based on decision variables of the cooling system size, the solar field size, the outlet temperature of the

Table 10.5 Decision parameters as inputs for optimization

Absorption chiller size (MW)	*HP ETC collector field (MW)*	*TFlue gas (°C)*	*Hot storage volume (m_3)*
5–50	10–70	900–1,100	500–2,000

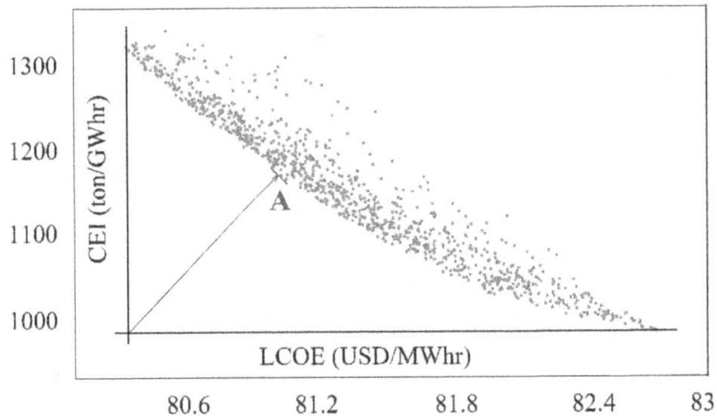

Figure 10.10 The Pareto frontier solution with the optimal point A.

Table 10.6 Optimization result and decision variables and objective functions for point A

	Decision parameter				*Objective function*	
Point	*Absorption chiller size (MW)*	*HP ETC collector field (MW)*	*TFlue gas (°C)*	*Hot storage volume (m3)*	*LCOE ($/MWhr)*	*CEI (ton/GWhr)*
A	10.2	14.2	1,100	500	81.04	1,158

flue gas, and the hot storage volume by implementing a multi-objective optimization genetic algorithm. These four decision variables are listed in Table 10.5 and the selected domain is specified for the solution process.

As mentioned before, two objective functions of LCOE and CO_2 emission index (CEI) are considered to compare the performance of the whole system based on the variation of decision variables. Figure 10.10 shows the Pareto frontier solution for LCOE and CO_2 emission index (CEI). The optimum point is a situation in which the LCOE and CEI are both at the minimum values. This point is not located on the Pareto frontier. Thus the nearest point on the Pareto frontier to the ideal point is considered as the best solution of the multi-objective optimization genetic algorithm. This point is point "A" which can be seen in Figure 10.10. Thus this point is considered as the base of the calculation and sizing. The optimum value of objective functions and main decision parameters at the selected point are tabulated in Table 10.6. The absorption chiller unit and HP ETC solar collector field are sized at 10.2 MW and 14.2 MW, respectively, and the flue gas temperature of 1,100 K is used to simulate the WI plant.

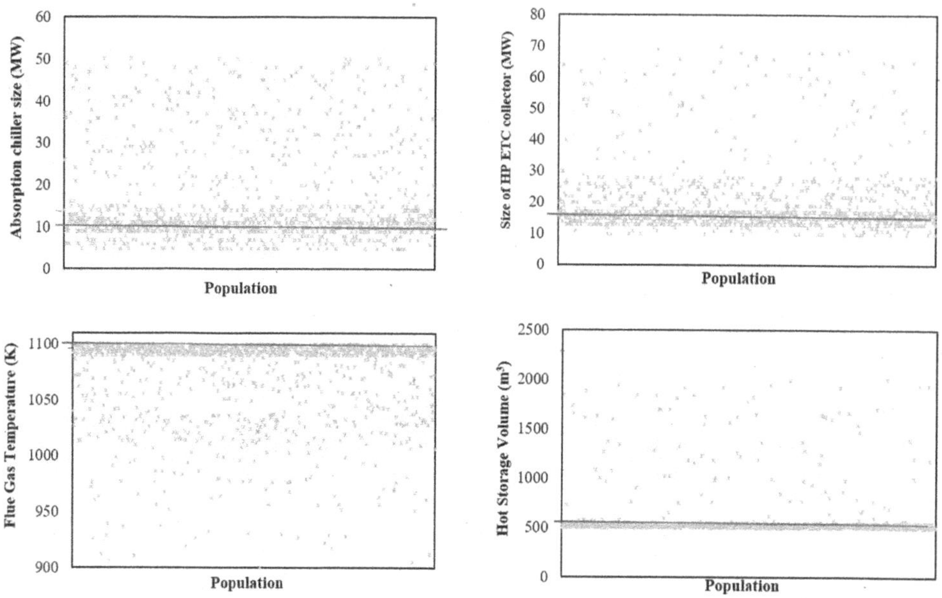

Figure 10.11 Scattered distribution of decision variables, (a) absorption chiller tonnage, (b) power of HP ETC filed, (c) flue gas temperature and (d) hot water volume.

To see the optimum domain of main decision variables, the distribution of data is presented in Figure 10.11. Based on Figure 10.11(a), the optimum tonnage of the absorption chiller would be considered as 10.2 MW or 2,900 TR. For the HP ETC solar field, the specified decision variable range is defined from 10 to 20 MW, and according to Figure 10.11(b), the best point is calculated around 14.2 MW. Based on Figures 10.11(c) and (d), decision variables of flue gas temperature and hot storage tank volume are considered in an appropriate range and the best size for these variables are found to be 1,100 K for temperature and 500 m^3 for volume.

The proposed system can now provide a portion of the cooling demand of the network in summer. For the cooling period, Figure 10.12 shows that there are 21 weeks of the year that the network requires cooling. The maximum cooling demand which occurs at week 4 is about 1.65 MWhr and the minimum cooling demand occurs in weeks 1 and 21 and is about 0.37 MWhr. The proposed absorption unit generally provides this demand partially; from at least 60% of demand occurring in weeks 4, 7, and 16 to completely occurring in the first week. As the cooling production of the absorption unit is constant, it should be mentioned that in months with lower cooling demand a higher percentage of the cooling load will be delivered by the cooling production unit; i.e. weeks 1, 20, and 21. Whenever the network requires 10 MW cooling or lower, the absorption unit will be ready to provide the complete demand and if the need is higher than 10 MW, then the extra amount of cooling should be provided by another cooling production unit.

Based on this cooling production which is represented in Figure 10.12 and because this demand is expected from the hybrid system, the total heating demand has changed specifically for summer when the heat demand was low and that was a challenge. In the proposed system,

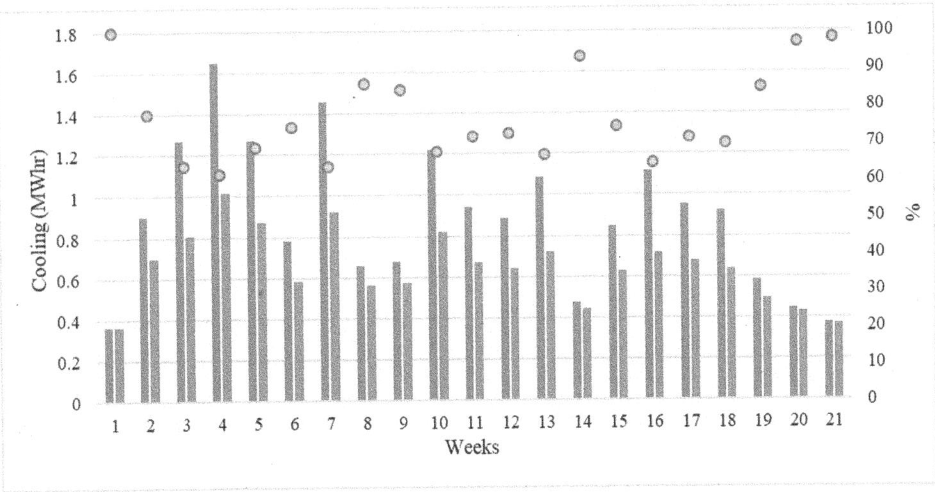

Figure 10.12 Demand of cooling load and production of the absorption unit.

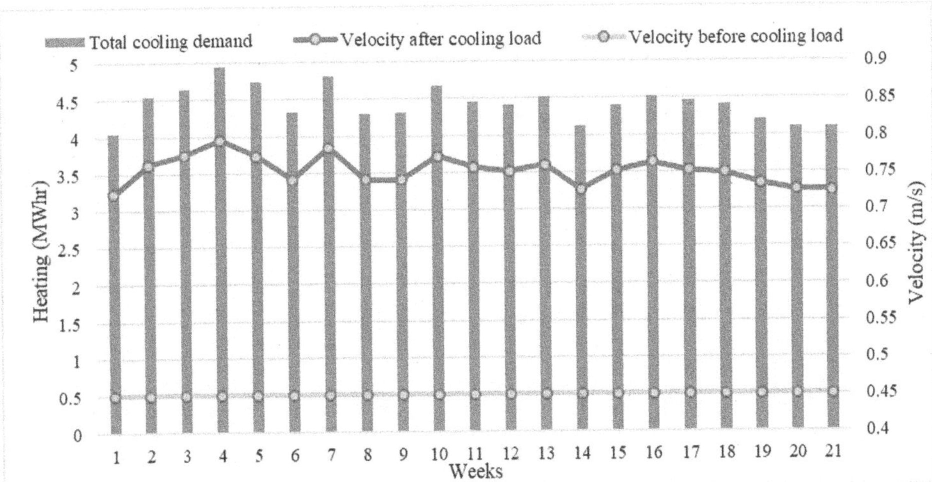

Figure 10.13 Summer heat demand of the district heating and the velocity.

to meet the heat demand of the absorption chiller, extra MSW will be burnt in the WI power plant and this heat will be provided for both production of cooling demand and supply of the domestic hot water of the network. Figure 10.13 shows the heat demand produced after cold production in the absorption chiller. This figure shows that there is a minor difference between the heat demand of each week and the changes are between 4 to 5 MWhr. This is because this amount of heat is obtained from the summation of two other heat sources; the heating load of the absorption chiller and heating load for the domestic hot water, although according to Figure 10.12, the fluctuations of the cooling demand are higher due to the variation of the

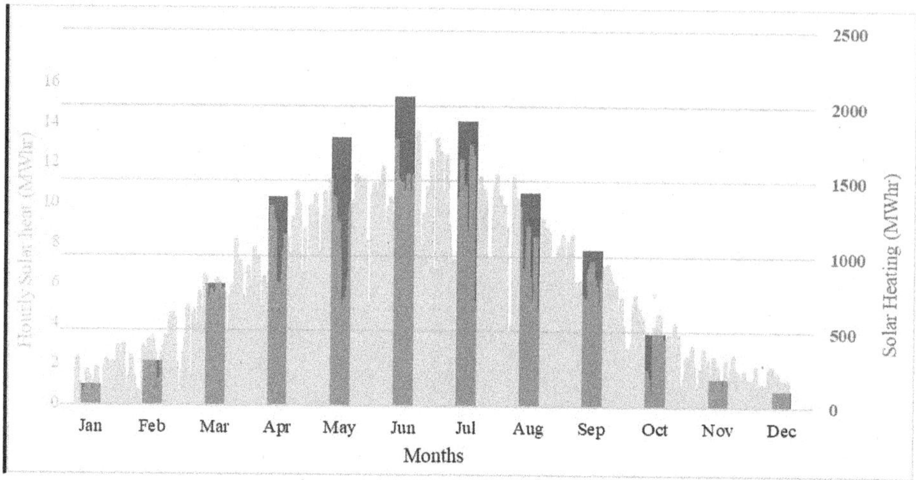

Figure 10.14 Amount of solar heat during the year.

ambient temperature. Also, in Figure 10.13 the velocity of the working fluid has been shown for 21 weeks in summer before and after adding the cooling load of the network. This figure shows the low velocity of the working fluid in the system before considering the cooling load of the network. After adding the absorption cooling unit into the hybrid system, the velocity increases up to 0.8 m/s which is an appropriate velocity inside a pipeline. In the optimum values of decision variables, the velocity increases about 90% compared to the velocity of the pipeline that just transferred the domestic hot water to the district heating system for summer.

In the integrated proposed system, a solar thermal field of HP ETCs is connected to the main heat exchanger of the cycle providing a portion of heat requirement during the year. This heat demand has two different applications based on the ambient temperature. In summer it is used for cooling production and in winter, it has application in heating the working fluid of the district heating. The criterion here for designing the solar field size is as follows; in the hottest hour of the year, the solar system would be able to provide the required heat demand of the absorption chiller. For this reason, the heat production of the solar field is represented in two sections specifying winter and summer. Figure 10.14 shows that in summer when there is the highest cooling demand, the highest amount of solar energy is achievable; around 14.2 MWhr at the optimum solar radiation hour and over 2,000 MWhr in June. This high amount of heating would be very beneficial for cooling production on hot days of summer.

The other problem of the system was the low rate of MSW combustion during summer. When the power plant has an accumulation of MSW, they ought to transfer the extra MSW towards another far away WI plant with a higher capacity of combustion, which for the case of Esbjerg is 300 km away from the city and causes extra expenses for the municipality. Figure 10.15 shows that the system, to provide the required domestic hot water for district heating, should burn 1.5 kg/s of MSW while when the partial supply of cooling load is assigned to the hybrid system, up to 2.5 kg/s of MSW will be burnt. It means that at each week 470,000 kg more MSW is combusted and it could be increased up to 600,000 kg for week 4.

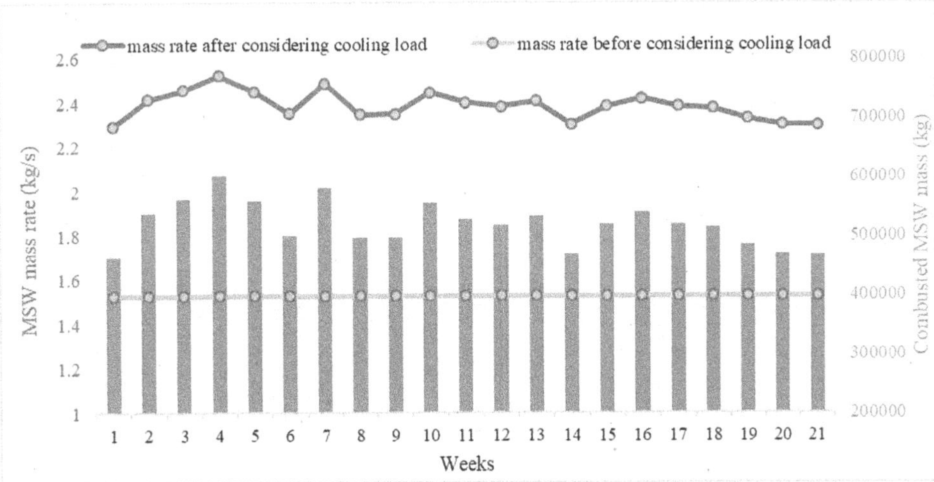

Figure 10.15 Daily consumption rate of MSW by the WI plant.

Table 10.7 Exergy destruction for all components

Components	*Exergy destruction (MW)*	*Relative exergy destruction (%)*
WI	78	49.6
SG	62.2	39.6
Turbine	8.6	5.5
Condenser	6.3	4.0
Pump	0.8	0.5
OFWH	1.2	0.8

The main scope of exergy analysis would be to find components with a high share of exergy destruction. This issue is useful to achieve a better operational design. The exergy of the MSW is about 178.25 MW and MSW produces 114.5 MW heat in the waste incinerator unit. From this amount of heat, 85.15 MW is used in the steam generator of the power plant and the other part of energy in the form of flue gas moves out the plant. In the middle of the way, 13 MW of energy is extracted from the flue gas and is used as the source of heating in the heat exchanger. The energy delivered into the steam generator is first used in the turbine to produce the electricity which is about 20 MW. Then the extra energy of the working fluid is removed out in the condenser and delivered to the heat exchanger supporting the district heating and absorption unit. Table 10.7 shows the exergy destruction of each component. Based on the table, the waste incinerator unit is the key source of exergy destruction, contributing 49.6%. The high exergy destruction in the incinerator is due to the partial combustion and the chemical reactions. Combustion processes, in general, have the most top exergy destruction due to their irreversible nature.

Table 10.8 Specification and properties of the absorption chiller

	h (kj/kg)	mass flow (kg/s)	P kPa	quality	T °C	X	s hJ/kg.K	Ex kW
c1	168.2	0.00458	7.4	0	40.1	0	-	-
c2	168.2	0.00458	0.67	0.064	1.8	0	-	-
c3	2504	0.00458	0.67	1	1.8	0	-	-
c4	86.66	0.05	0.67	0	31.8	0.5648	-	-
c5	86.66	0.05	7.4	-	31.8		-	-
c6	149.6	0.05	7.4	-	63.4		-	-
c7	2645	0.00458	7.4	-	77.1		-	-
c8	222.6	0.0455	7.4	0	88.8	0.6216	-	-
c9	154.8	0.0455	7.4	-	52		-	-
c10	154.8	0.0455	0.67	0.005	45	0.6216	-	-
c11	377.1	658.2	-	-	90		1.193	18134.1
c12	356	847	-	-	85		1.135	15431.5
c13	100.7	282	-	-	24		0.3532	−786.1
c14	150.9	282	-	-	36		0.5186	−296.0
c15	100.7	282	-	-	24		0.3532	−786.1
c16	142.5	282	-	-	34		0.4915	−425.7
c17	50.5	297.7	-	-	12		0.1806	−719.2
c18	16.91	297.7	-	-	4		0.0611	−295.4

Table 10.8 shows the specification of the absorption chiller as well as the properties of different states. The number of points corresponds exactly to the number represented in Figure 10.1, except that for several input/output streams of the generator, one input and output are considered as states c11 and c12. The cooling produced in the evaporator is considered as 10.2 MW. For this capacity, 14.4 MW heat is to be provided for the generator of the absorption chiller. The inlet flow (flow 11 of chiller) brings 248 MW energy into the absorption chiller which is equal to 18.1 MW exergy. From this amount of energy, 13.89 MW heat is extracted through the generator which is equal to 2.7 MW exergy. The exergy efficiency of the absorption chiller is defined as the ratio of the difference of exergy variation in the evaporator (which is the target of operation in the chiller) to the difference of exergy variation in the generator (which is the source of energy of operation in the absorption chiller). This parameter was found to be 0.16. Also, the amount of exergy destruction is calculated and is about 1.9 MW and the amount of output exergy at the evaporator is about 0.42 MW. Based on the data of this table and by considering the equation of Table 10.3, it could be concluded that the highest exergy destruction occurs in the absorber and generator.

For the economic evaluation, one would like to investigate how the proposed system can provide economic benefits for the policymakers and authorities. From two aspects, this issue is studied. First of all the challenge of extra money that should be paid for transferring and burning the excess MSW is considered. Then the levelized cost of energy production for this emerging technology is compared with the other well-established power generation technologies.

Figure 10.15 showed the amount of extra MSW that should be burnt after considering the cooling load of the network, one would know that in the former system this amount of MSW remained in the site of the power plant. Due to the place limitations and environmental considerations, this volume of MSW should be immediately transferred to be burnt

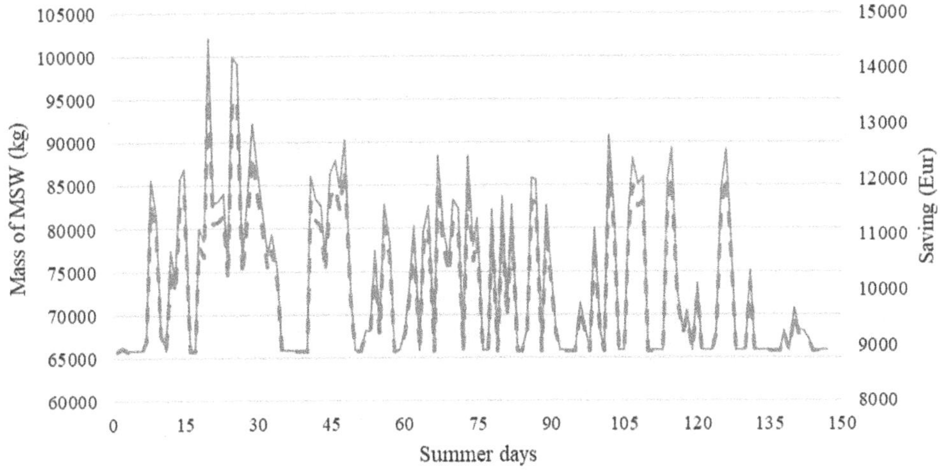

Figure 10.16 Saving due to increase in MSW consumption during summer.

or landfilled. When the proposed system governs the power plant and the burden of responsibility associated with cooling production falls on the shoulders of the hybrid system, then this excess mass of MSW will be burnt and a good saving will occur. Figure 10.16 shows the amount of money saved due to this change of strategy. This figure is represented for 147 days in the hot period of the year. It shows that this strategy saves between €9,000 to over €14,000 for specified days. And that would be a good result for the proposed hybrid system from an economic point of view. In addition, the figure shows the amount of mass if MSW will be burnt on each day of this period specifically for supplying the cooling unit. It could be seen that this amount would be between 65,000 kg to about 102,000 kg.

Figure 10.17 shows the LCOE of a few power generation technologies. A part of these mentioned power generation mechanisms is well-developed energy technologies that have less variability in LCOE due to established capital costs and operation and maintenance costs. As the proposed system would be considered as an emerging one, it could thus be subject to high uncertainties in both technical and economic performance, as expected for technologies in the early stages of development. Power generation with onshore wind and solar PV power plants represents the lowest LCOE which is in the range of 60 to 80 $/MWhr. Then, LCOE for waste incineration is estimated as 0.92 $/MWhr. This type of energy production includes electricity and heat production, which improves the cost of energy. In the proposed system again, one would see that electricity, heat, and cold production occur simultaneously and this modifies the cost of energy more compared to the former technology which was found to be 0.82 $/MWhr.

The environmental analysis could be considered as a process to identify elements that affect the performance of the system or compare the power generation technologies from a clean energy production point of view. The analysis entails assessing the level of threat or opportunity the factors might present. The environmental analysis is implemented to determine various environmental factors and their potential impact on a system. In this process, it is assumed that for both systems, the same amount of MSW will be burnt no matter if it is

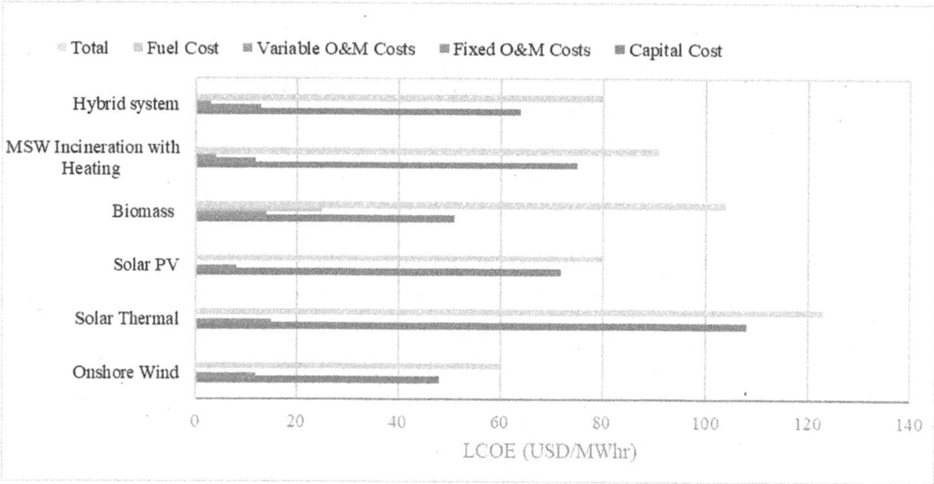

Figure 10.17 LCOE of the power generation technologies.

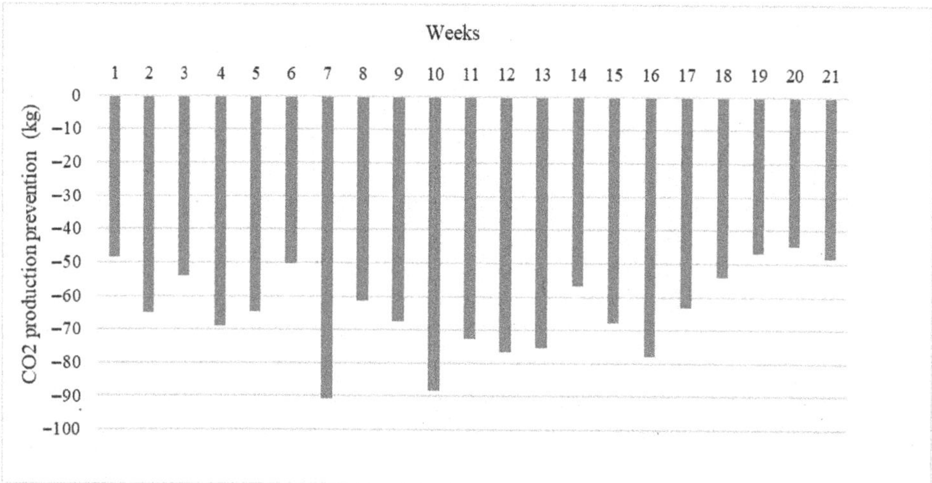

Figure 10.18 CO_2 production prevention by the use of solar energy.

in the power plant or the far away high-capacity waste incinerator. So from this point, there would be no difference between the current and proposed systems. While the application of solar energy means the implementation of clean energy and the heat generated in the solar field does not accompany CO_2 production. As much as the heat produced by solar energy and is used in the system, the environmental benefit is obtainable and can be quantified and weighed. Figure 10.18 shows the amount of CO_2 production prevention by the use of solar energy. For weeks with a higher amount of solar radiation, the system could use more from solar energy and this causes the potential to reduce CO_2 production. This figure shows that up to 90 kg CO_2 production prevention would occur during a week.

10.8 SUMMARY

With a look at the summer heat supply problem of district heating, a new solution has been presented in this work and the performance of the system before and after implementing the new approach was investigated. The challenges of the system would be considered as the low velocity of working fluid inside the feeding pipeline, decrease of the working fluid temperature at the destination, low consumption of MSW and accumulation of the waste at disposal point, and the expenses for the manipulation of the collected waste. To address this problem, the solution was considered to add a new section to the hybrid system to provide a cooling load for the system which is the need for the district network. For this, the size of the feeding pipe is calculated to minimize the expenses of the district heating unit. Then the whole system was considered to be optimized based on decision variables of absorption cooling unit, solar HP ETC field, the temperature of the flue gas, and volume of the hot storage. The values are found to be 10.2 MW, 14.2 MW, 1,100 K, and 500 m^3, respectively. In the optimum values of decision variables, the velocity increases about 90% with respect to when the pipeline just transferred the domestic hot water to the district heating system for summer. Whenever the network requires 10 MW cooling or lower, the absorption unit will be ready to provide the complete demand and if the need is higher than 10 MW, then the extra amount of cooling should be provided by another cooling production unit. The results showed that 0.73 of cooling demand responded by the proposed absorption unit. It means that at each week 470,000 kg more MSW is combusted and it could be increased up to 600,000 kg for week 4. Also, the amount of MSW would be burnt specifically for supplying the cooling unit found to be between 65,000 to about 102,000 kg per day. This strategy exhibited the potential to save between €9,000 to over €14,000 each day in summer and that would be a good result for the proposed hybrid system from an economic point of view. The other economic benefit of the proposed system is provided is a 10% LCOE reduction compared to the system with supplying the cooling load.

In summary, it should be said that the proposed system provides a 10 MW cooling load for the district network, furthermore, the summer challenges such as low velocity, accumulation of MSW, and the need for large cooling capacity of the power plant are removed.

REFERENCES

Abokersh, Mohamed Hany, Manel Vallès, Luisa F. Cabeza, & Dieter Boer. 2021. "Challenges associated with the construction and operation of seasonal storage for a small solar district heating system: A multi-objective optimization approach." In *14th International Renewable Energy Storage Conference 2020 (IRES 2020)*, pp. 150–160. Paris: Atlantis Press.

Abokersh, Mohamed Hany, Manel Vallès, Laureano Jiménez, & Dieter Boer. 2020. "Cost-effective processes of solar district heating system based on optimal artificial neural network." In Sauro Pierucci, Flavio Manenti, Giulia Luisa Bozzano, & Davide Manca (eds.), *30th European Symposium on Computer Aided Process Engineering*, pp. 403–408. New York: Elsevier. https://doi.org/10.1016/B978-0-12-823377-1.50068-9.

Achenbach, Elmar, & Ernst Riensche. 1994. "Methane/steam reforming kinetics for solid oxide fuel cells." *Journal of Power Sources* 52(2): 283–288. doi:10.1016/0378-7753(94)02146-5.

Apricus. 2006. "Evacuated tube solar collectors, solar collector." www.apricus.com/html/solar_collector.htm.

Arabkoohsar, Ahmad, & Meisam Sadi. 2020a. "A solar PTC powered absorption chiller design for co-supply of district heating and cooling systems in Denmark." *Energy* 193. doi:10.1016/j.energy.2019.116789.

Arabkoohsar, A., & M. Sadi. 2020b. "Thermodynamics, economic and environmental analyses of a hybrid waste–solar thermal power plant." *Journal of Thermal Analysis and Calorimetry* 144(3). doi:10.1007/s10973-020-09573-3.

Bai, Yakai, Ming Yang, Jianhua Fan, Xiaoxia Li, Longfei Chen, Guofeng Yuan, & Zhifeng Wang. 2021. "Influence of geometry on the thermal performance of water pit seasonal heat storages for solar district heating." *Building Simulation,* 14: 579–599.

Behzadi, Amirmohammad, Ehsan Houshfar, Ehsan Gholamian, Mehdi Ashjaee, & Ali Habibollahzade. 2018. "Multi-criteria optimization and comparative performance analysis of a power plant fed by municipal solid waste using a gasifier or digester." *Energy Conversion and Management* 171: 863–878. https://doi.org/10.1016/j.enconman.2018.06.014.

Burnett, Dougal, Edward Barbour, & Gareth P. Harrison. 2014. "The UK solar energy resource and the impact of climate change." *Renewable Energy* 71: 333–343.

Chen, Heng, Meiyan Zhang, Yunyun Wu, Gang Xu, Wenyi Liu, & Tong Liu. 2020. "Design and performance evaluation of a new waste incineration power system integrated with a supercritical CO_2 power cycle and a coal-fired power plant." *Energy Conversion and Management* 210: 112715. https://doi.org/10.1016/j.enconman.2020.112715.

Chen, Yuzhu, Jun Wang, & Peter D. Lund. 2020. "Sustainability evaluation and sensitivity analysis of district heating systems coupled to geothermal and solar resources." *Energy Conversion and Management* 220: 113084.

Chopra, Kapil, Atin K. Pathak, Vineet Veer Tyagi, Adarsh Kumar Pandey, Sanjeev Anand, & Ahmet Sari. 2020. "Thermal performance of phase change material integrated heat pipe evacuated tube solar collector system: An experimental assessment." *Energy Conversion and Management* 203: 112205. https://doi.org/10.1016/j.enconman.2019.112205.

da Silva, Leo Jaymee de Vilas Boas, Ivan Felipe Silva dos Santos, Johnson Herlich Roslee Mensah, Andriani Tavares Tenório Gonçalves, & Regina Mambeli Barros. 2020. "Incineration of municipal solid waste in Brazil: an analysis of the economically viable energy potential." *Renewable Energy* 149: 1386–1394.

EIA. 2017. "Construction costs for most power plant types have fallen in recent years." www.eia.gov/todayinenergy/detail.php?id=31912.

Energist. n.d. www.energnist.dk.

Fabricius, Mika, Daniel Øland Tarp, Thomas Wehl Rasmussen, & Ahmad Arabkoohsar. 2020. "Utilization of excess production of waste-fired CHP plants for district cooling supply, an effective solution for a serious challenge." *Energies* 13(13): 3319.

Hieminga, G. 2013. "Saving energy in the Netherlands: New EU directive takes on energy efficiency." Report, ING, May.

Hoornweg, Daniel, & Perinaz Bhada-Tata. 2012. *What a Waste: A Global Review of Solid Waste Management.* Washington, DC: World Bank.

Huang, Junpeng, Jianhua Fan, & Simon Furbo. 2020. "Demonstration and optimization of a solar district heating system with ground source heat pumps." *Solar Energy* 202: 171–189. https://doi.org/10.1016/j.solener.2020.03.097.

Janda, Kathryn B. 2009. "Worldwide status of energy standards for buildings: A 2009 update." In *Proceedings of the European Council for an Energy Efficient Economy (ECEEE) Summer Study*, pp. 1–6.

Kythavone, Latthaphonh, & Nattaporn Chaiyat. 2020. "Life cycle assessment of a very small organic Rankine cycle and municipal solid waste incinerator for infectious medical waste." *Thermal Science and Engineering Progress* 18: 100526. https://doi.org/10.1016/j.tsep.2020.100526.

Lewis, Nathan S. 2016. "Research opportunities to advance solar energy utilization." *Science* 351(6271).

Li, Zhanchao, Min Xu, Xiulan Huai, Caifeng Huang, & Kejian Wang. 2020. "Simulation and analysis of thermochemical seasonal solar energy storage for district heating applications in China." *International Journal of Energy Research* 45(1).

Liu, Jun, Xiaoyu Luo, Sheng Yao, Quangong Li, & Weishu Wang. 2020. "Influence of flue gas recirculation on the performance of incinerator-waste heat boiler and NOx emission in a 500 t/d waste-to-energy plant." *Waste Management* 105: 450–456. https://doi.org/10.1016/j.wasman.2020.02.040.

Liuzzo, Giuseppe, Nicola Verdone, & Marco Bravi. 2007. "The benefits of flue gas recirculation in waste incineration." *Waste Management* 27(1): 106–116.

LOGSTOR. n.d. "Profile: The most ambitious solutions in the energy sector." www.logstor.com/about-us/profile.

Ma, Chen, Bo Li, Dongmei Chen, Terrence Wenga, Wenchao Ma, Fawei Lin, & Guanyi Chen. 2019. "An investigation of an oxygen-enriched combustion of municipal solid waste on flue gas emission and combustion performance at a 8 MWth waste-to-energy plant." *Waste Management* 96: 47–56. https://doi.org/10.1016/j.wasman.2019.07.017.

Miao, Shanshan, Dehua Wang, Yaning Yin, Shengli Han, Ting Wang, Chong Li, Ying Huang, et al. 2021. "Feasibility analysis on the coupling power system of municipal solid waste incinerator and coal-fired unit." *IOP Conference Series: Earth and Environmental Science* 675(1): 12041. doi:10.1088/1755-1315/675/1/012041.

Mohammadi, Maryam, & Iiro Harjunkoski. 2020. "Performance analysis of waste-to-energy technologies for sustainable energy generation in integrated supply chains." *Computers & Chemical Engineering* 140: 106905. https://doi.org/10.1016/j.compchemeng.2020.106905.

Mondal, Pradip, Shambhunath Barman, & Samiran Samanta. 2021. "Integrated MSW to energy and hot water generation plant for Indian cities: Thermal performance prediction BT." In Shripad Revankar, Swarnendu Sen, & Debjyoti Sahu (eds.), *Proceedings of International Conference on Thermofluids*, pp. 569–578. Singapore: Springer Singapore.

PE Continuing Education. n.d. "Overview of Vapor Absorption Cooling Systems." www.cedengineering.com/courses/overview-of-vapor-absorption-cooling-systems.

Sadi, Meisam, & Ahmad Arabkoohsar. 2020a. "3E analysis of hybrid solar-waste driving CHP plant with flue gas recovery unit, a smart solution toward sustainable energy systems." In Farkhondeh Jabari, Behnam Mohammadi-Ivatloo, & Mousa Mohammadpourfard (eds.), *Integration of Clean and Sustainable Energy Resources and Storage in Multi-Generation Systems: Design, Modeling and Robust Optimization*, pp. 239–258. Cham: Springer International Publishing. doi:10.1007/978-3-030-42420-6_12.

Sadi, Meisam, & Ahmad Arabkoohsar. 2020b. "Techno-economic analysis of off-grid solar-driven cold storage systems for preventing the waste of agricultural products in hot and humid climates." *Journal of Cleaner Production* 275: 124143. https://doi.org/10.1016/j.jclepro.2020.124143.

Saloux, Etienne, & José A. Candanedo. 2020. "Optimal rule-based control for the management of thermal energy storage in a Canadian solar district heating system." *Solar Energy* 207: 1191–1201.

Shafieian, Abdellah, & Mehdi Khiadani. 2020. "Integration of heat pipe solar water heating systems with different residential households: An energy, environmental, and economic evaluation." *Case Studies in Thermal Engineering* 21: 100662. https://doi.org/10.1016/j.csite.2020.100662.

Song, Guohui, Laihong Shen, & Jun Xiao. 2011. "Estimating specific chemical exergy of biomass from basic analysis data." *Industrial and Engineering Chemistry Research* 50(16): 9758–9766. doi:10.1021/ie200534n.

Tian, Zhiyong, Shicong Zhang, Jie Deng, Jianhua Fan, Junpeng Huang, Weiqiang Kong, Bengt Perers, & Simon Furbo. 2019. "Large-scale solar district heating plants in Danish smart thermal grid: Developments and recent trends." *Energy Conversion and Management* 189: 67–80.

Chapter 11

Trigeneration system

Exergoeconomic and environmental analysis

Joy Nondy and Tapan Kumar Gogoi
Department of Mechanical Engineering, Tezpur University, Tezpur, India

11.1 INTRODUCTION

Waste heat recovery-based trigeneration systems have great potential for use in many practical applications due to their higher efficiency, cost-effectiveness, and lower environmental impact. One such trigeneration system that is already in use for the combined production of process heat, chilled water, and electricity is the gas turbine (GT) based combined cooling, heating, and power (CCHP) system. In the GT based CCHP system, the heat available at the GT exhaust gas is used for producing steam in a heat recovery steam generator (HRSG). The steam can be either used directly for process heat application or for operating a steam Rankine cycle for producing electricity. However, if the low-grade heat is available at the exhaust gas, then that can be used for producing electricity using an organic Rankine cycle (ORC). As reported by Cao et al. (2016), the GT-ORC plant gives better thermodynamic performance than the GT-ST plant for the same input parameters in the case of low-grade heat availability.

The waste heat from the GT exhaust can also be utilized for operating an absorption refrigeration system (ARS) to produce cooling. Moné, Chau, and Phelan (2001) demonstrated the economic viability of incorporating ARS into a GT-based combined heat and power system (CHP). They observed that the cooling capability obtained from the ARS is more a function of the gas exhaust flow rate than the temperature. Furthermore, the GT exhaust gas can also be used to generate power, cooling effect, and process heat all at the same time by combining HRSG, ST/ORC, and ARS to the GT plant to develop a CCHP system.

The feasibility of the CCHP or any other thermal system is determined primarily by using energy analysis that gives a quantitative measure of energy flow in the system. For instance, Mohammadi et al. (2017) proposed a combined system by integrating GT-ARS with an ORC. They observed that at the base case operating condition, the CCHP system gives a net power of 30 kW, a cooling capacity of 8 kW, and 7.2 tons of hot water with a thermal efficiency of 67.6%. However, for the qualitative study, exergy analysis is applied that pinpoints the location, cause, and the magnitude of irreversibility occurring in the system. Often researchers apply exergy analysis for assessing the feasibility of the CCHP systems. Khaliq (2009) analyzed a GT-HRSG-ARS integrated CCHP system based on exergy analysis and concluded that almost 80% of the overall exergy destruction occurs at the combustion chamber (CC) and the HRSG. Similarly, Sarkar et al. (2017) proposed a CCHP system that yields a net power of 33.60 MW, process heat of 40.78 MW, and a cooling load of 1 MW. By integrating ORC into the GT-ARS system, the thermal efficiency, and the exergy efficiency of the CCHP system have shown to be improved by 2.5% and 0.75%, respectively. The exergy analysis is not only crucial from the thermodynamics viewpoint but it is also a

DOI: 10.1201/9781003213741-11

prerequisite step for performing the exergoeconomic assessment that enables the designers to develop an efficient as well as cost-effective energy conversion system. Exergoeconomic analysis is the union of exergy and economic concepts that deal with the evaluation of the exergy and non-exergy related costs associated with the system. However, recently the focus has also shifted to the minimization of the environmental impact of a power system. Several studies related to exergoeconomic and environmental analyses considering various hybrid systems have been reported.

Ahmadi, Rosen, and Dincer (2011) provided an exergoeconomic and environmental analysis of a CCHP system where surplus heat from the GT cycle is utilized in the ST cycle and single effect ARS. They observed that due to the integration of the ARS into the system, both the energy and exergy efficiency increases. However, the share of ARS to the total exergy destruction and system cost is minimal. Anvari, Jafarmadar, and Khalilarya (2016) proposed a GT-ORC integrated CHP system and performed exergoeconomic and environmental assessment by adding the environmental cost of NOx and CO emission to the total system cost rate. They also performed the parametric study and concluded that increasing the pressure ratio and isentropic efficiency of the compressor and gas turbine increases the performance of the proposed CHP system.

According to a brief overview of the literature, several studies have suggested various integrations schemes for a GT-based CCHP system. However, the GT exhaust gas was only used for waste heat recovery in most of the papers. The proposed CCHP system in this chapter employs a dual approach to waste heat recovery, using both compressed air and GT exhaust gas. The novel aspects and the goals of the current chapter can be summarized as follows:

- Introducing a novel CCHP system that employs an intercooled-recuperative GT plant as the prime mover, with ARS to recover heat wasted during compressed air intercooling and HRSG and R-ORC to recover heat from exhaust gas for process heat and additional power.
- Investigating the exergoeconomic and environmental viability of adopting the proposed configuration to incorporate ARS, R-ORC, and HRSG into the IRGT plant.
- Conducting a parametric study to understand the impact of overall compressor pressure ratio, air compressor isentropic efficiency and gas turbine isentropic efficiency on the performance of the CCHP system.

11.2 SYSTEM DESCRIPTION

The schematic of the proposed CCHP system is shown in Figure 11.1. It includes an intercooled-recuperative gas turbine cycle and a heat recovery steam generator (IRGT-HRSG), a LiBr-H_2O operated single effect ARS and R-ORC. The ambient air first enters the IRGT unit through the double air compression stage (AC-I & AC-II) via a generator (GEN). Then GEN acts as the intercooler that extracts heat from the compressed air at the AC-I exit and lowers the air temperature for the second stage compression. The heat extracted at the GEN is utilized to drive the ARS for producing the cooling effect. Next, from the AC-II exit, compressed air passes through the air preheater (APH), gains some heat, and finally enters the CC. At the CC, the fuel in the presence of compressed air undergoes combustion to produce hot gases. The high-pressure combustion gases then expand in the GT and generate power. The combustion gases after doing work in the GT pass through the APH where

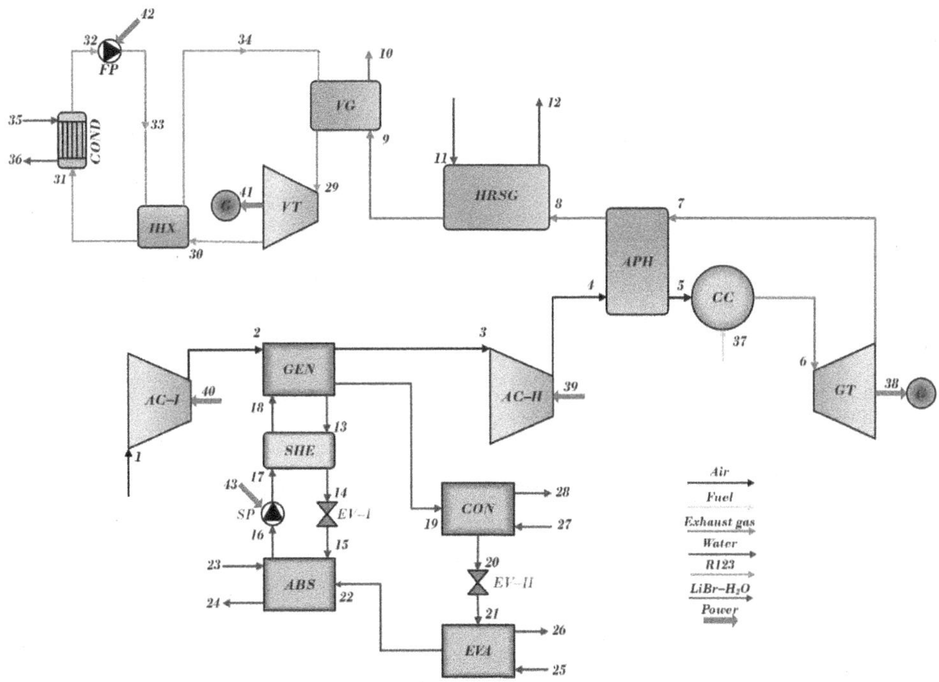

Figure 11.1 The proposed CCHP system is represented schematically.

it rejects some heat and then enters the HRSG. The HRSG utilizes the thermal energy of hot gas in producing saturated steam that is eventually used for providing the process heat. After exiting the HRSG, the flue gas passes through the vapor generator (VG) where the remaining low-grade heat is extracted for operating the R-ORC. The detailed working description of R-ORC and ARS can be found in the previous article of the author (Nondy & Gogoi 2020b).

11.3 MODELING

This section first discusses the assumptions considered for carrying out the study followed by the methodologies applied for modeling the proposed CCHP system.

11.3.1 Assumptions

The assumptions considered to simulate the proposed system are given as follows (Anvari, Jafarmadar, & Khalilarya 2016; Nondy & Gogoi 2019):

- Ambient temperature and pressure are 298.15 K and 101.3 kPa.
- Ambient air composition is 0.778 N_2, 0.2059 O_2, 0.0003 CO_2 and 0.019 H_2O.
- Methane with a molar lower heating value (LHV) of 802,361 kJ/kmol is used as fuel.
- 2% of the LHV is considered as heat loss from the CC.

Table 11.1 The base case working conditions of the CCHP system

Description	*Value*	*Description*	*Value*
Overall AC pressure ratio (r_p)	10	CON temperature	308.15 K
The isentropic efficiency of ACs	86 %	Fuel inlet temperature	298.15 K
The isentropic efficiency of GT	86 %	Fuel inlet pressure	1,200 kPa
APH outlet temperature (T_5)	850 K	Isentropic efficiency of VT	80 %
GT inlet temperature (T_6)	1,520 K	Isentropic efficiency of FP	85 %
Net power from IRGT cycle ($\dot{W}_{net,IRGT}$)	50 MW	Inlet temperature of ABS	298.15 K
PPTD of HRSG	5 K	Outlet temperature of ABS	303.15 K
VG outlet temperature (T_{29})	275 K	Inlet temperature of EVA	288.15 K
COND outlet temperature (T_{32})	302.8 K	Outlet temperature of EVA	283.15 K
PPTD of VG	5 K	Inlet temperature of CON	298.15 K
GEN temperature	373.15 K	Outlet temperature of CON	303.15 K
EVA temperature	278.15 K	Effectiveness of the SHX	75 %
ABS temperature	308.15 K	Isentropic efficiency SP	85 %

Sources: Nondy and Gogoi (2020a); Anvari, Taghavifar, and Parvishi (2017).

- The pressure drop in the HRSG's gas side is considered as 5%, while the pressure drop in the APH's air and gas sides is 3% and 5%.
- The inlet condition of the water at the HRSG is 298.15 K and 3,500 kPa.

Some other important parameters with their corresponding assumed values are given in Table 11.1.

11.4 ENERGY ANALYSIS

11.4.1 Modeling of IRGT cycle

The ideal gas mixture principle is used to determine the enthalpy and entropy of air and flue gas in the IRGT-HRSG cycle (Nondy & Gogoi 2021). The pressure at the outlet of AC-I and AC-II is evaluated using the following equations (Musharavati et al. 2021):

$$P_2 = r_{p1} \times P_1 \tag{11.1}$$

$$P_4 = r_{p2} \times P_3 \tag{11.2}$$

where r_{p1} and r_{p2} are the pressure ratio of AC-I and AC-II respectively, which are considered equal to the square root of the overall pressure ratio ($\sqrt{r_p}$).

Similarly, the temperatures at the outlet of AC-I and AC-II are evaluated by applying the ideal gas mixture principle considering isentropic compression. For a net IRGT power of 50 MW, the molar fuel-air ratio ($\underline{\lambda}$) is calculated by solving the energy balance and stoichiometry of the combustion reaction. For per mole basis, the combustion reaction can be given as follows (Nondy & Gogoi 2021):

$$\begin{aligned}\underline{\lambda} CH_4 + \left[x_{N_2} N_2 + x_{O_2} O_2 + x_{CO_2} CO_2 + x_{H_2O} H_2O\right] \rightarrow \\ (1+\underline{\lambda})\left[y_{N_2} N_2 + y_{O_2} O_2 + y_{CO_2} CO_2 + y_{H_2O} H_2O\right]\end{aligned} \tag{11.3}$$

where x is the mole fraction of ambient air and $y$is the mole fraction of the exhaust gas.

The $\underline{\lambda}$ is then evaluated solving the energy balance reaction at the CC given by Nondy and Gogoi (2021):

$$-0.02\underline{\lambda}\,\underline{LHV} + \underline{h}_5 + \underline{\lambda}\ \underline{h}_{37} = (1+\underline{\lambda})\ \underline{h}_6 \tag{11.4}$$

where $\underline{h}$ is the molar specific enthalpy.

The outlet condition of GT is calculated similarly to the ACs using the ideal gas mixture principle considering isentropic expansion.

11.4.2 Modeling of HRSG

The HRSG is designed in a single pressure mode that comprises of economizer (ECO) and an evaporator (EVA). The pinch point and approach point temperature difference (PPTD and APTD) for the modelling of the HRSG were assumed to be 5 K each. The amount of process heat ($\dot{Q}_{heating}$) generated from the HRSG is given by Nondy and Gogoi (2019):

$$\dot{Q}_{heating} = \dot{m}_w (h_{10} - h_9) \tag{11.5}$$

where $\dot{m}_w$ is the mass flow rate of water fed to the HRSG and h is the specific enthalpy.

The enthalpy and entropy of water at the inlet and outlet of the HRSG were calculated using REFPROP 9.0 (Lemmon, Huber, & McLinden 2010).

11.4.3 Modeling of ORC

In this study, R123 is chosen as the working medium for operating the R-ORC over other commonly used organic fluids since it has a lower value for environmental criteria such as global warming potential, ozone depletion potential, and atmospheric life (Tchanche et al. 2009). The properties of R123 at various state points of the ORC are evaluated using REFPROP 9.0 (Lemmon, Huber, & McLinden 2010). The heat exchange process between the GT exhaust gas and working fluid at VG is modeled assuming the stack temperature (T_{10}) as 380 K to avoid approaching the acid dew point temperature (ADPT). The ADPT is the temperature at which the vapor content of the exhaust gas condenses and causes corrosion in the HRSG components. Besides, the condition of working fluid at the inlet of VT and the outlet of COND is considered as saturated vapor and saturated liquid, respectively.

11.4.4 Modeling of ARS

The mass percentage (X) of LiBr in a strong solution (ss) and weak solution (ws) are evaluated using the equations given below (Salmi et al. 2017):

$$X_{ss} = \frac{49.04 + 1.125T_{GEN} - T_{CON}}{134.65 + 0.47T_{GEN}} \tag{11.6}$$

$$X_{ws} = \frac{49.04 + 1.125T_{ABS} - T_{EVA}}{134.65 + 0.47T_{ABS}} \tag{11.7}$$

where T_{GEN}, T_{ABS}, T_{EVA}, and T_{CON} are in °C.

The pressure level at which the various components of the ARS operates are evaluated using the analytical relation provided in Salmi et al. (2017). Besides, the thermodynamic properties of the LiBr-H_2O solution are calculated using the explicit functions of Pátek and Klomfar (2006) while REFPROP 9.0 (Lemmon, Huber, & McLinden 2010) is used for determining the properties of H_2O at vapor and liquid state. The ARS's detailed modeling can be referred from the previous article of the authors (Nondy & Gogoi 2020b).

11.5 EXERGY ANALYSIS

Exergy is the maximum work produced by a system during a reversible process from a given state to a reference state. It is a combined property of a system and the surrounding. Exergy can be broadly classified into physical and chemical components. The physical exergy (E_{ph}) is given by Nondy and Gogoi (2020a):

$$E_{ph} = \dot{m}\left[(h - h_0) - T_0(s - s_0)\right] \tag{11.8}$$

where $\dot{m}$ is the mass flow rate and h and s are the specific enthalpy and entropy, respectively.

The work potential achieved when the system's restricted reference state approaches the real reference state is known as chemical exergy. The chemical exergy of air, R123, and LiBr-H_2O solution are considered negligible while that of methane and water is estimated using standard chemical exergy data (Bejan, Tsatsaronis , & Moran 1995). Besides, the chemical exergy of flue gas (E_{fg}^{CH}) is computed as follows (Nondy & Gogoi 2020a):

$$E_{fg}^{CH} = \dot{n}_{fg} \times \left\{\sum_k x_k \left[\underline{e}_k^{CH} + \underline{R}T_0 \times ln\, ln(x_k)\right]\right\} \tag{11.9}$$

where, $\dot{n}_{fg}$ is the molar flow rate of the flue gas, x_k is the mole fraction of the flue gas component k, $\underline{R}$ is the universal gas constant and $\underline{e}_k^{CH}$ is the chemical exergy per mole of component k

.

The exergy balance equation for a control volume is given by Khandelwal et al. (2020):

$$\dot{E}_F = \dot{E}_P + \dot{E}_L + \dot{E}_D \tag{11.10}$$

where $\dot{E}_F$, $\dot{E}_P$, $\dot{E}_D$, and $\dot{E}_L$ denote the rate of fuel exergy, product exergy, exergy destroyed and the exergy loss, respectively.

The product exergy is the desired output given by a system, while the fuel exergy is the energy used to achieve that output. Exergy loss refers to the untapped exergy that flows out of the system, while exergy destruction refers to the exergy that is destroyed during the process due to the existence of irreversibilities. In an exergy analysis, the exergy efficiency (ε) and exergy destruction ratio (Y_d) are relevant criteria useful in comparing the true performance of all the components. These are computed as follows (Nondy & Gogoi 2020a):

$$\varepsilon = \frac{\dot{E}_P}{\dot{E}_F} \quad (11.11)$$

$$Y_d = \frac{\dot{E}_{d,k}}{\dot{E}_{d,tot}} \quad (11.12)$$

11.6 EXERGOECONOMIC ANALYSIS

The exergoeconomic analysis is the integration of exergy and economic principles that paves the way for developing a cost-effective energy conversion system. In this study, the specific exergy costing method (SEPCO) (Nondy & Gogoi 2021) is used to carry out the exergoeconomic analysis. In this method, the first step is to determine all the exergy flow rates crossing the control surface of all the components followed by the proper definition of fuel, product, and loss components of exergy. Then the cost rate is allocated to all the exergy streams applying the cost balance equation as given by Nondy and Gogoi (2021):

$$\sum\nolimits_0 \dot{C}_{o,k} + \dot{C}_{work,k} = \dot{C}_{heat,k} + \sum\nolimits_i \dot{C}_{i,k} + \dot{Z}_k \quad (11.13)$$

where $\dot{C}$ is the cost flow rate in $\$/h$, which can be defined in terms of exergy flow rate and cost per unit of exergy (c_i) as follows (Nondy & Gogoi 2021):

$$\dot{C}_i = c_i \times \dot{E}_i \quad (11.14)$$

In Equation (11.13), $\dot{Z}_k$ is the capital cost rate of a component k that also includes the operating and maintenance cost. $\dot{Z}_k$ is further calculated using the following relation (Nondy & Gogoi 2021):

$$\dot{Z}_k = \frac{PEC_k \times CRF \times \phi}{N} \quad (11.15)$$

where PEC stands for purchase equipment cost of the component k in US dollars, N is the annual service hours (7,446 hours), ϕ is the maintenance factor, considered as 1.06 (Nondy & Gogoi 2021), is CRF the capital recovery factor evaluated using the relation (Nondy & Gogoi 2021):

$$CRF = \frac{i \times (1+i)^n}{(1+i)^n - 1} \quad (11.16)$$

where i is the interest rate (12 %) and n is the service life of the components (20 years) (Nondy & Gogoi 2021).

Table 11.2 The relations applied for calculating the purchased equipment costs

Components	*Purchase equipment cost*
ACs	$\left(\frac{71.10\dot{m}_a}{0.9-\eta_{AC}}\right)\left(\frac{P_{out}}{P_{in}}\right)ln\left(\frac{P_{out}}{P_{in}}\right)$
APH	$\left[4122\frac{\dot{m}_g\left(h_7-h_8\right)}{18\times\Delta T_{lm,aph}}\right]^{0.6}$
CC	$\frac{46.08\dot{m}_a}{\left[0.995-\frac{P_6}{P_5}\right]}\left[1+exp\ exp\left(0.018T_6-26.4\right)\right]$
GT	$\left(\frac{479.34\dot{m}_g}{0.92-\eta_{GT}}\right)ln\left(\frac{P_6}{P_7}\right)\left[1+exp\ exp\left(0.036T_6-54.4\right)\right]$
HRSG	$6570\left[\left(\frac{\dot{Q}_{eco}}{LMTD_{eco}}\right)^{0.8}+\left(\frac{\dot{Q}_{eva}}{LMTD_{eva}}\right)^{0.8}\right]+21276\dot{m}_w+1184.4\dot{m}_g^{1.2}$
ARS	$0.322\left[30000+0.75\left(\frac{\dot{Q}_k}{0.5\times LMTD_k}\right)^{0.8}\right];k=GEN,ABS,EVA,CON,SHX$
VG	$309.143\left(A_{VG}\right)+231.915$
VT	$6,000\left(\dot{W}_{VT}^{0.7}\right)$
IHX	$1.3\left(190+310A_{IHX}\right)$
COND	$1,773\left(\dot{m}_{vapour}\right)$
FP	$3,540\left(\dot{W}_{FP}^{0.7}\right)$
SP	$3,540\left(\dot{W}_{SP}^{0.7}\right)$

Source: Anvari, Taghavifar, and Parvishi (2017).

The *PEC* for all the components of the CCHP system is determined by implementing the cost functions (Khaljani, Khoshbakhti Saray, & Bahlouli 2015; Anvari, Taghavifar, & Parvishi 2017) given in Table 11.2. These cost functions are the correlations involving various critical process parameters and constants. Because the *PEC* relations for different components are developed in different years, the Marshal and Swift index is used to transform them into the current year (Khaljani, Khoshbakhti Saray, & Bahlouli 2015). The *PEC* for IHX and VGs depends on the heat transfer area and in this study, they were determined following the procedures reported by Khaljani, Khoshbakhti Saray, and Bahlouli (2015).

The final step in the SEPCO method is to solve the cost balance equations for all the components with the aid of auxiliary equations. These equations are formulated based on the fuel and product approach. The cost balance and the auxiliary equations formulated for each of the components of the proposed CCHP system are given in Table 11.3. On solving these equations, the cost per unit of exergy (c_k) for each state points are determined and further used to evaluate various important exergoeconomic parameters *viz.* exergoeconomic factor (f),

Table 11.3 The cost equations formulated for each component of the proposed CCHP system

Components	*Cost balance equations*	*Auxiliary equations and assumptions*
AC-I	$\dot{C}_1 + \dot{C}_{40} + \dot{Z}_{AC-I} = \dot{C}_2$	$c_1 = 0$
AC-II	$\dot{C}_3 + \dot{C}_{39} + \dot{Z}_{AC-II} = \dot{C}_4$	–
APH	$\dot{C}_4 + \dot{C}_7 + \dot{Z}_{APH} = \dot{C}_5 + \dot{C}_8$	$c_7 = c_8$
CC	$\dot{C}_5 + \dot{C}_{37} + \dot{Z}_{CC} = \dot{C}_6$	$\dot{C}_{37} = c_f \dot{m}_f LHV$
GT	$\dot{C}_6 + \dot{Z}_{GT} = \dot{C}_7 + \dot{C}_{38} + \dot{C}_{39} + \dot{C}_{40}$	$c_6 = c_7, c_{38} = c_{39}, c_{38} = c_{40}$
HRSG	$\dot{C}_8 + \dot{C}_{11} + \dot{Z}_{HRSG} = \dot{C}_9 + \dot{C}_{12}$	$c_8 = c_9, c_{11} = 0$
VG	$\dot{C}_9 + \dot{C}_{34} + \dot{Z}_{VG} = \dot{C}_{10} + \dot{C}_{29}$	$c_9 = c_{10}$
VT	$\dot{C}_{29} + \dot{Z}_{VT} = \dot{C}_{30} + \dot{C}_{41} + \dot{C}_{42} + \dot{C}_{43}$	$c_{29} = c_{30}, c_{41} = c_{42}, c_{41} = c_{43}$
IHX	$\dot{C}_{30} + \dot{C}_{33} + \dot{Z}_{IHX} = \dot{C}_{31} + \dot{C}_{34}$	$c_{30} = c_{31}$
COND-II	$\dot{C}_{31} + \dot{C}_{35} + \dot{Z}_{COND-II} = \dot{C}_{32} + \dot{C}_{36}$	$c_{31} = c_{32}, c_{35} = 0$
FP	$\dot{C}_{32} + \dot{C}_{42} + \dot{Z}_{FP} = \dot{C}_{33}$	–
GEN	$\dot{C}_2 + \dot{C}_{18} + \dot{Z}_{GEN} = \dot{C}_3 + \dot{C}_{13} + \dot{C}_{19}$	$c_2 = c_3, \frac{(\dot{C}_{19} - \dot{C}_{18})}{(\dot{E}_{19} - \dot{E}_{18})} = \frac{(\dot{C}_{13} - \dot{C}_{18})}{(\dot{E}_{13} - \dot{E}_{18})}$
CON	$\dot{C}_{19} + \dot{C}_{27} + \dot{Z}_{CON} = \dot{C}_{20} + \dot{C}_{28}$	$c_{19} = c_{20}$
EVA	$\dot{C}_{21} + \dot{C}_{25} + \dot{Z}_{EVA} = \dot{C}_{22} + \dot{C}_{26}$	$c_{22} = c_{21}$
ABS	$\dot{C}_{22} + \dot{C}_{15} + \dot{C}_{23} + \dot{Z}_{ABS} = \dot{C}_{16} + \dot{C}_{24}$	$\frac{(\dot{C}_{22} + \dot{C}_{15})}{(\dot{E}_{22} + \dot{E}_{15})} = c_{16}$
SHX	$\dot{C}_{13} + \dot{C}_{17} + \dot{Z}_{SHX} = \dot{C}_{18} + \dot{C}_{14}$	$c_{13} = c_{14}$
SP	$\dot{C}_{16} + \dot{C}_{43} + \dot{Z}_{SP} = \dot{C}_{17}$	–
EV-II	$\dot{C}_{14} = \dot{C}_{15}$	–
EV-I	$\dot{C}_{20} = \dot{C}_{21}$	–

cost per unit of product exergy ($c_{p,k}$), cost per unit of fuel exergy ($c_{f,k}$), and the cost associated with the exergy destruction ($\dot{C}_{D,k}$) and the exergy loss ($\dot{C}_{L,k}$). The above-mentioned parameters were estimated as follows (Khaljani, Khoshbakhti Saray, & Bahlouli 2015):

$$c_{f,k} = \frac{\dot{C}_{f,k}}{\dot{E}_{f,k}} \tag{11.17}$$

$$c_{p,k} = \frac{\dot{C}_{p,k}}{\dot{E}_{p,k}} \tag{11.18}$$

$$\dot{C}_{D,k} = c_{f,k} \dot{E}_{D,k} \tag{11.19}$$

$$\dot{C}_{L,k} = c_{f,k} \dot{E}_{L,k} \tag{11.20}$$

$$f = \frac{\dot{Z}_k}{\left(\dot{Z}_k + \dot{C}_{D,k} + \dot{C}_{L,k}\right)} \tag{11.21}$$

11.7 ENVIRONMENTAL ANALYSIS

The polluting gases emitted from a fossil fuel-based energy generation system degrade the environment and lead to many environmental issues such as global warming and climate change. In environmental analysis, the impact on the environment is evaluated in terms of the environmental cost, which is charged directly to the total system cost as compensation or penalty for damaging the environment. Generally, the environmental cost is calculated considering carbon monoxide (CO) and nitrous oxide (NOx). However, in this study, the principal greenhouse gas; carbon dioxide (CO_2) is also considered. The mass flow rate of CO and NOx is calculated using the semi-analytical correlations reported by Ahmadi and Dincer (2011). Furthermore, the mass flow rate of CO_2 in kg/s is measured using the following correlation (Owebor et al. 2019):

$$\dot{m}_{CO_2} = y_{CO_2}\dot{m}_g\left(\frac{\underline{M}_{CO_2}}{\underline{M}_g}\right) \tag{11.22}$$

where $\underline{M}_g$ and $\dot{m}_g$ are the molar mass and mass flow rate of flue gas; $\underline{M}_{CO_2}$ and y_{CO_2} are molar mass and mass fraction of CO_2, respectively. The combustion stoichiometry is used to measure $\dot{m}_g$ and y_{CO_2}.

After determining the amount of CO, NOx, and CO_2, the final step is to compute the environmental cost rate ($\dot{C}_{env}$), which is calculated using the following equation (Ahmadi, Rosen, and Dincer 2011):

$$\dot{C}_{env} = \dot{m}_{CO_2} \times C_{CO_2} + \dot{m}_{CO} \times C_{CO} + \dot{m}_{NOx} \times C_{NOx} \tag{11.23}$$

In Equation (11.23), C_{Co_2}, C_{CO}, and C_{NOx} are the unit penalty costs corresponding to the respective gaseous species with the values of 0.02086 \$/kg, 6.853 \$/kg, and 0.024 \$/kg, respectively, as reported in Ahmadi, Rosen, and Dincer (2011).

11.8 OVERALL PERFORMANCE CRITERIA

The parameters considered for evaluating the overall performance of the CCHP system are defined as follows:

11.8.1 Total energy efficiency (η_{tot})

The energy efficiency of the overall system is given by (Shukla et al. 2019):

$$\eta_{tot} = \frac{\dot{W}_{net,IRGT} + \dot{W}_{net,ORC} + \dot{Q}_{heating} + \dot{Q}_{cooling}}{\dot{Q}_{CC}} \tag{11.24}$$

where $\dot{W}_{net,IRGT}$ and $\dot{W}_{net,ORC}$ are the net power obtained from the IRGT and ORC whereas $\dot{Q}_{heating}$ and $\dot{Q}_{cooling}$ are the process heat and cooling load obtained from the HRSG and the ARS. Further, $\dot{Q}_{CC}$ is the heat generated at the CC.

11.8.2 Total exergy efficiency (ε_{tot})

The exergy efficiency of the overall system is given by (Anvari, Taghavifar, & Parvishi 2017):

$$\varepsilon_{tot} = \frac{\dot{W}_{net,GT} + \dot{W}_{net,ORC} + \left(\dot{E}_{12} - \dot{E}_{11}\right) + \left(\dot{E}_{26} - \dot{E}_{25}\right)}{\dot{E}_{37}} \tag{11.25}$$

where the terms $\left(\dot{E}_{12} - \dot{E}_{11}\right)$ and $\left(\dot{E}_{26} - \dot{E}_{25}\right)$ are the net exergy transfer from the HRSG and the ARS for providing process heat and cooling effect, respectively and $\dot{E}x_{37}$ is the fuel exergy fed to the CC.

11.8.3 Total cost rate ($\dot{C}_{tot}$)

The total cost rate of the system is given by (Nondy & Gogoi 2020a):

$$\dot{C}_{tot} = \dot{C}_f + \sum_k \dot{Z}_k + \sum_k \dot{C}_{D,k} + \dot{C}_L + \dot{C}_{env} \tag{11.26}$$

where $\dot{C}_f$ is the fuel cost rate, $\dot{Z}$ is the capital cost rate, $\dot{C}_{env}$ is the environmental cost rate, $\dot{C}_D$ and $\dot{C}_L$ are the cost rate related to exergy destruction and exergy loss.

11.8.4 Specific CO_2 emission (S_{CO_2})

The environmental impact of CO_2 alone can also be measured using a parameter known as specific CO_2 emission (S_{CO_2}) which is given by (Owebor et al. 2019):

$$S_{CO_2} = \frac{\dot{m}_{CO_2}}{\dot{W}_{net,IRGT} + \dot{W}_{net,ORC} + \dot{Q}_{heating} + \dot{Q}_{cooling}} \tag{11.27}$$

where S_{CO_2} is the specific CO_2 emission in kg/MWh and $\dot{m}_{CO_2}$ is the emission rate of CO_2 in kg/h.

11.9 RESULTS AND DISCUSSION

11.9.1 Model validation

The proposed CCHP system is simulated using a MATLAB code, written following the mathematical formulations and principles of exergoeconomic and environmental analyses. The GT-HRSG and the ORC model adopted in the simulation has been separately validated in the previous articles of the authors (Nondy & Gogoi 2020b, 2021).

11.9.2 Energy results

For a fuel flow rate of 2.62 kg/s, the IRGT cycle generates a net power of 50 MW. It should also be noted that AC-II (24.27 MW) consumed more power than AC-I (19.93 MW). Since the air density is decreased in the second stage of compression due to the higher temperature, the compression work is increased. Further, the thermal energy of 48.22 MW is extracted from the GT exhaust gas at the HRSG, converting 64.43 ton/h of water into saturated steam for process heating. The heat generated at the combustion chamber is 131.2 MW, out of which 2.62 MW (2%) is considered as heat loss. The first law efficiency for the IRGT cycle is 38.89% and, after integrating HRSG, the thermal efficiency of the IRGT-HRSG cycle improves to 76.40%. Further with the inclusion of R-ORC, the thermal efficiency of the combined system (IRGT-HRSG/R-ORC) is increased to 77.13% with the additional ORC power of 937 kW. However, when the ARS is integrated into the IRGT-HRSG/R-ORC cycle, the efficiency of the CCHP system turns out to be 83.14%. Besides, the ARS generates 7.73 MW of cooling effect under the integrated setup with a coefficient of performance of 0.774.

11.9.3 Exergy results

The parameters of the exergy analysis described in Section 11.8.3 are compiled in Table 11.4. Exergy destruction occurs in a system due to irreversibilities caused by heat transfer, chemical reactions, and fluid friction. The chemical reaction, however, has the highest impact among the factors mentioned above. Therefore, CC, where the chemical reaction takes place in the form of combustion, has the highest exergy destruction accounting for 60.27% of the total exergy destruction in the CCHP system. The APH (10.34%) is the next big contributor, with the heat exchange between compressed air and the flue gas with a large temperature gradient being the cause of such high exergy destruction. HRSG is next in the order, with

Table 11.4 Summary of the findings from the CCHP system's exergy analysis

Components	$\dot{E}_D$*(MW)*	$\dot{E}_L$*(MW)*	*ε(%)*	Y_d	$\dot{C}_D$*($/h)*	$\dot{C}_L$*($/h)*	$\dot{Z}$ *($/h)*	*f(%)*
CC	38.82	-	80.85	60.27	754.89	-	12.68	1.65
APH	6.65	-	81.42	10.34	160.59	-	33.95	17.45
HRSG	5.57	-	75.86	8.65	134.36	-	64.16	32.32
GT	5.03	-	94.94	7.81	121.30	-	138.38	53.28
GEN	2.13	-	17.61	3.30	63.54	-	0.19	0.29
AC-II	1.95	-	91.96	3.03	52.47	-	22.18	29.71
AC-I	1.94	-	90.23	3.02	52.33	-	22.18	29.77
VG	1.3	2.32	56.03	1.54	23.94	86.51	0.43	1.73
ABS	0.52	-	13.10	0.80	7.93	-	0.19	2.28
EVA	0.21	-	60.91	0.35	5.79	-	0.19	3.11
CON	0.22	-	23.12	0.35	63.81	-	0.18	0.28
VT	0.2	-	82.45	0.31	4.93	-	13.88	60.37
SHX	0.08	-	48.05	0.13	53.86	-	0.19	0.34
COND	0.049	-	49.60	0.07	2.18	-	1.32	37.68
IHX	0.011	-	74.28	0.01	0.49	-	0.20	29.09
FP	0.001	-	85.25	0.00	0.19	-	0.39	76.96
CCHP	64.68	2.32	50.65	-	1502.6	86.51	310.70	16.36

a Y_d of 8.65%. However, when looking at the $\dot{E}_D$ from the perspective of subsystems, the IRGT-HRSG cycle accounts for 93% of the overall exergy destruction, while the R-ORC and the ARS account for 2% and 5%, respectively. The minimal $\dot{E}_D$ at the R-ORC and ARS can be explained by the fact that these cycles are not driven by combustion reactions, but rather by the IRGT-HRSG cycle's waste heat. In R-ORC, the most irreversible component is VG, while in ARS, the most irreversible component is GEN, with Y_d of 1.54% and 3.30%, respectively. In this study, the unused exergy (2.32 MW) carried away by the flue gas at the VG outlet (state 10) is considered as the exergy loss. Finally, the $\dot{E}_D$ for the entire CCHP system is estimated to be 64.4 MW.

Table 11.4 also shows that the GT is the most efficient component, with an exergy efficiency of 94.94%. In the case of R-ORC components, FP and VT have the highest exergy efficiency of 85.25% and 82.45%, respectively. In the case of ARS, EVA has the highest exergy efficiency of 60.91%. The IRGT cycle has an exergy efficiency of 36.82%, and when HRSG is introduced, it rises to 49.71%. Further, the exergy efficiency of the IRGT-HRSG/R-ORC increases to 50.40% when R-ORC is integrated. Finally, with the addition of ARS, the CCHP system's exergy efficiency is found to be 50.65%. Therefore, it can be concluded that with the addition of HRSG, R-ORC and ARS, the exergy efficiency of the combined system increases by 12.89%, 0.69%, and 0.25%, respectively.

11.9.4 Exergoeconomic results

The exergoeconomic parameters described in Section 11.8.4 are summarized in Table 11.4. The cost rate of exergy destruction is highest in the CC, accounting for 50.23% of the overall cost related to exergy destruction. It signifies that CC is the most critical component in the exergoeconomic viewpoint. However, the capital cost rate of CC is comparatively less, resulting in a lower value of exergoeconomic factor (1.65%). It shows that the exergy destruction cost rate accounts for about 98.35% of the total cost incurred by the CC. Therefore, the irreversibility losses in the CC need to be reduced even in the expanse of increasing capital cost. Thus, to reduce the exergy destruction cost rate in the CC, two key parameters *viz.* APH outlet temperature (T_5) and air flow rate ($\dot{m}_a$) could be adjusted or appropriately selected based on optimization.

The APH is next on the list accounting for 10.68% of the total exergy destruction cost rate. It is justified since, as already mentioned, APH is the site where heat is transferred between the compressed air and the GT exhaust with a wide temperature gradient. Moreover, the capital cost rate for the APH is 33.95 \$/h leading to an exergoeconomic factor of 17.45%. The overall AC pressure ratio (R_p), APH outlet temperature (T_5), and CC outlet temperature (T_6) can be adjusted to lower the exergy destruction cost rate at the APH.

The HRSG has the third-highest exergy destruction cost rate with an exergoeconomic factor of 32.32%. Besides, the GT has a relatively high exergoeconomic factor of 53.28%. It implies that in the case of GT, the capital cost predominates. It is also evident from Table 11.4, as the exergy destruction cost rate and capital cost rate for the GT are 121.30 \$/h and 138.38 \$/h, respectively. However, the capital cost rate of GT could be reduced by decreasing at least one of the variables *viz.* GT inlet temperature, GT expansion ratio (P_6/P_7) and $\eta_{s,GT}$ as they appear in the cost function of the GT. Also, it is seen that in R-ORC components, VG is responsible for the maximum exergy destruction cost rate with the exergoeconomic factor of 1.73%. However, in the case of VT, the exergy destruction cost rate is minimal while the capital cost

Table 11.5 Summary of overall exergoeconomic parameters for the CCHP system

Parameters	*Units*	*CCHP system*
$\dot{Z}_{tot}$	\$/h	310.70
$\dot{C}_{d,tot}$	\$/h	1,502.6
f_{tot}	%	16.36
$\dot{C}_L$	\$/h	86.51
$\dot{C}_f$	\$/h	1,889.8
$\dot{C}_{sys}$	\$/h	3,788.5

rate is relatively high (13.88 \$/h). CON, GEN, and SHX, however, have high exergy destruction cost rates of 63.81 \$/h, 63.54 \$/h, and 53.86 \$/h, respectively, among ARS components.

Table 11.5 includes the exergoeconomic parameters computed for the overall CCHP system. As can be observed, the fuel cost rate accounts for the largest fraction of the system cost rate, accounting for about 50% of the system cost rate. The exergy destruction cost rate, which accounts for 39.66% of the system cost rate, is the next major contributor. The capital cost rate, on the other hand, is quite low, at only 8.2%. Moreover, the exergoeconomic factor for the overall CCHP system is evaluated to be 16.36%. It implies that the irreversibility of the components in the proposed system, particularly the CC, APH, HRSG, GT, and ACs, should be reduced to lower the system cost rate.

11.9.5 Environmental results

The corresponding environmental cost rate for the CO, NOx, and CO_2 is 0.56 \$/h, 146.57 \$/h, and 627.13 \$/h. Therefore, the environmental cost rate for the CCHP system is computed to be 774.26 \$/h. The significance of CO_2 emissions in environmental analysis is evident by the fact that CO_2 emissions account for nearly 81% of the overall environmental cost rate. Finally, the total cost rate (system + environmental) for the proposed CCHP system is evaluated to be 4,562.8 \$/h. Furthermore, the CCHP system emits 244.45 kg of CO_2 per MWh of energy. When considering a standalone IRGT cycle, however, the specific CO_2 emission is 522.61 kg/MWh. It shows that introducing HRSG, R-ORC, and ARS to the IRGT cycle decreases specific CO_2 emissions by 53.22%.

11.10 PARAMETRIC RESULTS

The parametric analysis is performed to determine how key decision variables such as overall compressor ratio, AC isentropic efficiency, and GT isentropic efficiency impact overall system performance. It is carried out by varying the considered decision variables one at a time while keeping the remaining variables fixed at the base case.

11.10.1 Effect of overall compressor ratio

The effect of the overall compressor ratio on cooling capacity, process heat, and net ORC power is shown in Figure 11.2(a). It appears that cooling capacity increases with the increase in the overall compressor ratio. It is because as the overall compressor ratio increases, the

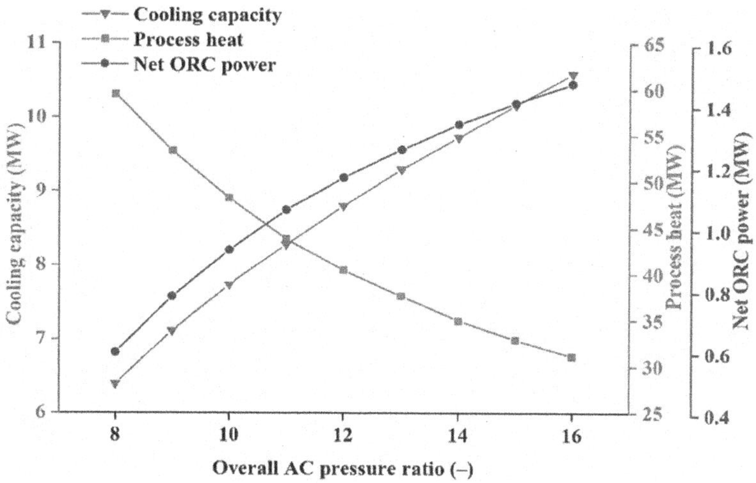

(a)

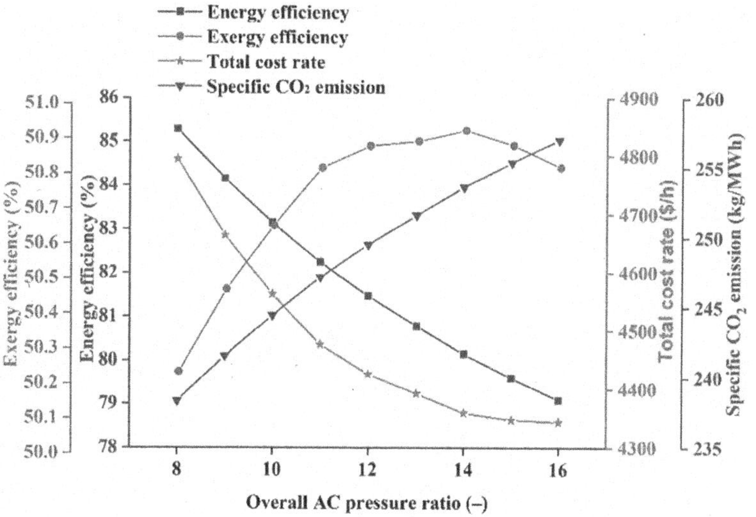

(b)

Figure 11.2 Effect of overall compressor ratio on the CCHP system's (a) output, (b) overall performance criteria ($\eta_{s,AC-I/II}$ = 86%, $\eta_{s,GT}$ = 86%).

individual compressor ratios of the ACs also increase. As a result, AC-I exit temperature raises leading to an increment in the heat availability at the GEN which in turn increases the cooling capacity. However, the process heat obtained from the HRSG shows a decreasing trend. It is because the net GT power is fixed (50 MW) and therefore with the rise in overall compressor ratio, the GT power and the power consumed by the ACs decreases, resulting in a decrease in the $\dot{m}_a$ and $\dot{m}_{fg}$. It eventually led to a drop in $\dot{m}_w$ which results in a decrease in process heat. Consequently, the heat availability at the VG increases since the HSRG exit temperature (T_9) shows an increment with the gain of the overall compressor pressure ratio while VG outlet temperature (T_{10}) is kept constant, hence the flow rate of R123 increases, resulting in a raise in ORC power.

Figure 11.2(b) illustrates the variation of energy efficiency, exergy efficiency, total system cost rate, and specific CO_2 emission with the change in overall compressor ratio. The plot shows that as the overall compressor ratio is varied from 8 to 16, the energy efficiency decreases from 85.28% to 79.09%. Both the net energy output and the CC heat generation rate ($\dot{Q}_{CC}$) decrease, as the overall compressor ratio increases. However, the rate of decrement of net energy output is higher compared to $\dot{Q}_{CC}$, leading to a reduction in energy efficiency. It is also worth noting that, unlike energy efficiency, the exergy efficiency of a CCHP system improves at the beginning then starts decreasing after reaching the peak value of 50.92%. The net exergy efficiency and the fuel exergy rate ($\dot{E}_{37}$) decrease as the overall compressor ratio rises. However, the rate of decrease in fuel exergy rate is higher than the rate of decrease in net exergy output, resulting in a decrease in exergy efficiency. Meanwhile, after reaching the peak, the rate of decrease in the net exergy output reduces leading to a decrease in exergy efficiency. The reason for the decrease in $\dot{Q}_{CC}$ and $\dot{E}_{37}$ is due to the reduction of fuel flow rate with the increase in the overall compressor ratio. Further, the total cost rate also appears to decrease, essentially due to a drop in fuel cost because of the reduction in fuel flow rate. However, the specific CO_2 emission increases due to a decrease in net energy output. The net energy output mainly reduces because as the overall compressor ratio increases the process heat load significantly reduces.

11.10.2 Effect of AC isentropic efficiency

Figure 11.3(a) shows that as the AC isentropic efficiency increases, the cooling capacity decreases. It is due to the drop in enthalpy difference across the GEN resulting fall in AC-I outlet temperature and the $\dot{m}_a$. Similarly, the process heat also decreases owing to a decrease in the $\dot{m}_w$ at the HRSG due to a drop in the $\dot{m}_a$. In contrast, the net ORC power increases with the increase in AC isentropic efficiency due to a rise in HRSG exit temperature (T_9) owning to rise in the enthalpy difference across the VG.

According to Figure 11.3(b), the energy efficiency decreases with the increase in isentropic efficiency of the ACs owing to the reduction in total net energy output. On the contrary, the exergy efficiency appears to increase linearly, essentially because of a drop in fuel exergy input to the CC. Meanwhile, the exergy destruction cost rate decreases while the capital cost rate of the ACs increases. Though the rate of increment of the capital cost rate is minimal at the beginning, it gradually increases and finally, when the isentropic efficiency reaches 88%, it raises exponentially. As a result, the total cost rate initially decreases then again it increases and reach the maximum value of 4614 \$/h. Lastly, due to the decrease in net energy output, the specific CO_2 emission increases, as observed in Figure 11.3(b).

11.10.3 Effect of GT isentropic efficiency

As the GT isentropic efficiency increases, the $\dot{m}_a$ decrease, resulting in a fall in the heat availability at the GEN. It causes the cooling capacity to reduce, as evident from Figure 11.4(a). Similarly, the process heat also decreases owing to a fall in the $\dot{m}_w$. However, the net ORC

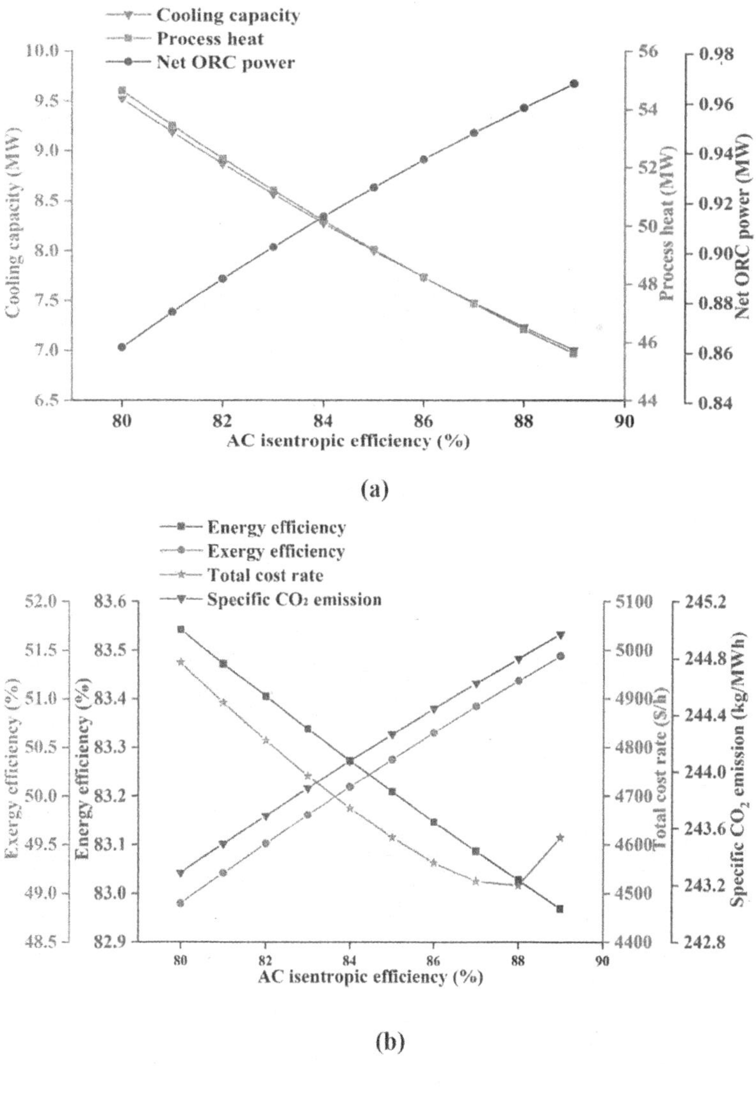

Figure 11.3 Effect of AC isentropic efficiency on the CCHP system's (a) output, (b) overall performance criteria (r_p = 10, $\eta_{s,GT}$ = 86%).

powershows increment due to a rise in HRSG exit temperature (T_9) that eventually led to an increase in enthalpy difference at the VG.

According to Figure 11.4(b), the energy efficiency decreases with the increase in GT isentropic efficiency due to the drop in net energy output. On the contrary, the exergy efficiency appears to increase linearly because of a decrease in fuel exergy. Consequently, the exergy destruction cost rate decreases while the capital cost rate rises. However, the rate of increase in the capital cost rate is minimal at the lower isentropic efficiency and when the isentropic

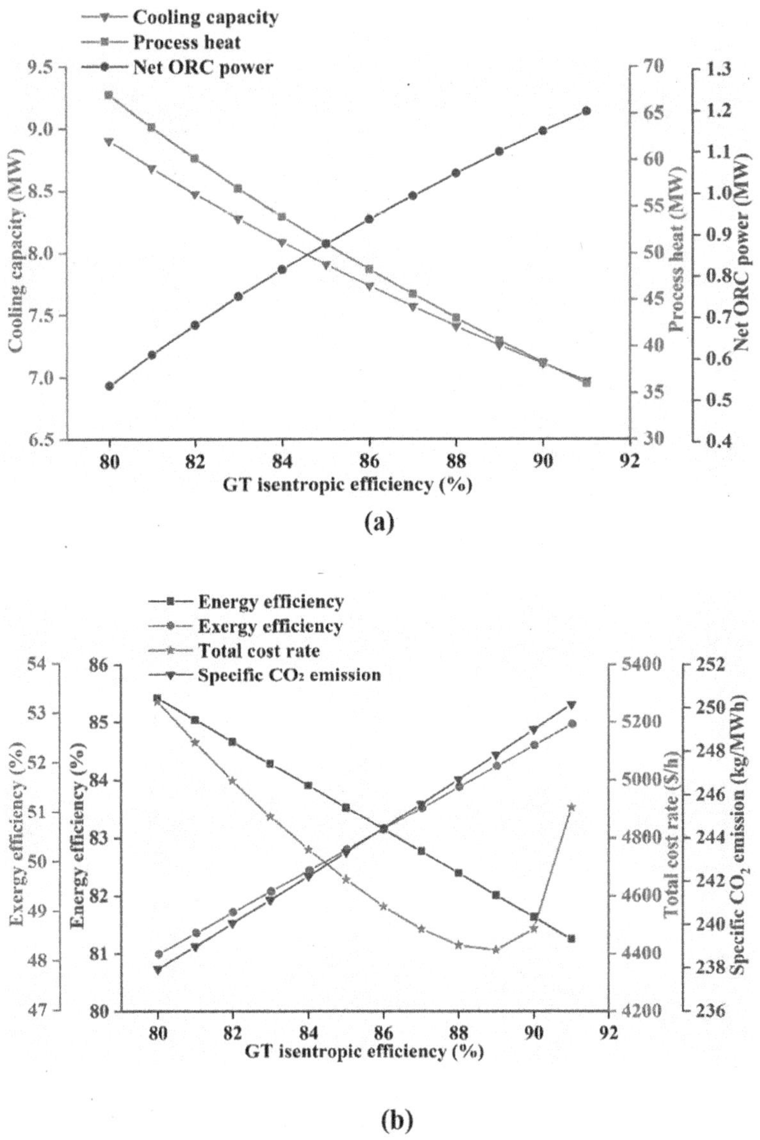

Figure 11.4 Effect of GT isentropic efficiency on the CCHP system's (a) output, (b) overall performance criteria (r_p = 10, $\eta_{s,AC-I/II}$=86%).

efficiency reaches 89%, it rises exponentially. Hence, the total cost rate initially decreases from 5,272.9 $/h to 4,412.8 $/h till 89%, then again increases to 4,905.6 $/h. Further, despite the increase in CO_2 emission rate caused by the rise in GT isentropic efficiency, the specific CO_2 emission increases due to the decrease in net energy output. The rate at which the total energy output decreases is higher than the rate of increment of CO_2 emission.

11.11 SUMMARY

In this chapter, a novel trigeneration system is proposed for the combined production of electricity, process heat and cooling. It includes IRGT cycle, HRSG, ARS, and R-ORC. The HRSG and R-ORC are operated using the excess heat present in the GT exhaust gas. However, aside from the traditional waste heat recovery technique using exhaust gas, the low-grade heat available in the air during intercooling is also used for energy conversion using an ARS. The main objective of this work is to address the thermodynamic, exergoeconomic, and environmental viability of ARS, R-ORC, and HRSG integration into the IRGT plant following the proposed setup. The key findings from this study can be summarized as follows:

- The IRGT cycle and the R-ORC yield a net power of 50 MW and 937 kW, respectively. The steam produced in the HRSG at a rate of 64.43 ton/h can meet a process heat load of 48.22 MW, while the ARS generates a 7.73 MW of cooling effect.
- Although the system cost rate with the inclusion of the above-mentioned subsystems increases by 11.67%, the energy efficiency and the exergy efficiency improve by 53.2% and 27.3%, respectively along with the reduction in specific CO_2 emission by 53.2%.
- Increasing the overall compressor pressure ratio lowers the total cost rate. However, it also reduces the system's thermodynamic and environmental performance.
- Increasing the AC isentropic efficiency reduces the cooling effect and the process heat. However, as AC isentropic efficiency increases, the system cost rate decreases at first and as it reaches 87%, the total cost rate rises abruptly. Therefore, it is recommended that AC isentropic efficiency be limited to 87%.
- Increasing GT isentropic efficiency decreases process heat to a substantial degree, resulting in a decrease in energy efficiency. The exergy efficiency, on the other hand, improves while the total cost rate decreases. Therefore, the GT isentropic efficiency should be limited to 88%, as the total cost rate rises exponentially above this limiting value.

Based on the findings, it can be concluded that while incorporating ARS, R-ORC, and HRSG into the IRGT plant would marginally increase the system cost rate, it would also significantly improve the thermodynamic, exergoeconomic, and environmental performance.

REFERENCES

Ahmadi, Pouria, & Ibrahim Dincer. 2011. "Thermodynamic and exergoenvironmental analyses, and multi-objective optimization of a gas turbine power plant." *Applied Thermal Engineering* 31(14–15): 2529–2540.

Ahmadi, Pouria, Marc A. Rosen, & Ibrahim Dincer. 2011. "Greenhouse gas emission and exergo-environmental analyses of a trigeneration energy system." *International Journal of Greenhouse Gas Control* 5(6): 1540–1549.

Anvari, Simin, Hadi Taghavifar, & Alireza Parvishi. 2017. "Thermo-economical consideration of regenerative organic Rankine cycle coupling with the absorption chiller systems incorporated in the trigeneration system." *Energy Conversion and Management* 148: 317–329.

Anvari, Simin, Samad Jafarmadar, & Shahram Khalilarya. 2016. "Proposal of a combined heat and power plant hybridized with regeneration organic Rankine cycle: Energy-exergy evaluation." *Energy Conversion and Management* 122: 357–365.

Bejan, Adrian, George Tsatsaronis, & Michael J. Moran. 1995. *Thermal Design and Optimization.* New York: John Wiley & Sons.

Cao, Yue, Yike Gao, Ya Zheng, & Yiping Dai. 2016. "Optimum design and thermodynamic analysis of a gas turbine and ORC combined cycle with recuperators." *Energy Conversion and Management* 116: 32–41.

Khaliq, Abdul. 2009. "Exergy analysis of gas turbine trigeneration system for combined production of power heat and refrigeration." *International Journal of Refrigeration* 32(3): 534–545.

Khaljani, Mansoureh, Rahim Khoshbakhti Saray, & Keyvan Bahlouli. 2015. "Comprehensive analysis of energy, exergy and exergo-economic of cogeneration of heat and power in a combined gas turbine and organic Rankine cycle." *Energy Conversion and Management* 97: 154–165.

Khandelwal, Neelam, Meeta Sharma, Onkar Singh, & Anoop Kumar Shukla. 2020. "Recent developments in integrated solar combined cycle power plants." *Journal of Thermal Science* 29(2): 298–322.

Lemmon, Eric, Huber, Marcia, & McLinden, Mark 2010. "NIST Standard Reference Database 23, Reference Fluid Thermodynamic and Transport Properties (REFPROP), Version 9.0, National Institute of Standards and Technology, R1234yf."

Mohammadi, Amin, Alibakhsh Kasaeian, Fathollah Pourfayaz, & Mohammad Hossein Ahmadi. 2017. "Thermodynamic analysis of a combined gas turbine, orc cycle and absorption refrigeration for a CCHP system." *Applied Thermal Engineering* 111: 397–406.

Moné, Christopher D., David S. Chau, & Patrick E. Phelan. 2001. "Economic feasibility of combined heat and power and absorption refrigeration with commercially available gas turbines." *Energy Conversion and Management* 42(13): 1559–1573.

Musharavati, Farayi, Shoaib Khanmohammadi, Mohammad Rahmani, & Saber Khanmohammadi. 2021. "Thermodynamic modeling and comparative analysis of a compressed air energy storage system boosted with thermoelectric unit." *Journal of Energy Storage* 33: 101888.

Nondy, Joy, & Tapan Kumar Gogoi. 2019. "Exergy analysis of a combined gas turbine and organic Rankine cycle based power and absorption cooling systems." In *ASME 2019 Gas Turbine India Conference, GTINDIA 2019*, Vol. 1. India: American Society of Mechanical Engineers (ASME).

Nondy, Joy, & Tapan Kumar Gogoi. 2020a. "A comparative study of metaheuristic techniques for the thermoenvironomic optimization of a gas turbine-based benchmark combined heat and power system." *Journal of Energy Resources Technology* 143(6): 062104.

Nondy, Joy, & Tapan Kumar Gogoi. 2020b. "Comparative performance analysis of four different combined power and cooling systems integrated with a topping gas turbine plant." *Energy Conversion and Management* 223: 113242.

Nondy, Joy, & Tapan Kumar Gogoi. 2021. "Performance comparison of multi-objective evolutionary algorithms for exergetic and exergoenvironomic optimization of a benchmark combined heat and power system." *Energy* 233: 121135.

Owebor, Kesiena, Chika Ogbonna Chima Oko, Ogheneruona E. Diemuodeke, & Oreva Joe Ogorure. 2019. "Thermo-environmental and economic analysis of an integrated municipal waste-to-energy solid oxide fuel cell, gas-, steam-, organic fluid- and absorption refrigeration cycle thermal power plants." *Applied Energy* 239: 1385–1401.

Pátek, Jaroslav, & Jaroslav Klomfar. 2006. "A computationally effective formulation of the thermodynamic properties of LiBr-H2O solutions from 273 to 500 K over full composition range." *International Journal of Refrigeration* 29(4): 566–578.

Salmi, Waltteri, Juha Vanttola, Mia Elg, Maunu Kuosa, & Risto Lahdelma. 2017. "Using waste heat of ship as energy source for an absorption refrigeration system." *Applied Thermal Engineering* 115: 501–516.

Sarkar, Jahar, Souvik Bhattacharyya, Pouria Ahmadi, Ibrahim Dincer, Marc A. Rosen, M. Khaljani, R. Khoshbakhti Saray, et al. 2017. "Exergetic evaluation of gas-turbine based combined cycle system with vapor absorption inlet cooling." *Applied Thermal Engineering* 122(1): 431–443.

Shukla, Anoop Kumar, Achintya Sharma, Meeta Sharma, & Gopal Nandan. 2019. "Thermodynamic investigation of solar energy-based triple combined power cycle." *Energy Sources, Part A: Recovery, Utilization and Environmental Effects* 41(10): 1161–1179.

Tchanche, Bertrand Fankam, George Papadakis, Gregory Lambrinos, & Antonios Frangoudakis. 2009. "Fluid selection for a low-temperature solar organic Rankine cycle." *Applied Thermal Engineering* 29(11–12): 2468–2476.

Chapter 12

Organic Rankine cycle integrated hybrid arrangement for power generation

Mohammad Bahrami and Fathollah Pourfayaz

Department of Renewable Energies and Environment, Faculty of New Sciences and Technologies, University of Tehran, Tehran, Iran

Ali Gheibi

Department of Mechanical Engineering, University of Kashan, Kashan, Iran

12.1 INTRODUCTION

The Rankine cycle or Rankine steam cycle is a set of closed processes that result in useful output work. Water is usually used as a working fluid in these cycles. Also, in some parts of the processes in this cycle, the fluid is in the form of vapor, and in other parts it is in the form of liquid. The Rankine cycle is mainly used to generate power in fossil fuel or nuclear power plants. In these power plants, the mentioned fuels are used to convert the water in the boiler into steam. Then, by passing the steam through the turbine, the desired work is produced.

The use of water as a heat transfer fluid in the Rankine cycle has the following problems (Macchi 2017):

1 Very low pressure is required to liquefy water vapor at ambient temperature. For example, 45°C condensation temperature requires about 0.1 bar pressure. This increases the row length of the last steam turbine blades and also increases the size of the condenser.
2 Due to the increase in the output pressure ratio of the steam turbine, a more complex turbine design and the use of multi-stage steam turbines are required.
3 To avoid water droplets in the final stages of the steam turbine, it is necessary to superheat the steam to higher temperatures. Higher temperatures affect the design and selection of turbines and heat exchangers.
4 Water has a high evaporation temperature. So it will need a heat source that can deliver a lot of heat energy at high-temperature levels.

Due to the problems of the water fluid, a more appropriate option is to use organic fluid. Waste heat recovery plays an important role in the management of energy resources. The organic Rankine cycle (ORC) can be used to recover low temperature dissipated heat. This cycle is similar to the Rankine cycle and uses only organic fluids as working fluids. The ORC works thermodynamically like a classical vapor cycle. The difference is that the heat transfer fluid in it is different. Organic fluids like R245Fa, toluene, pantane, and silicone oil can be used in ORC (Quoilin et al. 2013). The only difference between the ORC and the steam cycle is in the type of fluidization used in the cycle, but this small difference makes significant changes in the behavior and application of the cycle. It occurs much lower than water, so the temperature range of the organic Rankin cycle is lower than the steam Rankin cycle, and

DOI: 10.1201/9781003213741-12

the organic Rankin cycle can be used for heat recovery. There is a lot of work in the field of the ORC. Larjola (1995) investigated the generation of electricity from industrial heat loss. That work investigates a type of fossil fuel-free turbo generator pump for rapid power generation using an ORC. Hung (2001) investigated thermal recycling using an ORC with dry working fluid. Their obtained results show that irreversibility is a function of heat sources. Liu, Chien, and Wang (2004) investigated the effect of working fluid on the performance and efficiency of the ORC. Base on the results the critical temperature of the working fluid has little effect on the thermal efficiency of the cycle. Drescher and Brüggemann (2007) studied the selection of the working fluid for the ORC in biomass and thermal power plants. They achieved the best working fluid for high efficiency in biomass application. Mago et al. (2008) studied the ORC regarding the reluctance to use dry fluid. For this purpose, simultaneous analysis of the first law of thermodynamics was used. Other work (Dai, Wang, and Gao 2009) considered parametric optimization of ORC for heat recovery applications. Schuster et al. (2009), examined the ORC in terms of energy and economics. Efficiency, the rate of water production, and the prices of electricity generated are presented in this article. Heberle and Brüggemann (2010) investigated the selection of a suitable working fluid using exergy analysis for the cogeneration cycle. They also explored different concepts to achieve higher efficiency and selected the best ones. Bahrami, Hamidi, and Porkhial (2013) studied the combined Stirling-ORC power cycle in the aspects of thermodynamic and environmental investigation. Based on their obtained results, the combined cycle can be assisted by solar energy and an ORC used as an annular cold-side heat rejecter for a free-piston Stirling cycle. The power output efficiency of the ORC system increased about 4% to 8% compared to that of a Stirling standard cycle. The ORC plays an important role in using waste heat recovery. Chahartaghi and Babaei (2014) studied the effects of using two-component mixtures with different temperature gradients in the phase change process, on the performance of an ORC. They selected four two-component mixtures: n-pentane/n-butane, isopentane/isobutene, n-pentane/isobutene, and isopentane/n-butane. Their results show that using the two-component mixtures, in comparison with pure fluids, both energetic and exergetic efficiencies increase by approximately 9% and 14% for respectively simple and internal heat exchanger (IHE)- integrated configuration.

12.2 PLANT LAYOUT

Commercially there are two different types of ORC. Some of the plant layouts can be used on the market and other configurations are proposed in the research. The classification of the ORC is shown in Figure 12.1 (Macchi 2017).

As can be seen, ORC are divided into main systems, single pressure level cycle and multi-pressure level cycles, subcritical and supercritical/transcritical cycles are in the group of single pressure level and multi-pressure levels cycles. In addition, there are triangular and complete flash cycles that can be used for solar power applications (Macchi 2017). The thermodynamic properties consist of enthalpy, vapor density, and temperature of the working fluid, which have a great effect on the cycle performance. ORC cycles can be classified into subcritical and transcritical/supercritical cycles based on the operating pressures. The main differences are: (1) both heat addition and heat rejection at subcritical pressures for subcritical cycles, (2) both heat addition and heat rejection at supercritical pressures for supercritical cycles, and (3) heat addition at supercritical pressure and heat rejection at subcritical pressure

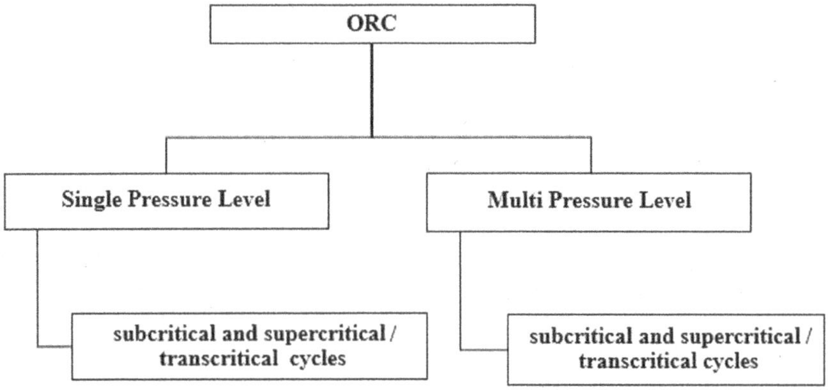

Figure 12.1 Classification of organic Rankine cycle.

for transcritical cycles. A simple ORC system and corresponding T-S diagram are shown in Figure 12.2. In a subcritical cycle, the working fluid at the evaporator outlet is in the saturated state and the superheated working fluid at the turbine condensed to a saturated liquid. In the transcritical cycle, the working fluid is compressed and heated to the supercritical state. So there is no change in the fluid state. The evaporator in the transcritical cycle is called a vapor generator (Lecompte et al. 2015).

Working fluids have a considerable role in low-grade heat recovery systems. Different working fluids in the ORC system are classified into three categories according to the slope of the saturation vapor line in a T-S diagram usually. There are three groups of working fluids in ORC. The dry fluid that has a positive gradient (dT/ds>0), a wet fluid has a negative gradient (dT/ds<0), whereas an isentropic fluid has an infinite gradient (Chahartaghi & Babaei 2014; Qiu et al. 2012; Gu and Sato 2002). Several working fluids were identified in ORC. In the selection of the best working fluid for a special application, safety factors such as being non-flammable, non-corrosive, non-toxic, and environmental impact (low ozone depletion potential and global warming potential) should be considered. In Figure 12.3, the classification of working fluids according to the operating cycle temperature is shown (Darvish et al. 2015).

The properties of the five working fluids which are the most used in the ORC systems are shown in Table 12.1. Table 12.1, $\xi = {ds}/{dt}$ is the inverse of the slope of the T-S diagram. Ozone depletion potential (ODP) and global warming potential (GWP) are two important factors that can be defined as:

- **ODP:** Potential of a molecule of refrigerant gas to degrade the ozone layer based on the CFC-11 (ODP=1.0). An ORC working fluid such as pentane has an ODP of zero so it will do no damage to the ozone layer.
- **GWP:** The effect of refrigerants on global warming is based on carbon dioxide as a base gas with a GWP of 1. The more GWP, the bigger its potential for global warming. A GWP is obtained over some time, usually about 100 years. CFC-11 has a GWP over 100 years of 4,250 while pentane has a GWP of less than 5.

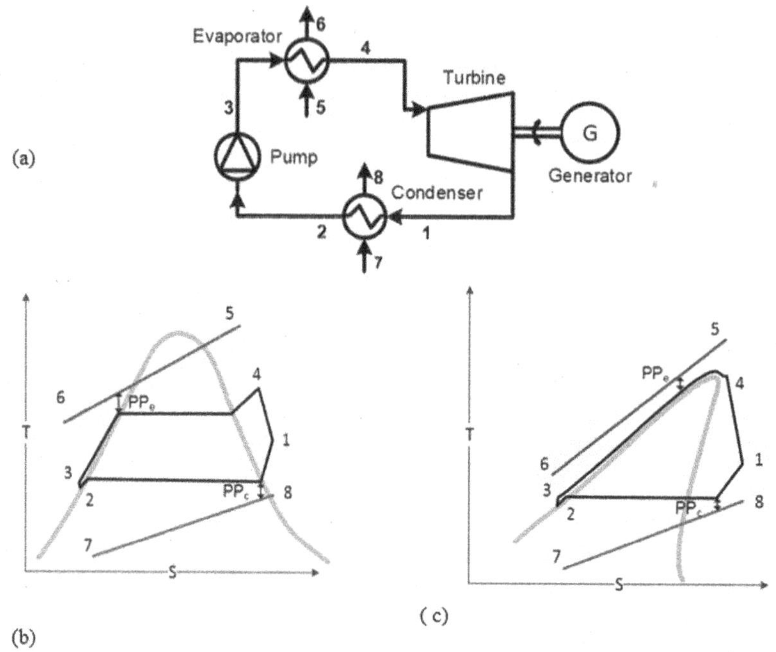

Figure 12.2 (a) Component layout in an ORC system, (b) T-S diagram in subcritical ORC, (c) T-S diagram in transcritical ORC. (Source: Lecompte et al. 2015.)

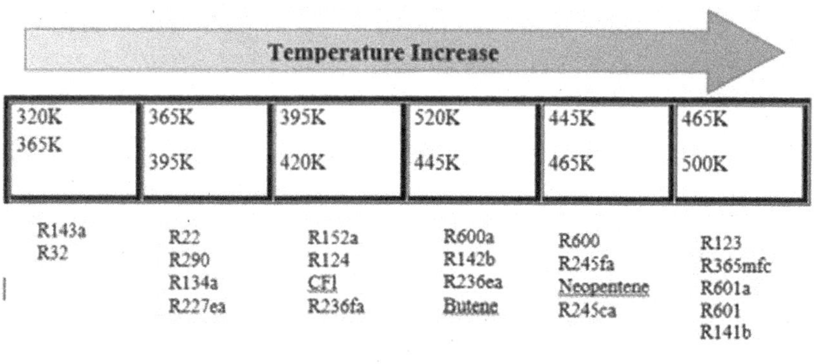

Figure 12.3 Classification of the working fluids based on the heat source temperature.

Figure 12.4 shows the influence of the different working fluids on the output work (Tchanche, Pétrissans, & Papadakis 2014). In the figure, the performance of three different fluid models in passing through the turbine is compared.

Organic fluid is a fluid that contains the element carbon in its formula. Saturation of organic fluid vapor for use in turbines occurs at much lower temperatures than water. Thus the temperature range of the organic Rankin cycle is lower than that of the steam Rankin cycle so

Table 12.1 The most used working fluid in ORC

Type	*Working fluid*	*Critical temperature (°C)*	*Critical pressure (MPa)*	*ξ (J/kg.K2)*
Wet	R-143a	72.71	3.76	−1.49
	CO_2	30.98	7.38	−8.27
Dry	R-218	71.87	2.64	0.45
	R123	183.7	3.66	0.26
Isentropic	R-142b	137.1	4.06	0

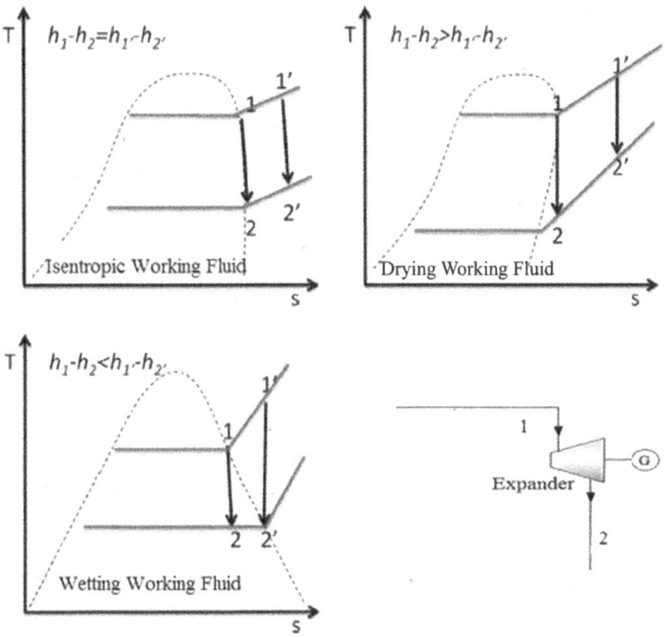

Figure 12.4 Expansion process in the turbine for different fluids. (Source: Turboden 2021.)

the ORC can be used in heat recovery applications. Important parameters in selecting the working fluid of the ORC are:

- **Isentropic saturation vapor:** Since the main aim of the cycle is to focus on low-temperature energy recovery, it is not appropriate to use a superheat approach like the traditional Rankin cycle. A small superheat at the operator output is always useful. This is the weakness of the wet fluid in the two-phase at the end of the expansion phase. In the case of dry fluid, a regenerator component must be used to increase the cycle efficiency.
- **High stability to temperature and low freezing point:** Organic working fluids usually undergo chemical erosion and decomposition at high temperatures. Therefore, the maximum high source temperature is limited based on the chemical stability of the working fluid. The lowest temperature in the cycle should be lower than the freezing point.

- **High density:** A fluid with latent heat and high density absorbs more energy from the operator source and thus reduces the required flow rate of the equipment size and energy consumption of the pump.
- **Low destructive effects on the environment:** In this case, two main parameters are ODP and GWP.
- **Safety:** The fluid must be non-flammable and non-toxic for a long lifetime.

12.3 SINGLE PRESSURE LEVEL ORC

The main component of the basic Rankine cycle is boiler, pump, condenser, and turbine. Water is used as the working fluid in Rankine cycles. In the organic Rankine cycle, the working fluid is an organic fluid such as refrigerants and hydrocarbons. Organic working fluids are in the temperature range of 150–300°C. A thermodynamic ORC is shown in Figure 12.5. As can be seen there are four main processes (Reddy et al. 2010).

State 1: Isentropic expansion: Superheated steam enters the turbine and thermal energy converts to mechanical energy. A generator is connected to the turbine which converts mechanical energy into electricity. The pressure and temperature of the steam is reduced in the turbine exit.

State 2: Isentropic heat rejection: The exit steam from the turbine enters the condenser and is condensed to liquid water.

State 3: Isentropic compression: The working fluid enters the pump.

State 4: Isentropic heat addition: Working fluid enters the boiler and by adding heat steam is generated and superheated within the boiler.

This group of cycles is simple and its simplicity and attainable efficiencies mean that this kind of cycle is the first option for different applications such as geothermal, solar energy, and waste heat recovery.

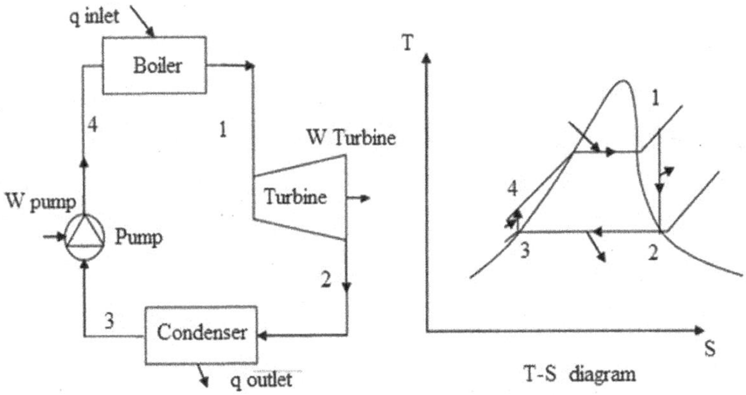

Figure 12.5 Schematic of Rankine cycle. (Source: Reddy et al. 2010.)

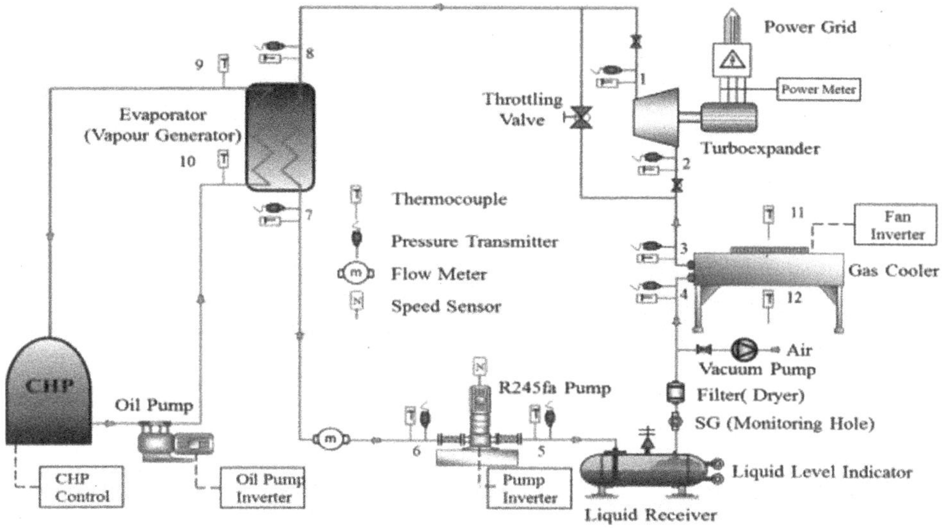

Figure 12.6 A real ORC system and facilities. (Source: Li et al. 2017.)

12.3.1 Subcritical ORC

A real schematic of the subcritical ORC can be seen in Figure 12.6. Using the dry fluids as the working fluid in the cycle allows the increasing of the efficiency of these cycles because the saturated vapor curve has a positive slope in the T-S diagram and these fluids do not need to be superheated. After all, the saturated vapor is placed in the superheated zone after expansion. In a subcritical ORC system, if the critical cycle temperature is higher than the maximum heat source temperature, using the cycle is very inefficient. In this case, the flow rate of the working fluid is reduced and as a result, the production of electrical energy will be reduced (Li et al. 2017).

12.3.2 Supercritical/transcritical ORC

Maximum pressure in a supercritical/transcritical ORC is higher than the critical one. Cycle efficiency can be influenced by fluid type. The superheating changes the amount of output work by the turbines. In the case of the dry working fluid, the thermodynamic state in the expansion process is in the superheated region. In the presence of the finite heat capacity, the efficiency of the supercritical ORC is higher than the subcritical cycle. A comparison of the subcritical and supercritical ORC is shown in Figure 12.7 (Becquin & Freund 2012). As can be seen in the subcritical cycle, a preheater is needed.

Also, some other ORC systems are used in industrial processes like Rankine cycles with internal heat exchangers and Rankine cycle with integrated feed liquid heaters which are shown in Figures 12.8 and 12.9 respectively. These cycles can be inserted in the group of single pressure levels of ORC.

As was said, isentropic and dry fluids are good choices for ORC (Figure 12.8). A regenerator could be used to increase the cycle thermal efficiency. Vanslambrouck et al. (2011)

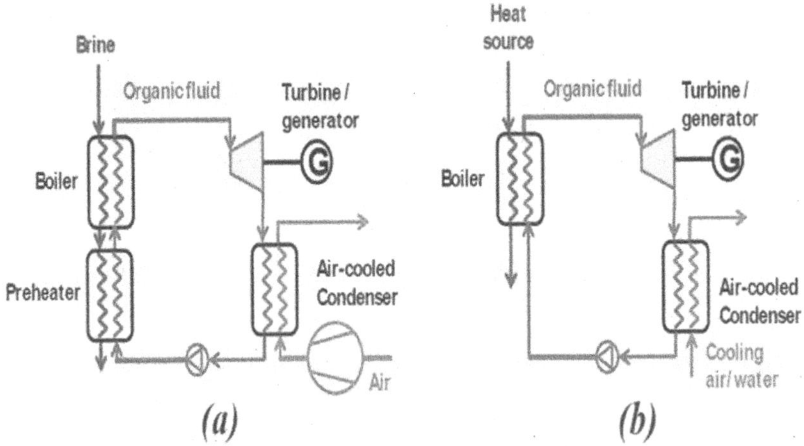

Figure 12.7 Single pressure ORC, (a) subcritical, (b) supercritical. (Source: Becquin & Freund 2012.)

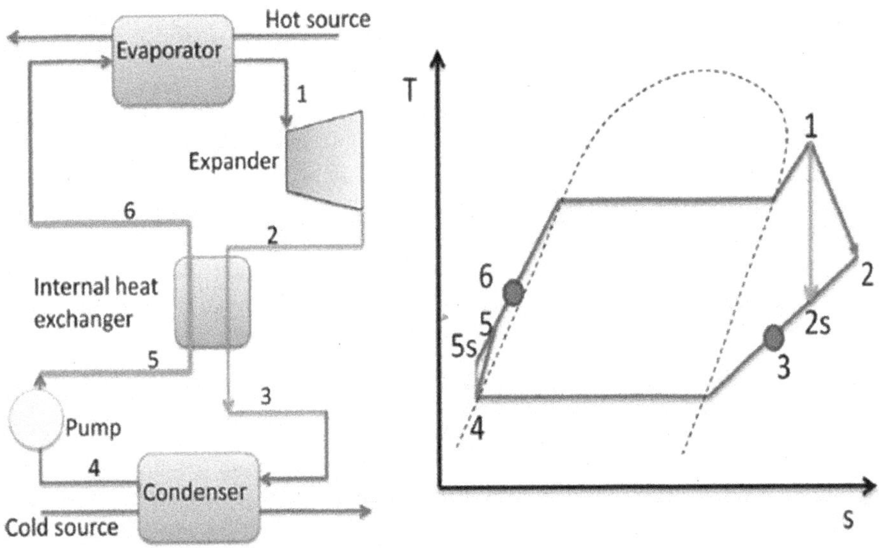

Figure 12.8 Rankine cycle with integrated regenerator. (Source: Tchanche, Pétrissans, & Papadakis 2014.)

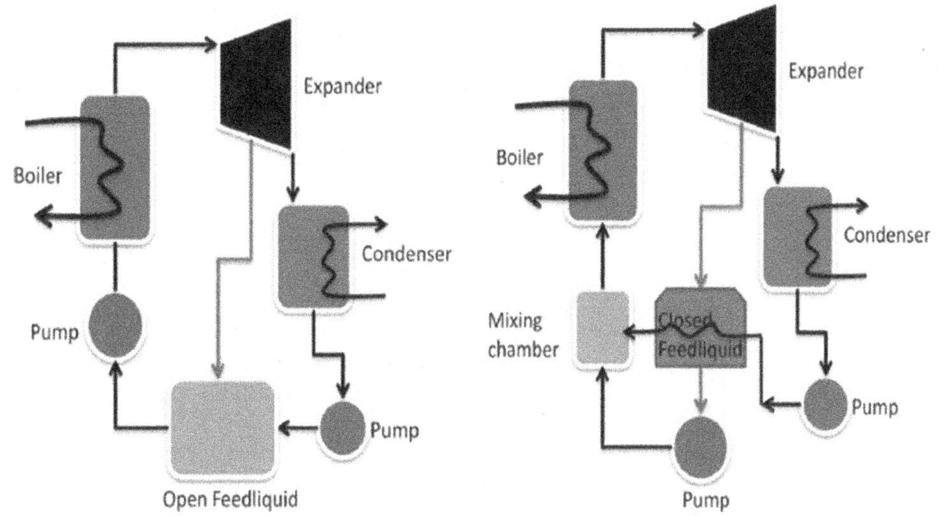

Figure 12.9 Rankine cycle with feed liquid heater. (Source: Tchanche, Pétrissans, & Papadakis 2014.)

presented a limit under which a regenerator is inoperative, this limitation is the temperature near 100°C.

In the cycle in an open feed liquid heater (Figure 12.9), two fluid streams from the second pump and the outlet fluid from the turbine are mixed up. This causes higher temperature fluid in the evaporator.

12.4 MULTI-PRESSURE LEVEL

12.4.1 Subcritical ORC multi-pressure level

Multi-pressure ORC systems have an extra pressure level. In these systems, an evaporation process occurs at two temperatures. The power cycle is divided into two different pressure levels (high-pressure and low-pressure loop). The loop high pressure increased as the high-temperature part in the heat source is increased. Dual-pressure level ORC are shown in Figures 12.10 and 12.11. Figure 12.10 shows a typical cycle that operates with wet fluid that may damage the turbine. So in these groups of cycles, a solution is to use a two-stage turbine. As can be seen from Figure 12.11, the amount of the working fluid which is required for the high-pressure loop preheating is sent to the high-pressure preheater, while the rest of the flow is used to evaporate a low-pressure loop. The dual-pressure ORC can work with the high-pressure level while using a high heat source (Guo et al. 2010).

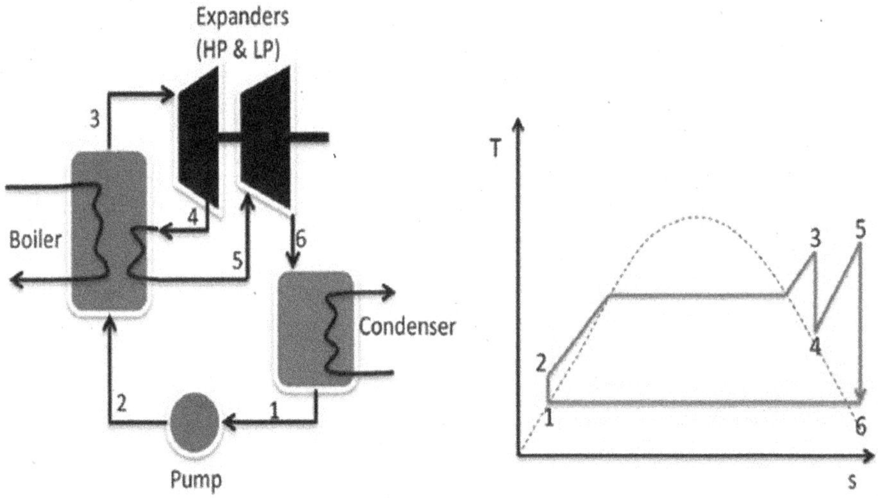

Figure 12.10 Rankine cycle with reheater. (Source: Tchanche, Pétrissans, & Papadakis 2014.)

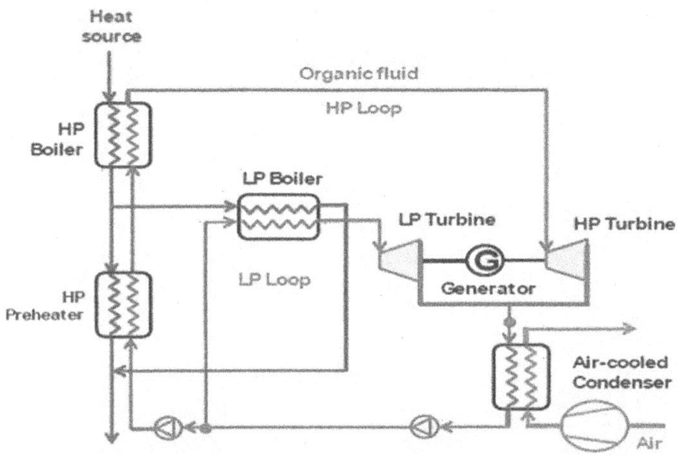

Figure 12.11 Subcritical-multi-pressure organic Rankine cycle (multi-pressure level ORC). (Source: Becquin & Freund 2012.)

12.4.2 Supercritical ORC multi-pressure level

Supercritical ORC multi-pressure levels can achieve higher efficiency compared to subcritical single level cycles and it performs similar to the supercritical cycles. These groups of cycles are more expensive because of using two turbines, larger heat transfer surfaces, etc., and a more complicated plan. Therefore these groups of cycles are proposed only in particular applications where the efficiency of power conversion is extremely important or the fixed cost of heat source greatly overcomes the cost of the power block.

12.5 ORC COMPONENTS

12.5.1 Turbine

There are three groups of expanders in ORC systems: turbines, positive displacement expanders, and ejectors. In the turbine, the pressure energy is transferred to kinetic energy. A turbine's geometry is divided into two main groups, radial, and axial turbines. The difference between these two groups is in their capacity (axial turbine capacity is greater than radial turbine capacity). Turbines used in the ORC system have no difference from steam turbines that are worked by the steam flow. It can be said that most power plants in the world use the method of generating electricity from steam turbines in thermal power plants (and this number is up to 90% in the United States). Steam turbines are usually divided into two categories: impact and reaction. As a hot, gaseous stream flows past through the turbine's spinning blades, the steam expands and cools, giving off most of the energy it contains. This steam revolves around the blades generally. The blades convert most of the steam's potential energy into kinetic energy (Chaibakhsh & Ghaffari 2008). Figures 12.12 and 12.13 show a turbine generator. The rotational force generated by the steam turbine is used to drive the generator and generate electrical energy (Explain That Stuff 2021).

The total work which is done by the turbine is obtained as shown in Equation (12.1).

$$w_t = (h_1 - h_{2s})\eta_t \tag{12.1}$$

Figure 12.12 Steam turbine. (Source: Explain That Stuff 2021.)

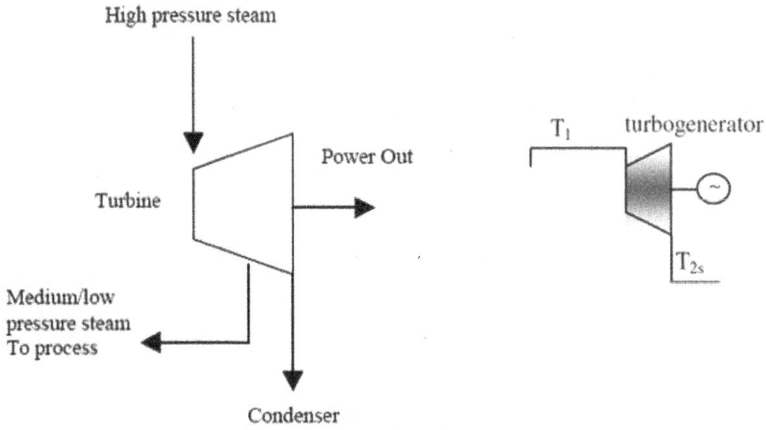

Figure 12.13 Fluid flow through the turbine.

In Equation (12.1), h_l in inlet fluid enthalpy and h_{2s} is the outlet fluid enthalpy by the assumption that this process is isentropic. η_t is turbine efficiency.

The energy transformation in the positive displacement expanders is based on the changing of the specific volume of the working fluid. There are four kinds of positive displacement expander: scroll expander, screw expander, rotary vane expander, and piston expander. An ejector as an expander device with any rotors can be used in ORC systems.

12.5.2 Condenser

A condenser is a device that is responsible for cooling the operating fluid and converting the vapor state to a liquid. This is done by cooling the material. During this process, the latent heat of the material is taken and transferred to the condenser cooler. Therefore, condensers can be considered as a kind of heat exchanger. Since heat transfer from the condenser surfaces is done by conduction, the condenser size (capacity) is a function of the heat transfer equation. The physical properties of the condenser depend on the size of the plant and the working fluid properties.

Depending on how the condenser exchanges heat with the environment and cools it, the condenser is divided into three main categories:

- **Cool air condenser:** In this condenser, air passes through the coils or pipes of the condenser, and through this, heat is transferred from the refrigerant to the environment. The air condenser is also divided into two categories of cooling with natural airflow or with forced airflow.
- **Coldwater condenser:** In a cold water condenser, the fluid that is responsible for transferring heat from the refrigerant to the outside is water.
- **Evaporative condenser for chillers and refrigerators:** Evaporative condenser (Figure 12.14) uses both water and air to receive heat from the refrigerant. An evaporative condenser combines the characteristics of a cooling tower with a water condenser.

The internal side of the heat exchanger is shown by indexed *i* and the outer side of the heat exchanger is shown by indexed *e*. The heat transfer in an elemental piece *dz* in the axial direction the local heat transfer rate can be expressed as Equation (12.2) (Guo et al. 2010).

$$dQ = U.dA.\left|(t_i - t_e)\right| \tag{12.2}$$

In Equation (12.2), A is the area of the heat transfer, t_i and t_e are the local bulk temperatures of the internal and the external fluid stream respectively, and U is the overall heat transfer coefficient referred to area A, which can be derived from Equation (12.3).

$$\frac{1}{U.dA} = \frac{1}{U_i.dA_i} = \frac{1}{U_e.dA_e} = \frac{1}{\Omega_i.h_i.dA_i} + \frac{r_{bw}}{dz} + \frac{1}{\Omega_e.h_e.dA_e} \tag{12.3}$$

In the constant value of h_i and h_e, total heat transfer coefficient is as Equation (12.4).

$$U = \left(\frac{A}{\Omega_i.h_i.A_i} + AR_{bw} + \frac{A}{\Omega_e.h_e.A_e}\right)^{-1} \tag{12.4}$$

h_i and h_e are internal and external convective heat transfer coefficients (fluid to the wall) respectively. Ω_i and Ω_e are the overall coefficients respectively of the finned-wall inner and outer surfaces also for a uniform surface $\Omega = 1$. R_{bw} is the conduction thermal resistance of the base wall. r_{bw} is the conduction thermal resistance per unit length of the base wall. d_i and d_e are the inner and outer diameter of the cylindrical prime wall.

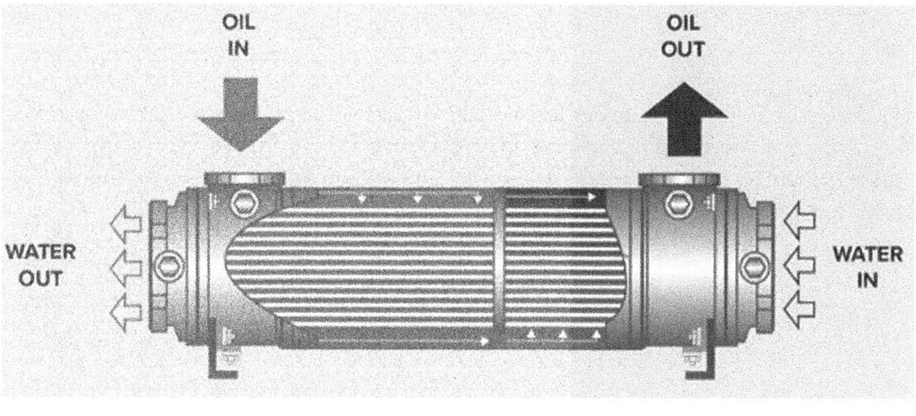

Figure 12.14 Condenser. (Source: Kelvion 2021.)

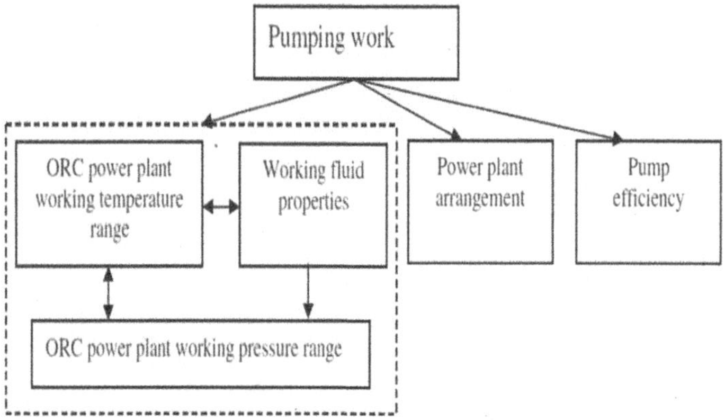

Figure 12.15 Important parameters affecting the pumping work value in ORC.

12.5.3 Pump

ORC pumps are usually variable speed multistage centrifugal pumps and their design is relatively common because of their wide use in both the chemistry and refinery fields. In the ORC system, pumps are handled to move the working fluid and raise its pressure from the value at the condenser to that required in the evaporator. Numerous factors affect the pumping work value. Some of the most important factors are shown in Figure 12.15 (Borsukiewicz-Gozdur 2013).

Pumps are one of the main energy consumers in the ORC systems so the first important point in the assessment of energy demand for the pumping work is to determine how that pumping work shall be calculated. For 1 kg of the working fluid, the specific pumping work (denoted by w_p) can be determined as Equation (12.5).

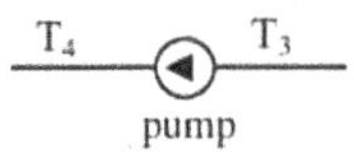

$$w_p = \frac{h_4 - h_3}{\eta_p} \tag{12.5}$$

In Equation (12.5), h_i is enthalpy at the inlet and outlet of the pump and η_p is the pump efficiency. The power decrease factor κ describes the portion of the turbine work output needed for the cycle pump supply is used for considerations on the pumping work demand and is obtained as Equation (12.6).

$$\kappa = \frac{w_p}{w_t}.100 \tag{12.6}$$

12.5.4 Boiler and evaporators

A boiler or steam generator is designed to convert a liquid to vapor. In a steam power plant, the thermal energy of the boiler is obtained by fossil fuels. Heat transmits from the combustion products to the working fluid. Nuclear energy can also be used as a heat source for generating high-pressure steam. In ORC systems, an evaporator is usually used instead of a boiler. An evaporator is a heat exchanger that has different types. In the ORC system, heat transfer between the working fluid and heat source occurs in the evaporator.

12.6 ORC APPLICATIONS

Waste heat recovery plays an important role in the management of energy resources. The organic cycle can be used for low-temperature dissipation heat recovery. Heat loss occurs at very low temperatures, where heat enters the environment through cooling towers or chimneys in industry. Some of these heat sources are used for district preheating and cooling. In cases where there is no specific use of this waste heat, it can be used to generate electricity by using an ORC system. One of the important sources of heat loss in the industry is related to the cement industry. Cement production is one of the most widely used industrial production processes in terms of energy consumption. In the application of an ORC system, the availability of an adequate heat source is important. Every heat-generating process in an industry such as obtaining heating energy from burning fuel can be taken as the heat source for an ORC. Figure 12.16, shows the percentage of the total output power produced for different heat source application types. As can be seen in Figure 12.16, much of the use of the ORC is related to the geothermal application (Rettig et al. 2011; Altun & Kilic 2020).

Waste heat recovery systems are widely used throughout many industries – and for good reason. The process of recycling expels heat and energy that would otherwise be wasted. This can have a surprisingly powerful impact, both on individual businesses and on industries as a whole.

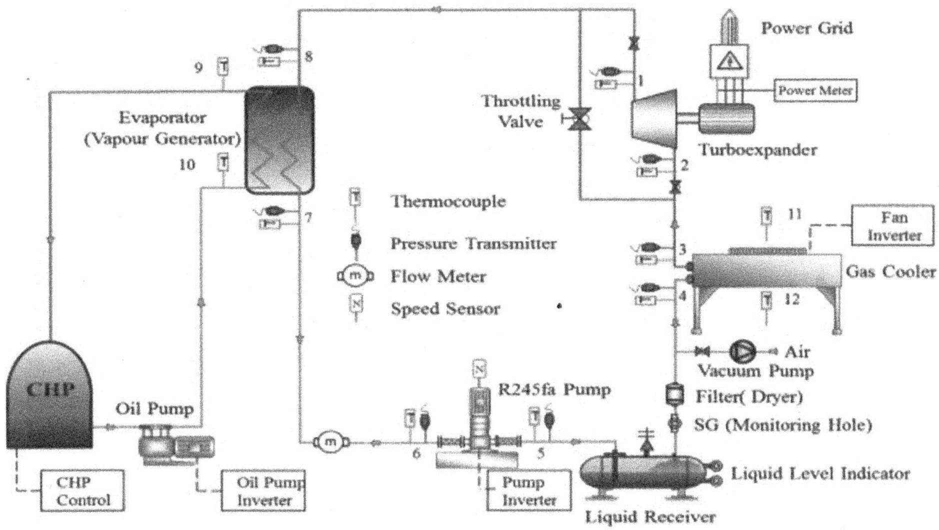

Figure 12.16 Percentage of the total power output produced for different heat source application type by ORC systems.

12.6.1 Geothermal

In general, "geothermal energy" is a type of thermal energy that is produced and stored on Earth. The heat energy in the Earth's crust is continuously generated due to the nature of the planet's formation and the radioactive decay of minerals. The geothermal gradient – the temperature difference between the core and the shell – results in a constant flow of heat from the core to the shell. Geothermal energy, which is classified as "alternative energy," is used for applications such as heating urban and industrial buildings or generating electricity. The term "geothermal power" refers to the same application of electricity generation. There are four main ways to use geothermal energy: geothermal power plants, geothermal heat pumps, direct use, and enhanced geothermal systems. This type of energy is available on a large scale, safely, and as a type of "renewable resource" that does not depend on climatic conditions. Regarding global warming, dependence on fossil fuels is reduced, and if the exploitation of the source is not more than the capacity, it leads to minimal damage to the environment. In addition, the development of technology has led to the possibility of using resources in a very large size and in a very secure way. Geothermal energy is generated from the initial heat of planet formation (about 20%) and the radioactive decay of minerals (about 80%). The main energy-producing isotopes are potassium-40, uranium-238, uranium-235, and thorium-232 (National Geographic 2021). Regardless of temperature changes in different seasons, the temperature gradient per kilometer from the depth of the Earth's crust is between 25–30°C, which leads to a thermal conductivity flux of 0.1 MW/m^2. In places where the crust is thinner, the heat flux is higher. This flux increases with the presence of magma movements, hot springs, water cycles, or a mixture of these factors. The first step in heat extraction is to enter the heat storage tank on the ground. There are three main ways to do this:

1 Power plants
2 Geothermal heat pump
3 Direct use of reinforced geothermal systems

It should be noted that the type of energy source determines which of the above methods should be used. The United States has the largest geothermal development in the world, the geysers north of San Francisco in California. In Iceland, many of the buildings and even swimming pools are heated with geothermal hot water. Geothermal power is economical, dependable, sustainable, and ecological, but has historically been limited to special areas of the Earth. ORC technology is particularly suitable for the exploitation of medium-to-low enthalpy sources. It is a cost-effective solution with power outputs reaching up to 40 MWe per single generator for sources with water temperature from 100°C to 200°C or higher. Figure 12.17 shows the geothermal ORC diagram (Altun & Kilic 2020).

In the ORC system, to achieve electrical energy from the cycle the geothermal fluid remains within a closed cycle without passing through the turbine and also it does no damage to the atmosphere. In this process, some of the produced heat which is not converted into electricity can have a thermal use, such as in room heating, etc. The application of the ORC system in geothermal is done with a binary power plant system. The binary power plant system is similar to the ORC system. Also, the ORC system can be used in geothermal power plants that use a flash system by handling waste brine from the separator. Also in this system, the ORC system should be selected based on some important parameters such as choice of the working fluid, cycle type, and selecting of the turbine technology. Selecting the suitable

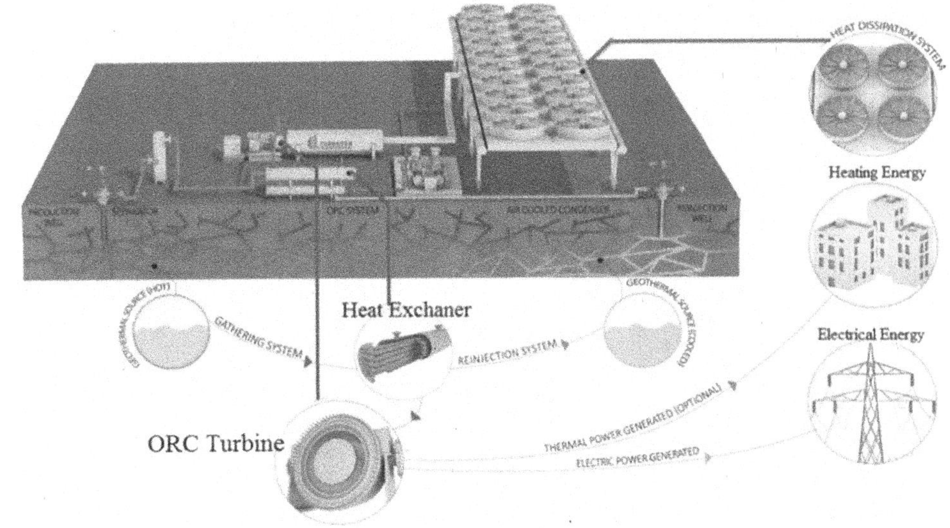

Figure 12.17 Binary geothermal ORC plant. (Courtesy of Turboden 2021.)

ORC system for the geothermal resource to obtain high efficiency using the proper working fluid and expander system is important.

Other advantages of using geothermal are:

- Renewable
- Huge potential
- Sustainable / stable.
- Heating and cooling.
- Reliable.
- No fuel required
- Rapid evolution.

An economic comparison of geothermal energy versus natural gas should mention that the geothermal heat pump uses electricity. In some parts of the country, natural gas costs are very low so it is much cheaper that a cycle that operates with natural gas. The temperature of the geothermal energy source is a range from 60°C to 350°C. In the temperature below 80°C overall system performance is reduced and the geothermal plant is not economical. There are two different categories for classifying geothermal energy resources:

1 Based on the temperature or enthalpy
2 Based on the exergy concept

In the classification based on the temperature or enthalpy, there are three groups: low, medium, and high enthalpy. Also in the case of the exergy, there are three groups: low, medium, and

Table 12.2 The capacity of the different countries in geothermal electricity generation

Country	*Installed Capacity (MW)*
United States	3.653
Indonesia	1.948
Philippines	1.868
Turkey	1.347
New Zealand	1.005
Mexico	951
Italy	944
Kenya	763
Iceland	755
Japan	459
Other	1.011

high exergy resources. The total capacity of the existed geothermal power plants is about 21.5 MW. The contribution of different countries in the use of geothermal energy is shown in Table 12.2.

12.6.2 Heat recovery

Heating energy is produced in many industrial activities in different factories. Most appliances that get their energy from burning fossil fuels producing high heat energy. We can change the heating energy to electrical energy by an ORC. These cycles can produce electricity by recovering heat from industrial processes and in combined cycles with reciprocating engines and gaș turbines. Figure 12.18 shows the ORC which uses heat recovery from production processes in a factory (Wei et al. 2007).

12.6.3 Biomass

Residues and substances derived from living organisms are called biomass. All forms of organic matter are considered biomass, such as plant and animal waste, municipal and factory waste. Important sources of biomass include energy production grains (grains produced for biofuel production in uncultivated lands), agricultural residues, forest residues, algae, microbial residues, wood industry residues, municipal wastes, and other sources. Lignocelluloses, which have the largest share in biomass production, include agricultural and wood industry wastes, forestry residues, and energy production grains. These three are among the most important sources of lignocellulose biomass production. For example, plant biomass including sugarcane, native plants, some weeds, corn, spruce, poplar, and various types of plants and residues of agricultural industries, paper, tree, and plant leaf residues, bran, stems, and roots are in these categories. Municipal waste, animal, and food waste are usually organic materials that are produced in solid or liquid form. Municipal solid waste includes municipal waste and rubbish. Paper, cardboard, and discarded food are some of these. Metal and plastic materials do not fall into this category. Animal waste also includes livestock waste, slaughterhouses, fishing and fish farming industries, dairy industries, manure, and

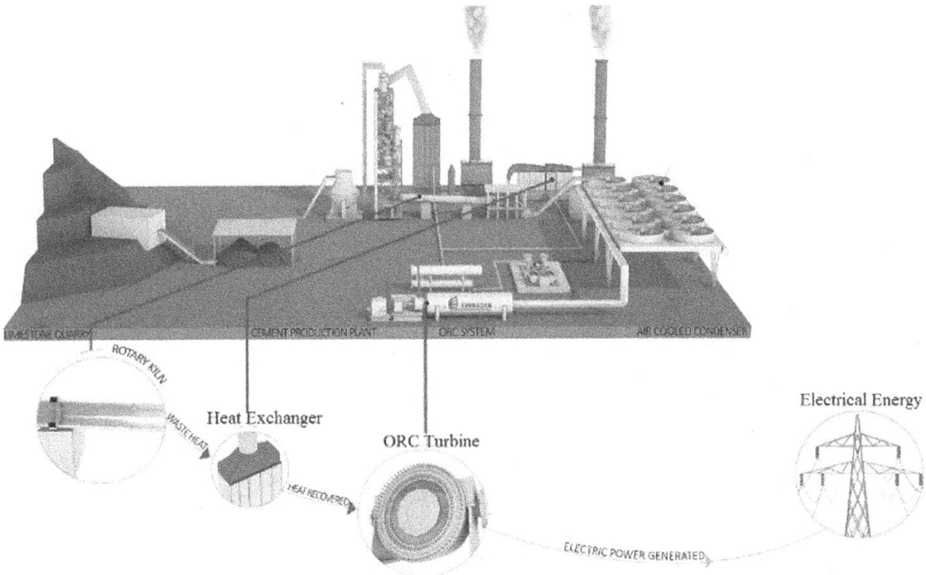

Figure 12.18 Heat recovery from industrial processes. (Courtesy of Turboden 2021.)

livestock sludge. It can be used directly as a fuel or processed into other formats. The carbon dioxide in the atmosphere is taken by the trees and converted into biomass. After they die, it is released back into the atmosphere. The chemical energy in the biomass is released by sunlight and the photosynthesis process. CO_2 and water in the environment with the sunlight energy are used to nurture the growth of organic material. Biomass energy is the stored solar energy that is converted into power with a null CO_2 balance. The most important parameters that most influences the process are the following:

Water content: The biomass must be completely dry or the biomass moisture in it must be reduced to a very low value (less than 10%) before entering the gasifier.

Ash content: Ashes are inorganic compounds that do not add heating value. Ash greatly reduces system efficiency due to the impurity of biomass. This reduction occurs when there is a significant percent of ash.

Heating value: Due to the calorific value of biomass fuel, it keeps its consumption constant at the output.

Composition: The chemical composition of the biomass (C and H and the chemical species prevailing in the biomass) greatly affects the final performance of the system.

Size: The biomass size is an important parameter in both plant and process. Too fine or dusty material does not allow the reactor bed to "breathe" properly, preventing the correct flow of the syngas. If the available biomass is too fine or dusty, the proper particle size of the material can be achieved through the production of pellets or briquette.

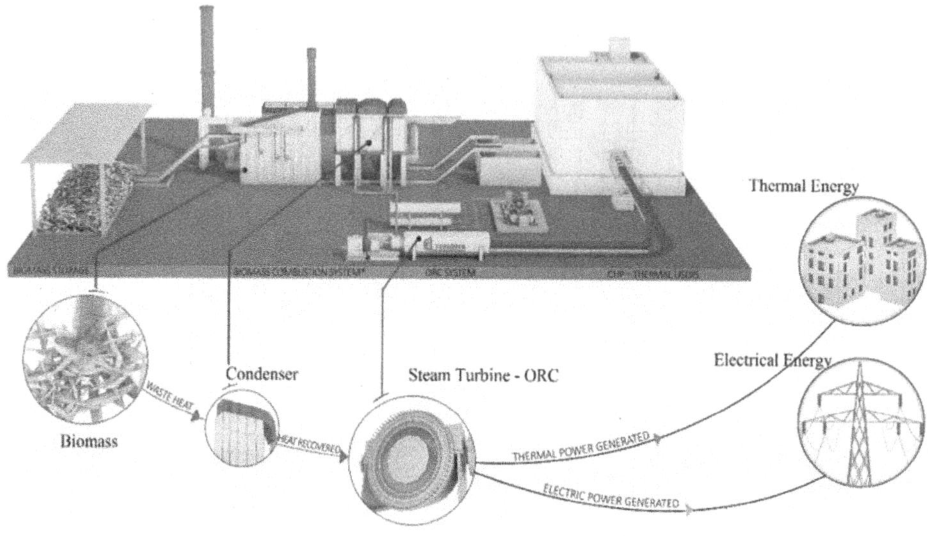

Figure 12.19 Biomass-ORC Plant. (Courtesy of Turboden 2021.)

An organic Rankine cycle (ORC) can produce electrical energy with high efficiency by using any kind of biomass process. Figure 12.19 Shows the ORC-biomass system (Drescher and Brüggemann 2007).

One of the most basic uses of biomass is to produce energy in the form of biofuels. Biomass can generate electricity or heat. Unlike fossil fuels, biomass resources are replaced easily and quickly. Also, by producing fuel from these sources, less carbon dioxide enters the atmosphere. Different biomass can be used to produce energy. Some biomass, such as energy-producing grains, can produce five to six times the energy used to produce them.

Heating energy obtained from biomass combustion is transferred to the working fluid in the ORC. The transfer process is done by an intermediate thermal oil circuit. This circuit transmits the heat energy of the biomass system to the ORC. In the systems where high temperatures are required, thermal oil is used. The technical specification of the thermal oil system is as follows.

12.6.4 Diathermic oil

This oil has a high heating capacity therefore it never burns off. The generated heat by the biomass system can be transferred to the ORC system by the presented thermodynamic chart as can be seen in Figure 12.20. There are many advantages to using an oil fluid in the middle cycle compared to water. In this state operating temperature is very important. Since hot oil has a higher boiling point than water, thermal oil heating systems can operate at higher temperatures.

An experimental study was done in Lienz (Obernberger, Thonhofer, & Reisenhofer 2002). Biomass combustion takes place in a thermal oil boiler followed by a thermal oil economizer with a 6.5 MW capacity and a hot water boiler with a nominal capacity of 7 MW (Table 12.3).

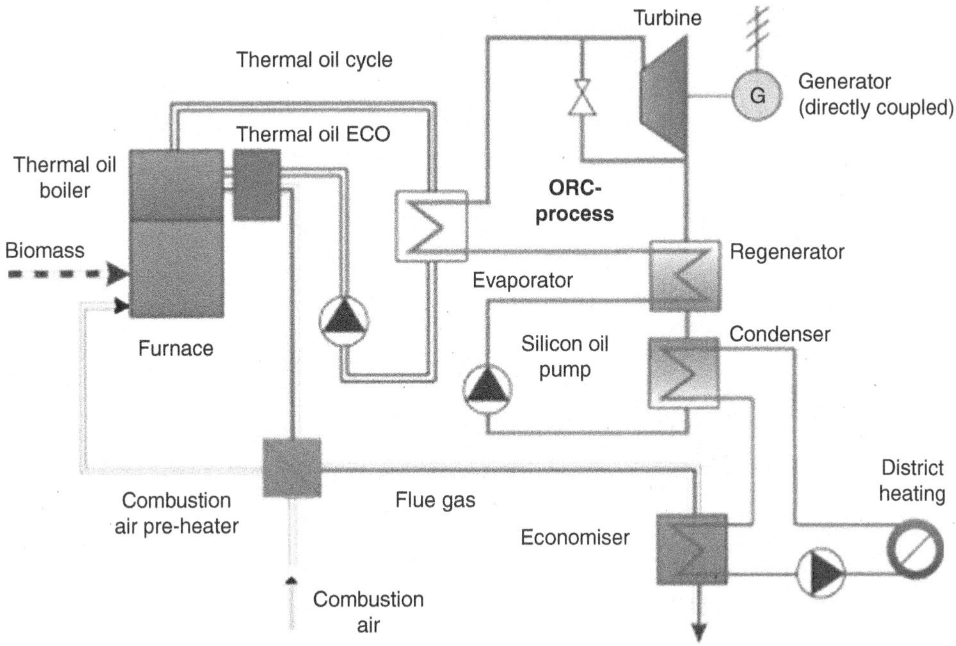

Figure 12.20 Transferred the heat energy from biomass source into the ORC. (Source: Obernberger, Thonhofer, & Reisenhofer 2021.)

Table 12.3 Technical data of the biomass CHP plant at Lienz

Roofed storage capacity	5,000 Srm
Open storage capacity	10,000 Srm
Solar thermal collector	630 m^2
Nominal power-thermal oil boiler	6,000 kW
Nominal power-thermal oil Eco	500 kW
Nominal power-hot water boiler	7,000 kW
Nominal power-hot water Eco	1,500 kW
Nominal power-oil boiler	11,000 kW
Maximum thermal power-solar collector	350 kW
Net electric power-ORC	1,000 kW
Production of heat from biomass	60,000 MWh/a
Production of heat from solar energy	250 MWh/a
Production of electrically from biomass	7,200 MWh/a

Source: Obernberger, Thonhofer, and Reisenhofer (2002).

12.6.5 Solar thermal

In solar systems, the impossibility of continuous access to energy and optimal use of energy received is one of the basic parameters and problems. Therefore, due to its high potential in energy conversion efficiency in the low-temperature range, considering the reduction of fossil fuel consumption and reduction of environmental problems, the organic

Rankine cycle is a suitable choice for use in solar systems. Given that global energy demand is growing rapidly and fossil fuels are no answer to this increase in demand, over the past 20 years the efficient use of low-temperature energy sources such as solar energy have received more attention, and research on them has become more important than ever. Concentrated solar power systems with ORC technology can be cost-effective in the range of up to 20 MW electric per single shaft. Figure 12.21 shows the solar ORC system (Turton et al. 2008).

The main concentrated solar power in the world can be seen in Table 12.4.

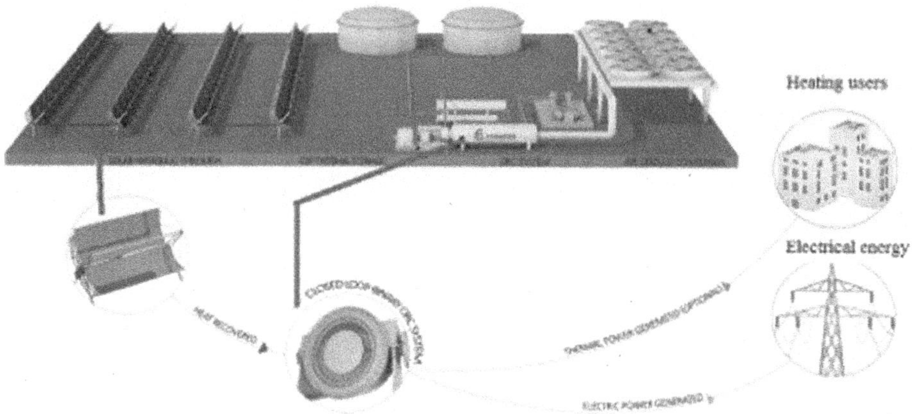

Figure 12.21 Solar ORC system. (Courtesy of Turboden 2021.)

Table 12.4 Concentrated solar power main references

ORC size	*Location*	*Customer*	*Status*	*Notes*
0.1 MWe	Lecce, Italy	Research project	Off	Electric power production
0.6 MWe	Ottana, Italy	ENAS (Ente Acque della Sardegna)	In operation since 2017	Electric power production from solar thermal energy (linear Fresnel collectors)
1 MWe	Melilli (SR), Italy	Archimede S.r.l.	Under construction	Electric power production from solar thermal energy (parabolic trough collectors)
2 MWe	Ait Baha, Morocco	Italcementi Group	In operation since June 2014	Electric power production from solar thermal energy (parabolic trough collectors) and waste heat from the cement production process (hybrid plant)
3.8 MWe	Brønderslev, Denmark	Brønderslev Forsyning	Under construction	Electric power and heat production from solar thermal energy (parabolic trough collectors) and biomass combustion (hybrid plant)

12.7 COMBINED HEAT AND POWER

Almost two-thirds of the total energy used by conventional electricity generation is wasted in the form of heating energy and enters the atmosphere. A large amount of the heating energy generated in the industrial process can be used by combined heat and power (CHP) technology which is one of the energy-efficient methods that generates both electricity and useful thermal energy – such as steam or hot water – that can be used for space heating, cooling, homemade hot water, and industrial processes. CHP can be located at an individual facility or building, or be a sub-region energy or utility resource and is located at areas where there is a need for both electricity and thermal energy.

CHP is used in over 4,400 conveniences nationwide, as:

- **Mercantile buildings**: hotels, hotels, hospitals
- **Residential**: condominiums, apartments, planned communities
- **Municipal**: regional energy systems, wastewater treatment, recycled equipment, schools
- **Manufacturers**: cement factory, ethanol, metal smelting plant, food processing, glass manufacturing

One of the advantages of the CHP systems is their decentralization. Unlike centrally generated power plant systems, these systems can be installed in different locations. Thus, due to the proximity of the power plant to the place of consumption, the losses due to the transmission of electricity will be very small, while the rate of loss in the transmission networks of national networks is about 20%. Therefore, in CHP power plants, due to the production of electricity locally and independently (decentralized) and the simultaneous use of wasted heat, the efficiency of power generators has increased significantly. The efficiency of electricity generation in gas power plants is about 30. With the increase of investment costs and being equipped with combined cycle equipment, this can finally be increased up to 55%. However, by using the technology of simultaneous production of electricity and heat, independently, the energy efficiency of these generators will reach about 75–95%, which is a high efficiency compared to gas turbines and combined cycle and create a great chance in fuel consumption. Meanwhile, in CHP systems, the amount of CO_2 produced per unit of electricity generation (kW) is much lower than other conventional methods, which will be very interesting in terms of reducing greenhouse gas emissions and CO_2 tax. Many European governments, the United States, and even some Asian countries such as Japan, have established policies and laws to encourage the use of CHP power plants. This shows the importance of CHP in reducing energy consumption.

Cogeneration systems use only one process to generate electricity, heat, or cold; therefore, the required heating/cooling capacities and electricity should be well estimated and the system equipment should be selected accordingly. Great care must be used when selecting cogeneration systems equipment for the required cooling, heating, and power. In any case, a cogeneration system will have four basic elements and components:

1. Propulsion generator: a mechanism that produces a mechanical force
2. Electric generator: a mechanism for generating electricity
3. Heat recovery system: lost heat recovery mechanism
4. Proper control system: management and control mechanism of all sensors and actuators

In general, due to the simultaneous production and consumption of electricity in cogeneration systems, it is necessary to set up such systems in places with the following special characteristics:

- Demand for energy consumption (electricity and heat) is constant.
- The demand for thermal energy is high.
- If possible, the ratio of electricity demand to heating should be balanced. In other words, the conditions for the operation of the system during the day and night and throughout the year should be such that continuous (permanent) operation of the system can be achieved.

Due to the dependence of the capacity of CHP systems on the heating needs of each complex and consequently their low capacity, CHP can be placed in the category of distributed power generation systems. In the current situation and despite the future potential until 2030, the types of CHP technologies in distributed electricity generation, considering natural gas as a primary energy source, are as follows:

- Reciprocating engines
- Microturbines
- Steam turbines
- Fuel cells

12.7.1 The importance of CHP in reducing energy consumption

Given the importance of energy-saving issues and the high share of energy consumption in industry, the use of effective methods in this regard will have significant effects on the progress and development of any country. Cogeneration of electricity and heat (CHP) as an efficient method to reduce energy consumption is currently on the agenda of many developed countries and a significant share of electricity and heat production using this method is provided (Arabkoohsar & Nami 2019). In CHP technologies, excess heat generated by power generation or mechanical power is recycled to reuse energy for various uses. The use of these technologies is due to a large number of losses when converting thermal energy into mechanical or electrical energy. These losses usually enter the chimney in the form of heat, its temperature is controlled and released into the atmosphere. By recycling some of the heat in the heat exchangers, the efficiency of the whole system increases significantly, and while electricity is generated, the heat required by commercial, industrial, and public centers is also provided. Simultaneous generation of heat, cold, and electricity (CCHP) systems also provide the required heat and cooling loads with the help of waste heat recovery in addition to generating electrical power. The increasing development of gas turbines and internal combustion engines of gas burners has made the use of production systems simultaneously with the primary propulsion of gas turbines as well as combustion engines. In these systems, by using an absorption chiller and also a heat recovery steam generator (HRSG), the excess heat released from the turbine (or combustion engine) can fuel the thermal and refrigeration load of the desired location without separate fuel consumption. This reduces fuel consumption and also reduces the pollutants emission. In

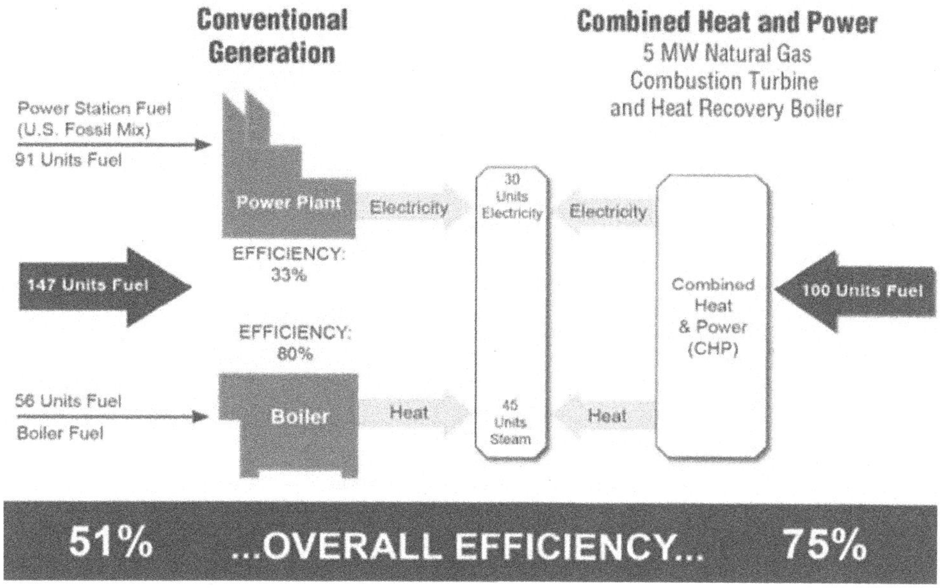

Figure 12.22 Comparing the efficiency of CHP (simultaneous generation of electricity and heat) and SHP (separate generation of electricity and heat) power plants. (Source: EPA 2021.)

Figure 12.22 a comparison between the CHP power plant with a conventional power plant is done (Reddy et al. 2010).

Financial constraints must be carefully considered in justifying CHP units. In each area, competing energies with concurrent production units need to be carefully compared. Cogeneration units usually require more investment than conventional energy conversion systems. But it should be noted that their energy consumption is much lower. In other words, the average conversion costs per unit of energy in CHP systems are lower than other methods.

12.8 ECONOMIC MODELING

The main components in an ORC which are used as a component capacity index in the cost estimation are the pump, turbine, and generator. A is needed for costing components and can be obtaiTned as Equation (12.7).

$$A = \frac{\dot{Q}}{U \cdot \Delta T} \tag{12.7}$$

In Equation (12.7), $\dot{Q}$ is the thermal duty, ΔT is log mean temperature difference (LMTD) between the cold and hot side of the heat exchanger, and U is the overall heat transfer coefficient which is related to the heat transfer coefficient of the shell and tube and the thermal resistance of the solid heat exchanger material. The heat transfer coefficient on shell and tube can be calculated as Equation (12.8) (Turton et al. 2008).

$$\alpha_s = \alpha_i \alpha_p \dot{m} Pr^{(-2/3)} (\frac{\mu_s}{\mu_w})^{0.14} j \tag{12.8}$$

In Equation (12.8), α_i is the ideal cross-flow heat transfer coefficient, μ_s and μ_w are viscosity and j is a correction factor. The heat transfer coefficient for the tube side in different states is obtained as Equations (12.9)–(12.12).

12.8.1 Single phase

$$\alpha_s = 0.023 \frac{\lambda}{d} \mathrm{Re}^{0.8} \mathrm{Pr}^{(n)} \tag{12.9}$$

In Equation (12.9), λ is the thermal conductivity and d is the diameter of the tube.

$$\alpha_t = \alpha_{NB} S_{chen} + \alpha_{DB} F_{chen} \tag{12.10}$$

$$\alpha_{NB} = 0.00122(\frac{\lambda_1^{0.45} c_{p,1}^{0.45} \rho_1^{0.45} \Delta T_{sat}^{0.24} \Delta P_{sat}^{0.75}}{\sigma^{0.5} \mu_1^{0.29} h_{1g}^{0.24} \rho_g^{0.24}}) \tag{12.11}$$

$$\alpha_{DB} = 0.023 \frac{\lambda_1}{d} \mathrm{Re}_1^{0.8} \mathrm{Pr}_1^{1/3} \tag{12.12}$$

α_{DB} is the nucleate boiling heat transfer coefficient, S_{chen} and F_{chen} are factors for the forced convection and nucleate boiling contributions respectively in overall heat transfer.

12.8.2 Two-phase

$$\alpha_t = 0.023 \frac{\lambda_1}{d} \mathrm{Re}_1^{0.8} \mathrm{Pr}_1^{0.3} [0.55 + 2.09(\frac{P_{cond}}{P_{crit}})^{-0.38}] \tag{12.13}$$

12.8.3 Supercritical

$$\alpha_t = \frac{f_r \mathrm{Re}_b \mathrm{Pr}_b}{1.07 + 12.7\sqrt{\frac{f_r}{8}}(\mathrm{Pr}_b^{2/3} - 1)} (\frac{\lambda}{d})(\frac{c_p}{c_{p,b}})^{0.65} \tag{12.14}$$

$$f_r = \frac{1}{(1.82 \log \mathrm{Re}_b - 1.64)^2} (\frac{\rho_w}{\rho_b})^{0.18} (\frac{\mu_w}{\mu_b})^{0.18} \tag{12.15}$$

Table 12.5 Coefficients for cost calculation of each component

Component	*K1, K2, K3*	*c1, c2, c3*	*B1, B2*	*FM*	*FBM*
Pump	K_1=3.3892 K_2=0.0536 K_3=0.1538	c_1=-0.3935 c_2=0.3957 c_3=-0.0023	B_1=1.89 B_2=1.35	1.0	-
Turbine	K_1=2.2476 K_2=1.4965 K_3=-0.1618	-	-	-	3.5
Heat exchanger	K_1=4.3247 K_2=-0.3030 K_3=0.1634	c_1=-0.0016 c_2=0.0063 c_3=0.0123	B_1=1.63 B_2=1.66	1.35	-
Generator	-	-	-	-	1.5

Source: Vant-Hull (2021).

The bare module cost (C_{BM}) of each component of the ORC system are as Equation (12.16). In this equation, *CEPCI* is a chemical engineering plant cost item (Turton et al. 2008).

$$C_j = C_{BM} \frac{CEPCI_{2017}}{CEPCI_{2007}} \tag{12.16}$$

$$C_{BM} = C_P^0 F_{BM} = C_P^0 (B_1 + B_2 F_M F_P) \tag{12.17}$$

$$\log(C_P^0) = K_1 + K_2 \log(X) + k_3[\log(X)]^2 \tag{12.18}$$

J introduced each component in the ORC system like pump, turbine, and condenser, etc. K_i, B_i, and F_i can be obtained as the values presented in Table 12.5. *X* is a corresponding component capacity index (power/hear exchange area). $CEPCI_{2001}$ = 397.0 and $CEPCI_{2017}$ = 567.5 (Vant-Hull 2021).

12.9 SUMMARY

In this chapter, the performance of the ORC cycles was investigated. The important topics of the investigation are:

1 Study on the ORC history
2 Working fluid which used in ORC system
3 Introduction of the different types of the ORC system
4 ORC economic modeling

ORC systems have a wide range of uses, especially in the case of low-temperature heat sources. In industry, heat loss at low temperatures occurs frequently and heat enters the environment through cooling towers or chimneys. Waste heat can be used to generate electricity by an ORC system. The only difference between the ORC and the steam cycle is in the type of fluid used in the cycle, but this slight difference makes significant changes in the behavior and application of the cycle. Vapor saturation of the organic working fluid for use in turbines at much lower temperatures occurs relative to water, thus the temperature range of the ORC is lower than that of the steam Rankin cycle. The choice of the working fluid in the parametric cycle is very effective and has a great impact on the performance of the system. Due to the low temperature of the system heat source, irreversibility may occur in heat exchangers, which can negatively affect the overall efficiency. This chapter also studies some of the important applications of the ORC. Paying attention to the practical aspects of the ORC system is always important as one of the important issues ahead.

REFERENCES

Altun, Ayşe Fidan, & Muhsin Kilic. 2020. "Thermodynamic performance evaluation of a geothermal ORC power plant." *Renewable Energy* 148: 261–274. doi:10.1016/J.RENENE.2019.12.034.

Arabkoohsar, Ahmad, & Hossein Nami. 2019. "Thermodynamic and economic analyses of a hybrid waste-driven CHP–ORC plant with exhaust heat recovery." *Energy Conversion and Management* 187: 512–522. doi:10.1016/j.enconman.2019.03.027.

Bahrami, Mohammad, Ali A Hamidi, & Soheil Porkhial. 2013. "Investigation of the effect of organic working fluids on thermodynamic performance of combined cycle Stirling-ORC." *International Journal of Energy and Environmental Engineering* 4(1): 1–9. doi:10.1186/2251-6832-4-12.

Becquin, Guillaume, & Sebastian Freund. 2012. "Comparative performance of advanced power cycles for low-temperature heat sources." In *The 25th International Conference on Efficiency, Cost, Optimization and Simulation of Energy Conversion Systems and Processes*, Vol. III. www.torrossa.com/gs/resourceProxy?an=2520328&publisher=FF3888#page=94.

Borsukiewicz-Gozdur, Aleksandra. 2013. "Pumping work in the organic Rankine cycle." *Applied Thermal Engineering* 51(1–2): 781–786. doi:10.1016/J.APPLTHERMALENG.2012.10.033.

Chahartaghi, Mahmood, & Mahdi Babaei. 2014. "Energy and exergy analysis of organic Rankine cycle with using two component working fluid in specified heat source conditions." *Modares Mechanical Engineering* 14(3): 145–156.

Chaibakhsh, Ali, & Ali Ghaffari. 2008. "Steam turbine model." *Simulation Modelling Practice and Theory* 16(9): 1145–1162. www.sciencedirect.com/science/article/pii/S1569190X08001196.

Dai, Yiping, Jiangfeng Wang, & Lin Gao. 2009. "Parametric optimization and comparative study of organic Rankine cycle (ORC) for low grade waste heat recovery." *Energy Conversion and Management* 50(3): 576–582. doi:10.1016/J.ENCONMAN.2008.10.018.

Darvish, Kamyar, Mehdi A. Ehyaei, Farideh Atabi, & Marc A. Rosen. 2015. "Selection of optimum working fluid for organic Rankine cycles by exergy and exergy-economic analyses." *Sustainability* 7(11): 15362–15383. doi:10.3390/SU71115362.

Drescher, Ulli, & Dieter Brüggemann. 2007. "Fluid selection for the organic Rankine cycle (ORC) in biomass power and heat plants." *Applied Thermal Engineering* 27(1): 223–228. doi:10.1016/J.APPLTHERMALENG.2006.04.024.

EPA. 2021. "Combined heat and power." www.epa.gov/energy/combined-heat-and-power.

Explain That Stuff. 2021. "How do steam turbines work?" www.explainthatstuff.com/steam-turbines.html.

Gu, Zhaolin, & Haruki Sato. 2002. "Performance of supercritical cycles for geothermal binary design." *Energy Conversion and Management* 43(7): 961–971. doi:10.1016/S0196-8904(01)00082-6.

Guo, Zengyuan Y., Xiongbin B. Liu, Wenquan Q. Tao, & Ramesh K. Shah. 2010. "Effectiveness–thermal resistance method for heat exchanger design and analysis." *International Journal of Heat and Mass Transfer* 53(13–14): 2877–2884. doi:10.1016/j.ijheatmasstransfer.2010.02.008.

Heberle, Florian, & Dieter Brüggemann. 2010. "Exergy based fluid selection for a geothermal organic Rankine cycle for combined heat and power generation." *Applied Thermal Engineering* 30 (11–12): 1326–1332. doi:10.1016/J.applthermaleng.2010.02.012.

Hung, Tzu Chen. 2001. "Waste heat recovery of organic Rankine cycle using dry fluids." *Energy Conversion and Management* 42(5): 539–553. doi:10.1016/S0196-8904(00)00081-9.

Kelvion. 2021. "Products." www.kelvion.com/products/product/shell-tube-single-comfin/.

Larjola, Jaakko 1995. "Electricity from industrial waste heat using high-speed organic Rankine cycle (ORC)." *International Journal of Production Economics* 41(1–3): 227–235. doi:10.1016/0925-5273(94)00098-0.

Lecompte, Steven, Sanne Lemmens, Henk Huisseune, Martijn Van den Broek, & Michel De Paepe. 2015. "Multi-objective thermo-economic optimization strategy for orcs applied to subcritical and transcritical cycles for waste heat recovery." Energies 8(4): 2714–2741. www.mdpi.com/95796.

Li, Liang, Yunting T. Ge, Xiang Luo, & Savvas A. Tassou. 2017. "Experimental investigations into power generation with low grade waste heat and R245fa organic Rankine cycles (ORCs)." *Applied Thermal Engineering* 115: 815–824. doi:10.1016/j.applthermaleng.2017.01.024.

Liu, Bo Tau, Kuo Hsiang Chien, & Chi Chuan Wang. 2004. "Effect of working fluids on organic Rankine cycle for waste heat recovery." *Energy* 29(8): 1207–1217. doi:10.1016/j.energy.2004.01.004.

Macchi, Ennio. 2017. "Theoretical basis of the organic Rankine cycle." In *Organic Rankine Cycle (ORC) Power Systems: Technologies and Applications*. Duxford: Woodhead Publishing, pp. 3–24. doi:10.1016/B978-0-08-100510-1.00001-6.

Mago, Pedro J., Louay M. Chamra, Kalyan Srinivasan, & Chandramohan Somayaji. 2008. "An examination of regenerative organic Rankine cycles using dry fluids." *Applied Thermal Engineering* 28(8–9): 998–1007. doi:10.1016/j.applthermaleng.2007.06.025.

National Geographic. 2021. "Geothermal energy." www.nationalgeographic.org/encyclopedia/geothermal-energy/.

Obernberger, Ingwald, Peter Thonhofer, & Erwin Reisenhofer. 2002. "Description and evaluation of the new 1,000 KWel organic Rankine cycle process integrated in the biomass CHP plant in Lienz, Austria." *Euroheat & Power* 10. https://pure.tue.nl/ws/files/3237054/Metis220459.pdf.

Qiu, Guoquan, Yingjuan Shao, Jinxing Li, Hao Liu, & Saffa B. Riffat. 2012. "Experimental investigation of a biomass-fired ORC-based micro-CHP for domestic applications." *Fuel* 96: 374–382. doi:10.1016/j.fuel.2012.01.028.

Quoilin, Sylvain, Martijn Van Den Broek, Sébastien Declaye, Pierre Dewallef, & Vincent Lemort. 2013. "Techno-economic survey of organic Rankine cycle (ORC) systems." *Renewable and Sustainable Energy Reviews* 22: 168–186. doi:10.1016/j.rser.2013.01.028.

Reddy, Vundela Siva, Subash Chndra Kaushik, Sudhir Kumar Tyagi, & Narayanlal Panwar. 2010. "An approach to analyse energy and exergy analysis of thermal power plants: A review." *Smart Grid and Renewable Energy* 2010(3): 143–152. doi:10.4236/SGRE.2010.13019.

Rettig, Adrian, Martin Lagler, Thomas Lamare, Shuai Li, Varnabhye Mahadea, Sean McCallion, & Julia Chernushevich. 2011. "Application of organic Rankine cycles (ORC)." *World Engineers' Convention*. www.zhaw.ch/storage/engineering/institute-zentren/iefe/PDFs/orc-final-paper-wec2011-2011-07-30.pdf.

Schuster, Andreas, Sotirios Karellas, Emmanouil C. Kakaras, & Hartmut Spliethoff. 2009. "Energetic and economic investigation of organic Rankine cycle applications." *Applied Thermal Engineering* 29(8–9): 1809–1817. doi:10.1016/j.applthermaleng.2008.08.016.

Tchanche, Bertrand Fankam, Mathieu Pétrissans, & Georg Papadakis. 2014. "Heat resources and organic Rankine cycle machines." *Renewable and Sustainable Energy Reviews* 39: 1185–1199. doi:10.1016/J.RSER.2014.07.139.

Turboden. 2021. www.turboden.com.

Turton, Richard, Richard C. Bailie, Wallace B. Whiting, & Joseph A. Shaeiwitz. 2008. *Analysis, Synthesis and Design of Chemical Processes*. New Jersey: Prentice Hall.

Vanslambrouck, Bruno et al. 2011. "Turn waste heat into electricity by using an organic Rankine cycle." www.researchgate.net/publication/257569971_Turn_waste_heat_into_electricity_by_using_an_Organic_Rankine_Cycle_part_2.

Vant-Hull, Lorin L. 2021. "Central tower concentrating solar power systems." *Concentrating Solar Power Technology*. Duxford: Woodhead Publishing, pp. 267–310. doi:10.1016/B978-0-12-819970-1.00019-0.

Wei, Donghong, Xuesheng Lu, Zhen Lu, & Jianming Gu. 2007. "Performance analysis and optimization of organic Rankine cycle (ORC) for waste heat recovery." *Energy Conversion and Management* 48(4): 1113–1119. doi:10.1016/J.ENCONMAN.2006.10.020.

Chapter 13

Power-to-fuel

A new energy storage technique

Alper Can Ince

Gebze Technical University, Faculty of Engineering, Mechanical Engineering Department, Gebze, Kocaeli, Turkey

Can Ozgur Colpan

Dokuz Eylul University, Faculty of Engineering, Mechanical Engineering Department, Buca, Izmir, Turkey

Mustafa Fazıl Serincan

Gebze Technical University, Faculty of Engineering, Mechanical Engineering Department, Gebze, Kocaeli, Turkey

13.1 INTRODUCTION

The increment in fossil fuels depletion and greenhouse gas (GHG) emissions have become a major issue faced by our world in recent decades. Currently, 86% of the overall key energy supply is provided by fossil fuels (Momeni et al. 2021). Therefore, the total CO_2 emissions measured above 30 Gt in 2020 (IEA 2020a). The primary energy sources which cause CO2 emission are coal (e.g. 14.5 Gt), oil (e.g. 11.37 Gt), natural gas (e.g. 6.7 Gt), and others (IEA 2020b). In this regard, several environmental meetings and protocols have been arranged to reduce GHG emissions and global warmings such as the Kyoto Protocol in 1997 (Böhringer 2003), the Copenhagen Accord in 2009 (Lau, Lee, & Mohamed 2012), the "Energy Roadmap 2050" in 2011 (EU 2011), and the Paris Agreement in 2015 (Savaresi 2016). One of the most important aims from these events is to provide almost 100% electricity generation from renewable energies and fossil-based energy systems integrated with carbon capture and storage (CCS) (Hübler & Löschel 2013). Therefore, the share of renewable energy sources in the total primary energy supply for power and heat generation has remarkably increased and is expected to be accelerated (Giglio et al. 2018). In this context, a 100% renewable power system is technically and economically achievable in Europe and the USA by 2050 (Jacobson et al. 2015; Zappa, Junginger, & van den Broek 2019). However, a significant challenge of renewable energy resources is to produce intermittent electricity that leads to balancing problems between power supply and demand in the grid network (Morales et al. 2014). Electrical energy storage (EES) systems bring a potential solution to store surplus electricity and, thus, variations in the grid can be smoothed for stability and consistency. Different techniques in a variety of scales are developed to store this electricity. As displayed in Figure 13.1, for example, high-power superconductors are capable of storing kilowatts of power in short periods (seconds), while pump-hydro or chemical storage has the capability of storing gigawatts of power for long periods (months and years) (Valera-Medina et al. 2021).

In this regard, power-to-fuel (P-t-F) is a promising chemical energy storage technique that serves two main purposes.

DOI: 10.1201/9781003213741-13

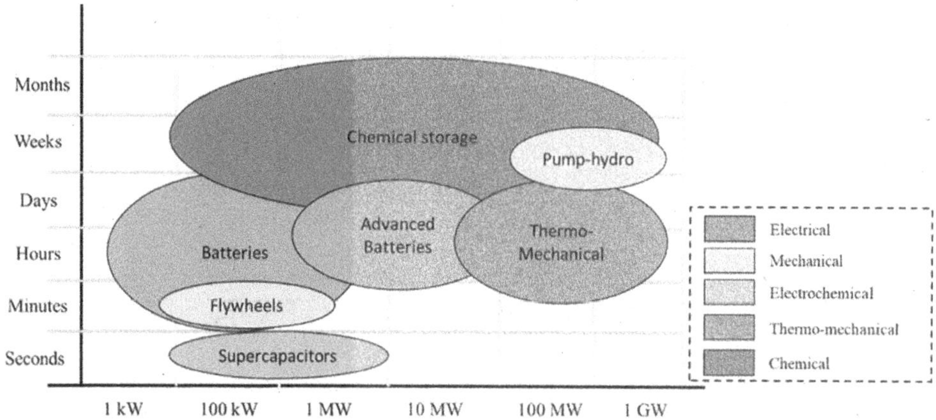

Figure 13.1 Various energy storage techniques. (Source: Adapted from Valera-Medina et al. 2021.)

1 Through the electrochemical process (e.g., water electrolysis, CO_2 electrolysis), the excess power is utilized to produce directly fuel or produce hydrogen and syngas that are further sourced for fuel production. Therefore, the excess power from renewables can be stored as fuel.
2 The P-t-F system includes CO_2 capture systems for hydrocarbon-based fuel production. Therefore, the CO_2 can be captured from the flue gas of the heat and power plant and can be utilized for fuel production (Adnan & Kibria 2020, p. 2; Hidalgo & Martín-Marroquín 2020; Peters et al. 2019; Schemme et al. 2017).

Even though there are numerous pathways in a P-t-F system as discussed in Section 13.3, a typical pathway of the P-t-F system to produce hydrocarbon-based fuels starts with converting renewable power into hydrogen through water electrolysis. CO_2 is captured from different carbon sources (e.g., flue gas, biogas, air) through various separation techniques and processes. Finally, CO_2 combines with hydrogen in a reactor through the appropriate catalyst and operating conditions resulting in produce storable liquid and gaseous fuel (e.g., methane, methanol), as displayed in Figure 13.2.

This chapter presents the fundamental operating principle of the main pathways employed in P-t-F such as direct electrochemical reduction pathway (D-ERP), syngas followed by catalytic step (syn-CS), and water electrolysis followed by catalytic step (H2-CS). Moreover, sub-processes in the P-t-F system from renewable power production to fuel upgrading are described briefly. The key contribution of this chapter is to present a thermodynamic assessment of available P-t-F studies and projects. Hence, the chapter aims to be a research guide for prospective novel P-t-F designs.

13.2 KEY SUB-PROCESS OF P-T-F PATHWAYS

13.2.1 Renewable power production

Wind, solar, or other renewable energy resources (e.g., hydropower, nuclear) ensure zero-carbon power for the P-t-F system. In the world, 2,537 GW of global renewable generation

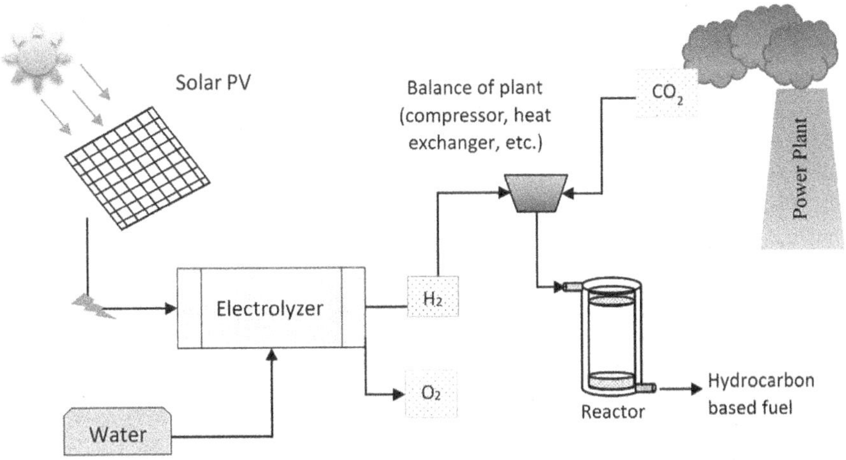

Figure 13.2 Typical P-t-F system process configuration.

Table 13.1 A basic comparison of solar and wind renewable energy resources

Energy resources	*Shares of primary energy (2018)*	*Shares of primary energy (2050)*	*Cost of producing power ($/kWh)*	*Efficiency (for hydrogen production)*	*Present cost (for hydrogen production)*	*Future cost projection for the year of 2050 (hydrogen production)*
Solar	1%	69 %	14-30	20 %	7–11 $/kg$_{H2}$	3–4 $/kg$_{H2}$
Wind	1%	18 %	4–18	21 %	10–30 $/kg$_{H2}$	3–4 $/kg$_{H2}$

Sources: Manage et al. (2011); Ram et al. (2019).

was installed at the beginning of 2020 (Mohn 2020). In this section, the current economic situations, major developments and forecasting future perspectives in power generation from solar and wind that are generally used for P-t-F systems are compared in Table 13.1 and discussed in the following paragraphs.

The power produced from wind turbines has a considerable growing trend in the renewable energy industry with 1,273,409 GWh (shares 1% of total primary energy) of total electricity in 2018 (IEA 2020b). However, some issues such as loading, intermittency and integration with the grid are still a significant challenge. The use of integrated electrolyzer with wind power generators can be an alternative solution for these problems with following two routes. In the first route, H_2 is produced during higher power generation and low electricity demand, and the second route provides continuous H_2 production (Jørgensen & Ropenus 2008). The cost of producing power plays a critical role in determining the economic feasibility of P-t-F systems. Here, the wind speed is a key factor to determine the cost of electricity and hydrogen production. For instance, when the velocity of wind is 6.7 m/s, the cost of electricity and cost of producing hydrogen (with current density and voltage of 0.1 A/cm^2 and 1.60 V, respectively) are calculated as 4.5 $/kWh and 26.3 $/GJ, respectively. However, when the wind velocity increases to 8.9 m/s, the cost of electricity and the cost of producing hydrogen is decreased to 1.89 \$/kWh and 15.73 \$/GJ, respectively (Bockris & Veziroglu 2007).

Solar PVs contribute to the electricity production with a value of 554,382 GWh (shares about 1% of total primary energy) of total electricity in 2018 (IEA 2020a). The power production from solar PVs suffers from low efficiencies and high costs. However, a significant cost reduction has been experienced in recent years, and the trend of this decrement is predicted to continue. Currently, the capital cost of the utility scale of PV plants (including the cost of module, invertors, installation work, and others) is around 100 €/kW (Ikäheimo et al. 2018). A significant factor that affects the solar PV potential is the area of installation (open space and rooftop). For example, Bossavy, Girard and Kariniotakis (2016) reported that the average open space energy potential is calculated annually as 179 Wh/m^2 in Brittany (France). In other words, the capacity is 0.16 MW/km^2.

13.2.2 Water electrolyzer (for hydrogen production)

Water electrolysis has become a considerable process to produce hydrogen over conventional methods (e.g., steam reformation, gasification) because the water electrolysis provides purified hydrogen production environmentally with higher conversion efficiency. However, the hydrogen production through this process is still around 5% due to the cost and durability issues for the electrolyzer technology as well as the cost of electricity that provides the required power. The water electrolyzer is classified according to the ion conductor material (alkaline, polymer, and solid) and working temperature (low-temperature and high-temperature). The comparison of different electrolyzer technologies is presented in Table 13.2. The low-temperature electrolyzer group (60–80°C) includes alkaline and PEM electrolyzers. Alkaline electrolyzers are more mature (i.e., technology readiness level, TRL is 9) and economically feasible technology (non-precious catalysts are used) over other electrolyzers. A high hydrogen production capacity and high durability are important benefits of alkaline electrolyzers, while low partial load range and limited current density are major drawbacks (Amores, Rodríguez, & Carreras 2014; Barco-Burgos et al. 2020; Demirdelen et al. 2020; Zhang & Zeng 2016). Polymer electrolyte membrane electrolyzer (or proton exchange membrane) is another low-temperature water electrolyzer based on a solid polymer electrolyte (e.g., Nafion, Fumapem) that provides high proton conductivity and low gas crossover. Unlike alkaline electrolyzers, PEM electrolyzers can be operated at a higher current density (0.8–2 A/cm^2). Moreover, the PEM electrolyzer provides a fast response to fluctuating power. However, the use of precious catalyst and durability is a significant challenge in terms of commercialization (Bessarabov, Wang, & Zhao 2016; Carmo et al. 2013; Shiva Kumar & Himabindu 2019). On the other hand, high-temperature electrolyzers are superior compared to low-temperature electrolyzers since high temperature provides an increase in the ionic conductivity and reduction in the activation overpotential. Moreover, water management is not an important issue as liquid water is not fed (Toghyani et al. 2018). Therefore, the essential power that split water for hydrogen production is decreased significantly. A high-temperature PEM electrolyzer (HT-PEM) is an alternative and relatively newer application of high-temperature electrolyzers (elevated at 170°C). Here, phosphoric acid doped Nafion that prevents membrane dehydration at high temperatures can be used in the HT-PEM electrolyzer (Aili et al. 2011; Hansen et al. 2012). However, solid oxide electrolyzers (SOEs) are very promising in thermodynamic considerations because a higher operating temperature (700–1,000°C) can be achieved. Therefore,

Table 13.2 Comparison of water electrolyzer technologies

		Low-temperature		*High-temperature*	
Properties/electrolyzers	*Unit*	*Alkaline*	*PEM*	*SOE*	*H-SOE*
Electrolyte material	-	Liquid Potassium hydroxide (KOH) solution % 25–30 wt	Polymer electrolyte (e.g., Nafion)	YSZ	Perovskite-type ceramics e.g., $BaCe_{1-x}Zr_xO_3$
Reactions	Anode	$2H_2O_{(l)} + 2e^- \rightarrow H_{2(g)} + 2OH^-$	$H_2O_{(l)} \rightarrow 2H^+ + 1/2O_{2(g)} + 2e^-$	$O^{2-} \rightarrow 2e^- + 1/2O_2$	$H_2O_{(g)} \rightarrow 2H^+ + \frac{1}{2}O_{2(g)} + 2e^-$
	Cathode	$2OH^-_{(aq)} \rightarrow 0.5O_{2(g)} + H_2O + 2e^-$	$2H^+ + 2e^- \rightarrow H_{2(g)}$	$H_2O_{(g)} + 2e^- \rightarrow H_{2(g)} + O^{2-}$	$2H^+ + 2e^- \rightarrow H_2$
Ion transfer	-	OH^-	H^+	O^{2-}	H^+
Current density	A/cm^2	0.2–0.4	0.8–2	0.3–1	0.5
Temperature	°C	60–90	70–80	650–850	550–750
Pressure	bar	10–30	20–50	1–15	1
Efficiency	%	65	>70	>80	-
Capacity (hydrogen production per stack)	Nm^3/h	<1,400	<400	<10	-
Durability (efficiency degradation)	%/a	0.25–1.5	0.5–2.5	3–50	-
Flexibility (cold start-up time)	min	15	<15	>60	-
Economic feasibility (investment costs)	€/kW	800–1,500	1,400–2,100	>2,100	-

Sources: Buttler and Spliethoff (2018); Grigoriev et al. (2020); Hu et al. (2018); Schwabe et al. (2019).

the highest efficiency can be obtained (>70%) when SOE is chosen over other electrolyzer types. However, high cell degradation and low hydrogen production yield in SOE are significant drawbacks related to other electrolyzers (Bi, Boulfrad, & Traversa 2014; Costa et al. 2019; Lenser et al. 2020). Even though a typical oxygen ion conductive electrolyte (e.g., YSZ) is used in SOE cell, the use of proton conductive material for electrolyte (e.g., perovskite-type ceramic $BaCe_{1-x}Zr_xO_3$ based materials) has gained more attention in the last few years. The reason can be explained that proton conductive SOE (H-SOE) technology offers the following advantages: lower activation energy and simple gas separation (Ding, Wu, & Ding 2019; Lei et al. 2019; Tucker 2020).

13.2.2.1 CO2 capture (for hydrocarbon-based fuels) and N2 production (for ammonia fuels) techniques

The integration of the CO_2 capture process into a plant becomes a significant environmental and engineering task to reduce CO_2 emission that comes from flue gas. In fact, the plant integrated with the capture process leads to producing more CO_2 per kWh when compared to the power station without the capture process. However, more than 85% of the CO_2 can be captured (Kanniche et al. 2010). The CO_2 capture strategies from flue gas can be classified

Table 13.3 Comparison of main CO_2 separation techniques

Techniques	*Energy requirement*	*Advantages*	*Disadvantages*
Absorption	4–6 MJ/kg_{CO2}	Rapid response to variations, high CO_2 recovery, matured technology for the post-combustion CO_2 capture process	High energy requirement, solvent losses, corrosion, low efficiency of solvent regeneration
Adsorption	2–3 MJ/kg_{CO2}	Fast kinetic, high thermal stability	Low CO_2 selectivity, solvent losses, needs for solvent regeneration
Membrane	0.5–6 MJ/kg_{CO2}	Low energy requirement, low control requirement	High-pressure difference requirement, fouling, high manufacturing cost, high membrane surface area
Cryogenic	6–10 MJ/kg_{CO2}	High CO_2 recovery, producing liquid CO_2 without separation unit	High energy requirement, suitable for high CO_2 concentration content (>50 %), longer response to variations
Chemical looping	-	Inherent process, low cost, high CO_2 capture efficiency	High investment of installation cost, thermal stability, oxygen transport capacity

Sources: Lyngfelt (2020); Mondal, Balsora, and Varshney (2012); Song et al. (2018).

according to the location of conversion where the carbon source (e.g., fuel, raw material) converts the CO_2 rich gas such as post-conversion, pre-conversion, and oxy-fuel combustion (Mondal, Balsora, & Varshney 2012). The CO_2 capture process is followed by separation techniques such as absorption, membrane, cryogenic separation, adsorption, and chemical looping to obtain pure CO_2 (Song et al. 2018). The key comparison for these techniques is presented in Table 13.3. One of the important disadvantages of CO_2 capture techniques is the requirement of high energy and thus, total plant energy efficiency is reduced significantly. Here, the membrane separation technique offers lower energy requirements, while the cryogenic separation technique is challenging in terms of high energy penalty. Moreover, the membrane technique suffers from high capacity and stability. In the absorption technique, solvent degradation, corrosion, and high cost are the main problems. One of the most mature processes is the pre-combustion process where the CO_2 is captured after following two different routes: steam reforming or partial oxidation and water-gas shift reaction. Then, CO_2 is separated through the solvents. Another important process is the post-combustion process, where CO_2 is captured after the combustion of fossil fuels. In the oxy-fuel combustion process, the combustion of fuel is done with pure oxygen to produce high CO_2 rich gas. The process is followed by condensation to obtain CO_2. Several reviews studies that address CO_2 capture process such as pre-combustion (Babu et al. 2015; García et al. 2011; Jansen et al. 2015), post-combustion (Samanta et al. 2012; Y. Wang et al. 2017), and oxy-fuel combustion (Habib et al. 2011; Wu et al. 2018) have been published in the last decade.

The CO_2 can also be captured from biogas and air. CO_2 capture from the air is a relatively new and innovative technology and is generally called direct air capture (DAC) (Koytsoumpa, Bergins, & Kakaras 2018). Here, the CO_2 is captured from the air through the mostly chemical sorbents (e.g., calcium carbonate precipitators) and adsorption process (e.g., temperature vacuum swing and moisture swing adsorption) (Fasihi, Efimova, & Breyer. 2019). There is no uncertainty that the DAC enables the reduction of CO_2 concentration directly from the air. However, CO_2 separation from the air requires more energy and cost when compared to CO_2 separation from flue gas (Wilcox, Psarras, & Liguori 2017).

The production of N_2 also necessitates high energy, and thus it leads to a decrease in the energy efficiency of the renewable ammonia synthesis process. N_2 is generally obtained from air by a cryogenic technique. Here, the air is compressed through the many stages of the compressor. After the air is cooled by many heat exchangers, high purity N_2 is used for ammonia synthesis (Fu et al. 2016; Wijayanta & Aziz 2019). The enhancement of the N_2 production process improves the performance of the synthetic ammonia production process significantly. For this purpose, the N_2 production unit proposed by Aziz et al. (2017) enables a decrease in energy consumption of about 43% when compared to other N_2 production processes.

13.3 THE OPERATING WINDOW OF VARIOUS P-T-F PATHWAYS

Three main pathways can be considered for the P-t-F process depending upon whether the process involves purely electrochemical or electrochemical followed by chemical processes. In the first pathway, the fuel can be produced through the electrochemical process directly decomposing CO_2 and/or H_2O or N_2 and/or H_2O (D-ERP). Another pathway, which is also the most typical process, covers water electrolysis for hydrogen production and is followed by a catalytic step through the chemical reactor (H2-CS). In some cases, syngas is produced through electrochemical reduction by replacing steam methane reformer which harms the environment. Subsequently, the syngas is used in the catalytic step for hydrocarbon-based fuel synthesis. This pathway is called syn-CS in this chapter. In the following subsections, these pathways are discussed in terms of their basic operating window, advantages, and drawbacks.

13.3.1 Direct electrochemical reduction pathway

Direct electrochemical reduction pathway (D-ERP) offers to produce fuel directly in a stack and compact electrochemical reactors. One of the most applications of these pathways is the CO_2 electrolyzer (CO_2 reduction reactor, CO_2-RR), as shown in Figure 13.3a. Different electrolyzer types from alkaline to solid oxide cells can be tailored for the CO_2-RR. In other words, the CO_2-RR can be worked at both high and low operating temperatures (Song et al. 2019). The working principle behind CO_2-RR is based on the electrochemical reduction of CO_2 through numerous electron (from 2 e- to 18 e-) reduction (Arquer et al. 2020; Kuhl et al. 2012). Therefore, the CO_2-RR is capable of producing liquid (e.g., HCOOH, CH_3OH, C_2H_5OH) and gases (e.g., CO, CH_4, C_2H_4). Here, since CO cannot be considered as a fuel, CO is considered as a tailored gas to further use for the catalytic step where hydrogen reacts with CO to produce synthetic fuel. Since the CO_2 electrolyzer is a relatively novel component, the studies on the CO_2-RR mechanism are generally considered for material, achievements, and future directions (Huang et al. 2020; Lee et al. 2020; Liu et al. 2018; Zhou et al. 2018).

This pathway is also used for the electrochemical synthesis of ammonia in an electrochemical Haber-Bosch reactor stack (HB-RR) based on the electro-reduction of nitrogen. Recent progress of electrochemical ammonia synthesis is presented in the literature (Kyriakou

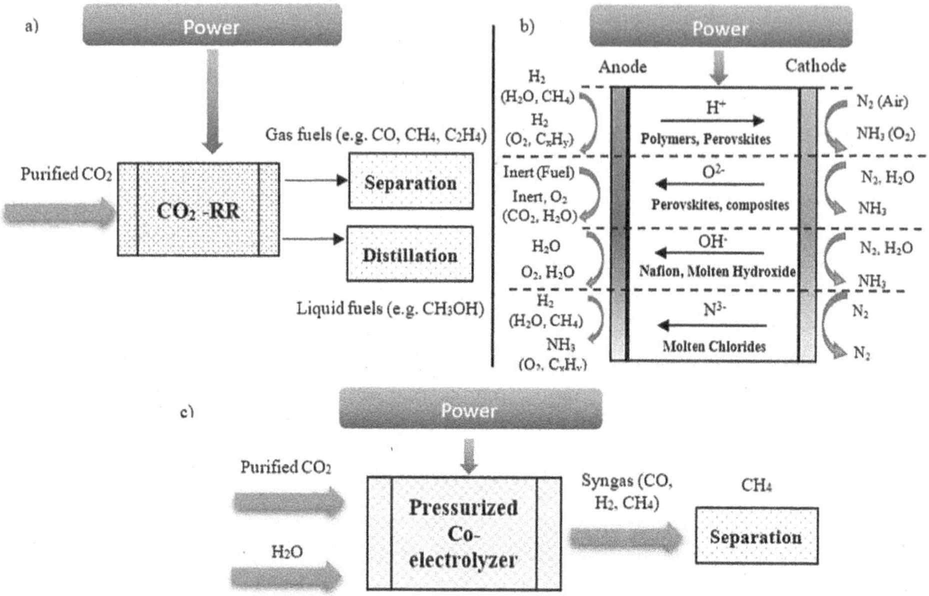

Figure 13.3 (a) Basic working principle of CO_2-RR; (b) HB-RR; and (c) Co-SOE. (Sources: Jouny, Luc, & Jiao 2018; Kyriakou et al. 2017; Luo et al. 2018.)

et al. 2017; Shipman & Symes 2017; J. Wang et al. 2020). HB-RR can be operated in different electrolyzer types considering various charge carriers and electrolytes as displayed in Figure 13.3(b). Material advances of HB-RR is currently a hot topic in improving the design of electrode/catalyst/electrolyte for the enhancement of selectivity, overpotentials, and energy efficiency.

In the D-ERP pathway, direct methane synthesis through the co-electrolyzer in which steam and CO_2 are decomposed simultaneously through the Ni catalyst (Co-SOE) can be provided as shown in Figure 13.3(c) (Chen, Chen, & Xia 2014). However, methane formation is not preferred thermodynamically. Li et al. (2013) reported that the methane generation is around 0.2% when the co-electrolyzer is used. Similar results are found repeatedly in the literature (e.g., Lei et al. 2017; Xie et al. 2011; Xu et al. 2016). However, the methane formation from a co-electrolyzer unitary electrochemical reactor highly depends on the operating window of the co-electrolyzer. For example, when the operating pressure of co-electrolyzer and reactant utilization increase, the methane fraction at the outlet of co-electrolyzer increases significantly (Luo et al. 2018; Ligang Wang et al. 2019). On the other hand, one of the most promising and feasible applications of co-electrolyzer is to produce syngas that can be further used in the catalytic step to produce synthetic fuel (Deka et al. 2019; Ni 2012).

One of the significant contributions of D-ERP is to be more compact and unitary, and thus, the pathway does not require many chemical processes. Moreover, some applications such as CO and HCOOH production through the CO_2 electrolyzer is currently more feasible, while some application of this pathway such as ammonia production through the HB-RR is projected to be more feasible when the electricity price is lower over other pathways

(Hochman et al. 2020; M. Wang 2020) (Suryanto et al. 2019; Wan, Xu, & Lv 2019; X. Wang et al. 2020). However, some issues on the catalyst (poisoning, carbon deposition risk), slower charge transfer kinetics, and lower product yield are the main drawbacks for D-ERP among other pathways (Duboviks et al. 2014; Qi et al. 2014; Skafte et al. 2018).

13.3.2 Water electrolyzer followed by catalytic step

The typical pathway of the P-t-F concept is based on the water electrolyzer followed by catalytic step (H2-CS), as shown in Figure 13.2. The working principle behind this pathway starts with the production of hydrogen from power through water electrolysis. In the second step, CO_2 is obtained from mainly three different carbon sources (e.g., air, biogas, and flue gas) for hydrocarbon-based fuels. Alternatively, for ammonia production, nitrogen (N_2) is mainly obtained from air. The final step of this pathway is the fuel upgrading section (e.g., Sabatier reactor, methanol reactor, FT reactor, HB reactor, and membrane reactor) where the desired fuel (e.g., methane, methanol, diesel) is produced from a combination of H_2 with CO or CO_2 or N_2. One of the most important benefits of this pathway is to use more mature, simply and economic technology in the fuel upgrading section when compared to D-ERP (Bailera et al. 2017; Giglio et al. 2018; Hidalgo & Martín-Marroquín 2020). Different routes can be used in accordance with CO_2 conversion, as shown in Figure 13.4. In the first route, CO_2 can be combined with hydrogen directly in the catalytic step (CO_2 hydrogenation). In the second route, CO_2 can be converted to CO through the conventional reverse water gas shift reaction (Ham et al. 2019; Küngas 2020). CO is then combined with hydrogen in the catalytic step. In the third route, CO_2 can be converted into CO electrochemically through the CO_2-RR to be further used in the catalytic step. Here, the role of the coupled CO_2 and water electrolyzers system is to replace the steam- methane reforming process. In the first and second routes, the CO_2 conversion is provided with chemical methods through the non-precious and prevalent catalysts (Atsonios, Panopoulos, & Kakaras 2016; Huang et al. 2021). Therefore, carbon deposition and catalyst poisoning are not a significant problem; furthermore, the product yield and selectivity shows relatively better performance when compared to the third route. However, hydrogen management (e.g., production, storage) is a significant drawback for all routes of this pathway when compared to other pathways. Moreover, the system generally requires many chemical processes such as separation, purification, distillation, flash, and absorber.

13.3.3 Syngas followed by a catalytic step (case: co-electrolyzer step)

The syngas followed by the catalytic step (Syn-CS) provides fuel production through the two steps. One of the most significant advantages of this pathway is the lack of requirement of a separate hydrogen production section and its easy management (e.g., storage) when compared to other pathways. In the conventional route, syngas is produced through the steam methane reforming of methane. Subsequently, syngas is converted into fuel in the catalytic step. However, this system damages the environment significantly. In an innovative route, syngas is produced by co-electrolysis process, as displayed in Figure 13.5. Then, it is converted to fuel in the catalytic step through the chemical reactor. Here, as CO_2 is converted to syngas with the electrochemical process, some catalyst issues such as carbon deposition

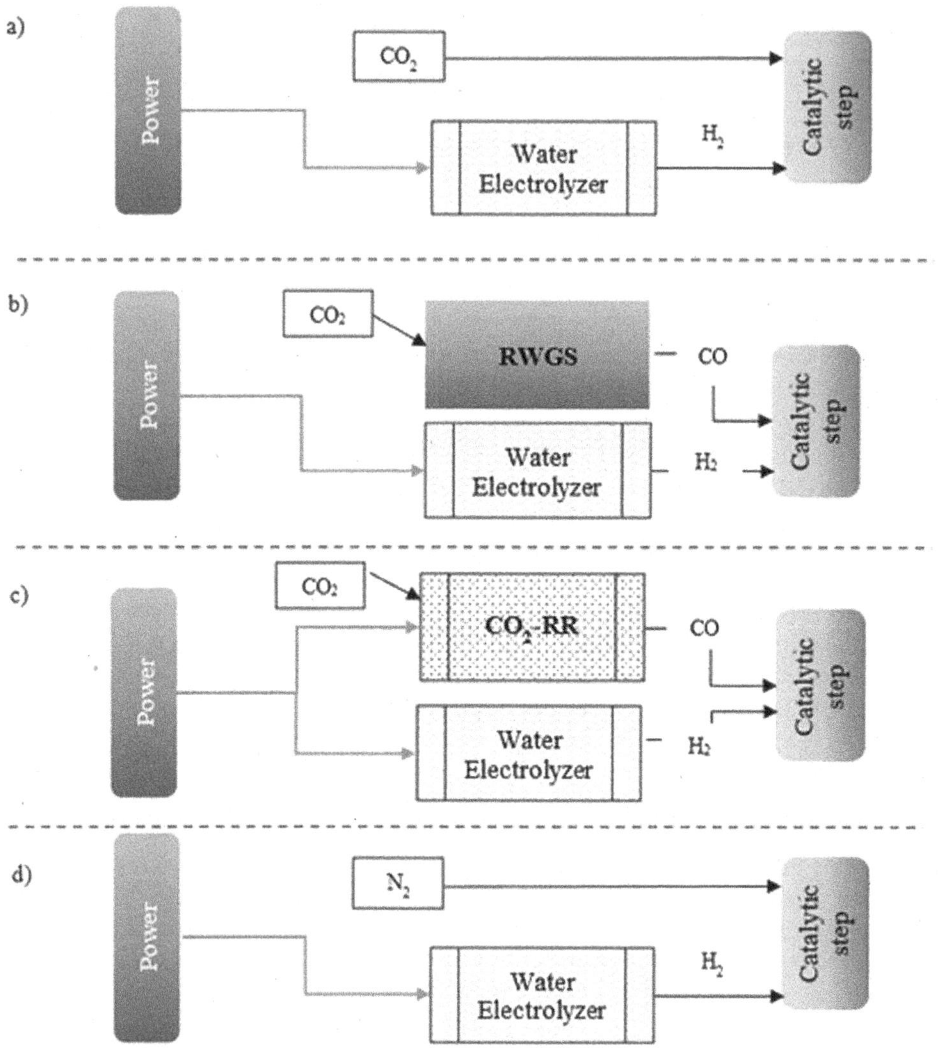

Figure 13.4 Various routes for H2-CS.

and catalyst poisoning can occur. However, a significant benefit of this route is to provide syngas production environmentally when compared to the conventional route. Moreover, the surplus power can be utilized to produce syngas in this route.

13.4 THERMODYNAMIC ASSESSMENT

Thermodynamic assessment of the system is a significant tool to understand the feasibility of the system. The principle of thermodynamic assessment is based on mass, energy, and exergy balances applied to the control volume of system components. The energy and exergy

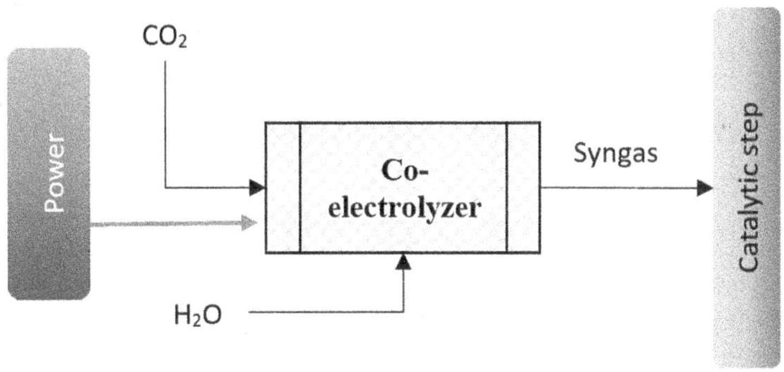

Figure 13.5 Syngas followed by catalytic step.

loss within the system can be found through the thermodynamic assessment. Moreover, the power and heat transfer between the system and the environment can be obtained. Therefore, the system thermodynamic efficiency (energy and exergy) can be revealed using two main equations (Equations (13.1) and (13.2), respectively) as follows. Here, x represents the desired fuel. LHV and HHV are the low and high heating value of desired fuel. $\dot{W}_{p,in}$ is the total power requirement. Ex_{heat} is the exergy transfer accompanying heat transfer.

$$\sum_x \frac{\dot{m}_x \cdot LHV_X \left(or\ HHV\right)}{\dot{W}_{p,in}} \tag{13.1}$$

$$\frac{\dot{E}x_{flow,x}}{\dot{E}x_{heat} + \dot{W}_{p,in} + \sum \dot{E}x_{flow,in} - \sum \dot{E}x_{flow,out}} \tag{13.2}$$

Different route-based pathways aforementioned can be used to produce synthetic fuel from power. In this chapter, thermodynamic performance assessment for Power-to-Methanol and Power-to Ammonia processes with their alternative routes is presented and discussed.

13.4.1 Case 1: Power-to-methanol

Since methanol which is miscible with gasoline has a high energy density (22.7 MJ/kg, 18 MJ/L HHV) and higher octane number (110–112), the combustion efficiency is very high and thus, the methanol can be used in diesel and Otto motors (Dieterich et al. 2020). Another important benefit of methanol is being liquid at ambient conditions. Therefore, methanol can be stored and transported easily. Moreover, methanol can also be converted to power through the methanol fuel cell (openly or with reformer) with high efficiency (Ince et al. 2019; Karaoglan et al. 2021). The history of methanol synthesis on an industrial scale back to 1923 (by BASF facilities). In that process, ZnO/Cr_2O_3 catalysts were used at the operation of high pressure and temperature (Elvers & Bellussi 2011). Today, methanol is mainly produced commercially in the following process; after conversion of methane into syngas through the steam methane reforming (SMR) reaction, the syngas is upgraded to methanol in the catalytic step. This mature technology currently dominates to meet 90% of methanol production demand. However, it heavily depends on fossil fuels, which leads to a high amount of 670–790 g CO_2 eq greenhouse gas emission (GHG) to produce 1 kg methanol (Adnan & Kibria 2020). In this regard, global methanol production is almost 100 million metric tons per year, through mainly coal and natural gas as primary feedstock (Kaithal, Werlé, & Leitner 2021). On the other hand, the development of renewable methanol synthesis from renewable energy has received considerable traction in recent decades. In Section 13.3, the P-t-X pathways were described. Several routes of these pathways can be modified for renewable methanol synthesis. Excluding the direct methanol synthesis through the CO_2 electrolyzer, the methanol synthesis for other pathways is provided by the catalytic reactor (catalytic step). Three typical reactors can be used for fuel syntheses such as fixed bed reactor, fluidized bed reactor, and liquid phase reactor. The overview that addresses the comparison of these reactor designs were presented by Dieterich et al. (2020). The bubble formation is a vital issue for the fluidized bed reactor, while the high mechanical stress of the catalyst is challenging for the liquid-phase reactor. The methanol synthesis in the reactor is imparted at the temperature of 250–300°C and pressure of 50–100 bar through the catalyst in different compositions of CuO/ZnO/Al_2O_3. Various reactor configurations (e.g., Lurgi tubular reactor, Linde Variobar, Toyo MRF, Mitsubishi Superconverter, Haldor Topsøe adiabatic reactor, Lurgi MegaMethanol, and Air Products LP-MEOH) have been developed for methanol synthesis. Quasi-isothermal steam-raising fixed bed reactors (SRC) are preferable for today's reactor design. The Lurgi tubular reactor (as displayed in Figure 13.6a) that is the first representation of SRC is capable of producing 1,500–2,200 tons per day. On the other hand, the Toyo MRF type reactor

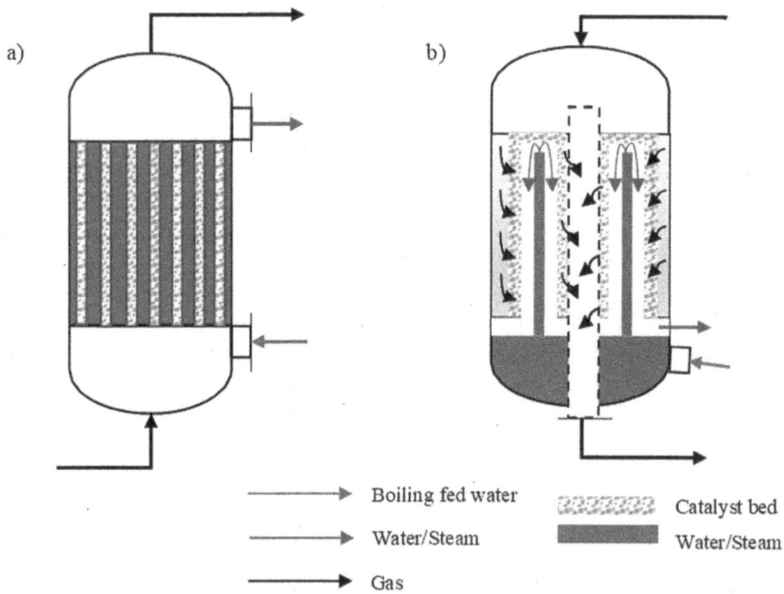

Figure 13.6 Methanol synthesis reactors: (a) Lurgi tubular reactor; (b) Toyo MRF. (Sources: Adapted from Buttler 2018; Bozzano & Manenti 2016.)

(as displayed in Figure 13.6b) is capable of producing 5,000 tons per day (Dahl et al. 2014; Dieterich et al. 2020).

There are four main routes in the literature depending on the feed gas options (direct CO_2, CO_2+H_2, CO+H_2, and syngas) to produce renewable methanol. In the first route, the CO_2 conversion directly into methanol can be provided in an electrochemical reactor, CO_2-RR (Figure 13.2(a)). Since the development of CO_2-RR is still at material-level and a system including balance of plant has not been developed yet, the number of thermodynamic assessment on CO_2-RR is limited and the component-level models are generally developed to find the electrochemical performance of CO_2-RR, and the selectivity and yield of methanol under various operational conditions (Jouny et al. 2018; Kotb et al. 2017). Therefore, electrochemical assessment is a decisive performance parameter. In Table 13.4, different studies that addressed CO_2-RR for direct renewable methanol production are compared in terms of electrochemical performance under different operating conditions. An important challenge of CO_2 electrochemical reduction to methanol is the six-electron transfer. Therefore, different catalysts have been developed to make the electrochemical reduction mechanism more efficient and increase Faradaic efficiency. Many studies suffer from poor selectivity of methanol and low Faradaic efficiency (e.g., Low et al. 2019). On the other hand, Huang and Yuan, (2020) reported that the Faradaic efficiency can reach 59.9% with their in-house heterogeneous catalyst C-Py-Sn-Zn. Similarly, Liwen Wang et al. (2021) formed a novel catalyst (CuO/CN/Ni) to promote the selectivity of methanol and Faradaic efficiency. They calculated the Faradaic efficiency as 70.7% at an applied potential of –0.7 V (vs. RHE).

In Table 13.5, recent studies that address the renewable methanol production (P-t-Methanol) process and the thermodynamic assessment of its system are presented; furthermore, they are discussed below. Other routes are abundant in the literature. In the first route (as displayed in Figure 13.3(a)), CO_2 can be converted into methanol directly through some heterogeneous

Table 13.4 Basic electrochemical performance comparison for CO_2-RR

Study	*Catalyst*	*Applied potential*	*Current density*	*Product yield*	*Maximum Faraday efficiency*
Huang and Yuan (2020)	C-Py-Sn-Zn	−0.5 V	N.A.	N.A.	59.9 %
Irfan Malik et al. (2016)	Cu_2O catalyst supported on MWCNTs	-0.8 V	N.A.	N.A.	38 %
Hazarika and Manna (2019)	Cu_2O catalyst	-2 V	7.8 mA/cm^2	50 μmol/l	47.5 %
Low et al. (2019)	PD-Zn/Ag foam	-1.38 V	2.7 mA/cm^2	4.9 μmol/m^{2}s	10.5 %
Albo et al. (2017)	Cu_2O/ZnO	-0.26 V	1 mA/cm^2	2.91 μmol/m^{2}s	25.6 %
Yuan et al. (2018)	Cu/TiO_2-NG	-0.20 V	0.061 mA/cm^2	0.21 μmol/m^{2}s	19.5 %
Liwen Wang et al. (2021)	CoO/CN/Ni	-0.7 V	15 mA/cm^2	N.A.	70.7%

Note: MWCNTs: Multi Wall Carbon Nanotubes

Table 13.5 Basic comparison of performance of P-t-Methanol systems

Study	Pathway	Route scheme	Energy sources	CO2 source	Hydrogen production	Reactor type	Thermal integration	Thermodynamic assessment				
								Software	Product yield	Total energy/ power requirement	Energy efficiency	Exergy efficiency
Schwabe et al. (2019)	H2-CS	Fig 13.4a	N/A	N/A	H-SOE	N/A	✘	N/A	N/A	617.2 kJ/mol	N/A	✘
Leonzio, Zondervan, and Foscolo (2019)			N/A	N/A.	N/A	Lurgi type	✓	Aspen Plus + Matlab	55.1 ton/h	N/A	N/A	✘
Bos et al. (2020)			Wind	Air	Alkaline	Liquid-Out Gas-In Concept	✓	N/A	33 kton/y	100 MW	50 %	✘
Battaglia et al. (2021)			N/A	Flue gas	Alkaline	Horizontal isothermal plug-flow reactor	✓	Aspen Plus	788 kg/h	15 MW	37.22 %	✘
Kotowicz, Węcel, and Brzęczek (2021)			Wind	Flue gas	N/A	Fixed bed adiabatic reactor	✓	N/A	4.643 kg/kgH_2	40 MW	45–53 %	✘
Atsonios, Panopoulos, and Kakaras (2016)		Fig 13.4b	N/A	Flue gas	Alkaline	Zeolite membrane reactor	✘	Aspen Plus	98,343 tons/d	0.46 MWe	85 %	✘
Smith et al. (2019)		Fig 13.4c	Solar	Air	N/A	N/A	✘	N/A	10,000 tons/d	1148 kJ/ mol_{CO2}	50 %	✘

Al-Kalbani et al. (2016)	Syn-CS	Fig 13.5	N/A	Flue gas	✘	Lurgi type	✘	Aspen HYSYS	1500 ton/d	40-50 GJ_{th}/ $ton_{methanol}$	41 %	✘
Zhang and Desideri (2020)			N/A	Flue gas	✘	N/A	✓	Aspen Plus	100 kton/ year	110 MW	72 %	76 %
Chaniago et al. (2019)			N/A	N/A	✘	N/A	✓	Aspen HYSYS	940.23 kmol/h	220 MW	N/A	N.A.
Rivera-Tinoco et al. (2016)			N/A	N/A	✘	Isothermal plug-flow reactor	✓	Aspen Plus	19,200 ton/ year	25 MW	54.8 %	✘

catalysts (With the composition of $CuO/ZnO/Al_2O_3$ wt%) in a methanol reactor. For example, Bos, Kersten, and Brilman (2020) performed an energy assessment of a P-t-Methanol system that incorporated an alkaline electrolyzer and a direct air capture system. The system was capable of producing 33 kton/year methanol in a Liquid-Out Gas-In Concept reactor (introduced in the study of Bos & Brilman 2015) with an electrical efficiency of 50%. To do this, total power was provided as 100 MW from the wind park. Kotowicz, Węcel, and Brzęczek (2021) presented a renewable synthetic methanol production system through CO_2 hydrogenation. CO_2 is captured from the exhaust gas of the power plant (flue gas). The system produces 4.643 kg methanol from 1 kg hydrogen and 7.277 kg CO_2 in a fixed bed adiabatic reactor with a system electrical efficiency range of 45–52%. The required power was supplied by a wind park with 40 MW. The direct CO_2 fed system is different through heterogeneous catalysts as discussed above, there are also CO fed systems for renewable methanol production. In this regard, the second route presents a CO fed system through the chemical (e.g., RWGS reactor) or electrochemical conversion of CO_2 (e.g., CO_2 electrolyzer). Subsequently, CO reacts with H_2 in a chemical reactor for the methanol synthesis. For instance, Smith et al. (2019) conducted an energy analysis for the P-t- Methanol system that couples the CO2 electrolyzer (for CO production) and water electrolyzer (for hydrogen production). The system has the capability for 10,000 tons of renewable methanol per day with an electrical efficiency of 50%. In that process, CO_2 is captured from the air through the hollow fiber gas-liquid membrane contactors with an energy consumption of 13 kJ per 1 mole CO_2. The total energy requirement was found as 1,148 kJ per 1 mole CO_2. In the third route (Figure 13.4), the renewable methanol can be produced from syngas followed by the catalytic step. Elsernagawy et al. (2020) presented a CO fed system for renewable methanol synthesis. For this purpose, the CO_2 captured from the air is converted to CO through an external RWGS reactor. The thermal integration was provided for the system through the pinch analysis. They found that the temperature of higher RWGS is to provide a great advantage for heat integration. In a third route, through the co-electrolysis process, the syngas can be obtained electrochemically from CO_2 and H_2O by replacing the steam methane reforming process (Andika et al. 2018). According to the best knowledge of the authors, the first energy analysis on a co-electrolysis based P-t-Methanol system was performed by Al-Kalbani et al. (2016). They compared the syngas fed system through the co-electrolysis (the third route) and the CO_2 hydrogenation-based system (the first route). The electrical efficiency of the co-electrolysis based system was calculated as 41%, while the renewable methanol through the CO_2 hydrogenation-based was produced with an efficiency of 20%. Another co-electrolysis-based P-t-Methanol system energy performance assessment was performed by Rivera-Tinoco et al. (2016). The system provides renewable methanol synthesis of 19,200 tons annually with consuming energy of 25 MW. They found the system electrical efficiency as 54.8%. The dramatic increment in the system efficiency was reported by Zhang and Desideri (2020). They presented the co-electrolysis-based P-t-Methanol with a heat recovery system. The presented system produces 100 kton methanol per year with an electrical efficiency of 72%. For this purpose, the system consumes water and CO_2 of 146.7 tons and 163.2 tons per year, respectively; furthermore, the total power requirement was calculated as 110 MW in that process.

13.4.2 Case 2: Power-to-ammonia

Ammonia is one of the most produced chemicals in the industry (c. 180–200 million metric tons per year and its predicted to increase to c. 270 million metric tons per year in

2050) because ammonia is used with a variety of applications such as fertilizers, polymers, nitric acids, refrigerants, and neutralizers (FAO 2017; Hochman et al. 2020; Iqbal Cheema & Krewer 2018). Moreover, since ammonia has higher volumetric (22.5 MJ/kg) and energy density (13.6 GJ/m^3) when compared to other candidates (e.g., hydrogen and methanol) and it is also competitive with some fossil sources (gasoline, diesel, natural gas), ammonia can be utilized as a fuel for ignition engines, gas turbines, and fuel cells (Dimitriou & Javaid 2020; Jeerh, Zhang, & Tao 2021; Valera-Medina et al. 2021; Yapicioglu & Dincer 2019). Basically, ammonia is produced through the combination of hydrogen and nitrogen in the Haber-Bosch reactor system through the Fe- or Ru-based catalysts. Ammonia production is currently dominated by fossil fuels (e.g., steam reforming of natural gas, 72%) because hydrogen that is used as a feedstock for ammonia production comes from fossil fuels. In this regard, about 1.6% of fossil fuels (e.g., natural gas, coal) are used for ammonia production in the world (Iqbal Cheema & Krewer 2018). In this case, 1.87 tons of the greenhouse gas CO_2 can be released for producing 1 ton of ammonia (Foster et al. 2018). The renewable-based ammonia production pathways have therefore gained significant attraction. Two alternative pathways can be followed. The first pathway is based on the electrochemical reduction of nitrogen for the direct production of ammonia. The technology is relatively new, and the research is still on a material level. Therefore, different catalysts are used to promote the selectivity of ammonia and Faradaic efficiency (Zhao et al. 2019). Some recent studies that address the electrochemical performance of HB-RR are presented in Table 13.6. According to this, a couple of Ag-Pd catalyst is the most preferable for electrochemical ammonia synthesis. Au-based materials, which are stable and metallic, are another significant catalyst because of low hydrogen evolution reaction activity. However, theoretical calculations show that the catalytic activity of Au is low. Therefore, Nazemi, Panikkanvalappil, and El-Sayed (2018) promoted the catalytic activity with modifying morphology (i.e., using hollow Au nanocages) in order to enhance electrochemical ammonia synthesis yield. They obtained a relatively higher ammonia yield. Mo, which is non-noble metal, is an efficient electrocatalyst for the electrochemical ammonia synthesis under ambient conditions. For example, Yang, Chen, and Wang (2017) reported that a Faradaic efficiency of 72 % can be achieved at the applied voltage of 0.29 V through

Table 13.6 Basic electrochemical performance comparison for HB-RR

Study	*Catalyst*		*Applied potential*	*Reactants*		*Product yield*	*Faraday efficiency*
	Anode	*Cathode*		*Anode*	*Cathode*		
W. Wang et al. (2010)	Ag-Pd	Ag-Pd	N/A	H_2	N_2	3.50×10^{-9} mol/s·cm^2	52 %
Nazemi, Panikkanvalappil, and El-Sayed (2018)	N/A	Au-based	−0.4 V	H_2O	N_2	3.9×10^{-9} μg/ cm^2·h	30.2 %
Yang, Chen, and Wang (2017)	N/A	Mo-based	−0.49 V	H_2	N_2	3.09×10^{-9} mol/s·cm^2	72 % (−0.29 V)
Zhang et al. (2018)	N/A	TiO_2-rGO	−0.90 V	H_2O	N_2	15.13 μg/ mg_{cat}·h	3.3 %
Yang et al. (2019)	N/A	TiO_2 (defective)	−0.15 V	H_2O	N_2	7.59 μg/ cm^2·h	9.17 %

Mo-based catalyst. However, the Mo-based catalyst shows poor performance in terms of the selectivity. Zhang et al. (2018) investigated electrocatalyst of TiO_2 nanoparticle for the selectivity enhancement. They found the product yield as 15.13 μg/mgcat·h at the applied voltage of 0.90 V. In this case, the Faradaic efficiency was calculated as 3.3%. Similar catalyst was investigated by Yang et al. (2019). They used a defective TiO_2 catalyst for the electrochemical reduction of nitrogen. The study contributed to the increase of the product yield from 1.04 μg/ cm^2·h to 7.59 μg/ cm^2·h.

In the second pathway, as shown in Figure 13.4(d), the hydrogen produced by the water electrolyzer is combined with N_2 that is generally obtained from air. The reaction occurs in a chemical reactor with a Fe- or Ru-based catalyst. There are only a few studies conducted to understand the thermodynamic feasibility of the P-t-Ammonia process, as presented in Table 13.7. For example, Cinti et al. (2017) compared low-temperature and high-temperature water electrolyzer based P-t-Ammonia with a conventional ammonia production system. The highest energy storage efficiency was found as 62.41% when the high-temperature electrolyzer is integrated into the ammonia production system. On the other hand, there is a minor difference between a low-temperature electrolyzer based system and a conventional system in terms of storage efficiency. Zhang et al. (2020) presented trade-offs between thermodynamic performance for three ammonia production systems: 1) SOE integrated P-t-Ammonia system, 2) biomass to ammonia, and 3) conventional system. The highest electrical efficiency was achieved by the SOE integrated ammonia production system. Here, the electrical efficiency for these three cases was calculated as 74% (for case 1), 44% (for case 2), and 61% (for case 3). Siddiqui and Dincer (2019) performed a thermodynamic analysis (including energy and exergy performance) for solar and wind integrated green ammonia production systems. The PEM electrolyzer was used for renewable hydrogen production. The maximum ammonia production was found as 107.7 mol/s. The maximum energy was calculated as 53.3%, while the maximum exergy efficiency was 41.5%. The produced ammonia is then used for fuel

Table 13.7 Basic comparison of the performance of P-t-Ammonia systems

			Thermodynamic assessment			
Study	*Energy sources*	*Hydrogen production*	*Product yield*	*Power requirement*	*Energy efficiency*	*Exergy efficiency*
Cinti et al. (2017)	N/A	SOE PEM	N/A	SOE:8.30 kWh/kg$_{NH3}$ PEM:14.25 kWh/kg$_{NH3}$	SOE:62.41% PEM: 36.35%	✗
Zhang et al. (2020)	N/A	SOE	50 kton/year	48 MW	74 %	N/A
Siddiqui et al. (2020)	Solar	PEM	0.016 mmol/s	601.3 W	1.6%	1.8%
Siddiqui and Dincer (2019)	Wind	PEM	107.7 mol/s	N/A	53.5 %	41.5 %
Ozturk and Dincer (2021)	Solar	PEM	0.22 kg/s	3410 kW	26.08 %	30.17%,

for solid oxide fuel cell. Ozturk and Dincer (2021) presented a thermodynamic assessment (including energy and exergy performance) for the solar-based P-t-Ammonia process where hydrogen is produced through the PEM electrolyzer. The system was capable of producing 0.22 kg/s ammonia from hydrogen amount of 0.04 kg/s and nitrogen amount of 0.18 kg/s. In this case, the energy and exergy efficiencies were calculated as 26.08% and 30.17%, respectively.

13.5 CONCLUSION

The synthetic fuel production from renewables (power-to-fuel) has gained significant consideration in energy storage technologies in the last decades since this concept provides storage of electricity in chemicals (e.g., ammonia, methane, methanol) for long-term periods. Moreover, since hydrogen is produced from water in an eco-friendly way, the fuel synthesis through the P-t-F concept leads to less greenhouse gas emission and reduces fossil fuel depletion when compared to conventional fuel production systems. There is a considerable effort to develop different pathways of the P-t-F system in order to find a more accurate combination in terms of thermodynamic, economic, and environmental aspects. The functional window of these pathways is described and introduced in this chapter. The recent studies that address the thermodynamic assessment of the P-t-F system that employs these pathways are reviewed for two desired fuel cases such as P-t-Methanol and P-t-Ammonia. According to the literature survey conducted, the remarkable conclusions are derived as follows:

- Thermal integration plays a critical role for the enhancement of the thermodynamic performance of the P-t-F system. In this context, for example, the high-temperature electrolyzer integrated P-t-Ammonia system shows better performance than the low-temperature electrolyzer.
- The co-electrolysis process is a promising technology to be integrated into P-t-F systems since the co-electrolysis process eliminates the steam methane reforming that damages the environment significantly. Moreover, many studies show that co-electrolysis-based P-t-Methanol systems present a more efficient way in terms of thermodynamics.
- There is some clue that one of the most feasible ways to use CO_2-RR is to produce CO that can be then used for methanol synthesis in a chemical reactor. The reason can be explained by a low number of electron transfers. For more electron transfers, increment of product selectivity and yield is currently the main research task in CO_2-RR as well as increment of Faradaic efficiency.
- The electrochemical HB reactor (HB-RR) is a promising technology for ammonia synthesis when compared to other ammonia production technologies since the HB-RR is compact and unitary and requires fewer chemical processes. However, like CO_2-RR, the low Faradaic efficiency and ammonia selectivity are the main challenges in terms of performance assessment. In order to increase Faradaic efficiency, different catalysts are still performed, and the most preferable catalysts selected are Ag-Pd, Ru, Au-based, Mo-based, and TiO_2.

In the future, the direct electrochemical reduction pathway could be dominated for fuel synthesis in the industry. The cell architecture, stack development, and balance of plant around the compact electrochemical reactor could be significant research tasks hierarchically to be addressed for prospective studies. On the other hand, from the system perspective of P-t-F, the effect of integration of CO_2-RR that provides CO and co-electrolyzer that provides tailored syngas into the P-t-Methanol system is very limited in the literature. Moreover, the thermodynamic optimization for P-t-Methanol and P-t-Ammonia systems to find the right combination of sub-process should be carried out.

REFERENCES

Adnan, Muflih Arisa, & Md Golam Kibria. 2020. "Comparative techno-economic and life-cycle assessment of power-to-methanol synthesis pathways." *Applied Energy* 278: 115614. https://doi.org/10.1016/j.apenergy.2020.115614.

Aili, D., M. K. Hansen, C. Pan, Q. Li, E. Christensen, J. O. Jensen, & N. J. Bjerrum. 2011. "Phosphoric acid doped membranes based on Nafion®, PBI and their blends: Membrane preparation, characterization and steam electrolysis testing." *International Journal of Hydrogen Energy* 36(12): 6985–6993. https://doi.org/10.1016/j.ijhydene.2011.03.058.

Albo, J., G. Beobide, P. Castaño, & A. Irabien. 2017. "Methanol electrosynthesis from CO_2 at Cu_2O/ZnO prompted by pyridine-based aqueous solutions." *Journal of CO_2 Utilization* 18: 164–172. https://doi.org/10.1016/j.jcou.2017.02.003.

Al-Kalbani, H., J. Xuan, S. García, & H. Wang. 2016. "Comparative energetic assessment of methanol production from CO_2: Chemical versus electrochemical process." *Applied Energy* 165: 1–13. https://doi.org/10.1016/j.apenergy.2015.12.027.

Amores, E., J. Rodríguez, & C. Carreras. 2014. "Influence of operation parameters in the modeling of alkaline water electrolyzers for hydrogen production." *International Journal of Hydrogen Energy* 39(25): 13063–13078. https://doi.org/10.1016/j.ijhydene.2014.07.001.

Andika, R., A. B. D. Nandiyanto, Z. A. Putra, M. R. Bilad, Y. Kim, C. M. Yun, & M. Lee. 2018. "Co-electrolysis for power-to-methanol applications." *Renewable and Sustainable Energy Reviews* 95: 227–241. https://doi.org/10.1016/j.rser.2018.07.030.

Arquer, F. P. G. de et al. 2020. "CO_2 electrolysis to multicarbon products at activities greater than 1 A cm^{-2}." *Science* 367(6478): 661– 666. https://doi.org/10.1126/science.aay4217.

Atsonios, K., K. D. Panopoulos, & E. Kakaras. 2016. "Thermocatalytic CO_2 hydrogenation for methanol and ethanol production: Process improvements." *International Journal of Hydrogen Energy* 41(2): 792–806. https://doi.org/10.1016/j.ijhydene.2015.12.001.

Aziz, M., A. Putranto, M. K. Biddinika, & A. T. Wijayanta. 2017. "Energy-saving combination of N2 production, NH3 synthesis, and power generation." *International Journal of Hydrogen Energy* 42(44): 27174–27183. https://doi.org/10.1016/j.ijhydene.2017.09.079.

Babu, P., P. Linga, R. Kumar, & P. Englezos. 2015. "A review of the hydrate based gas separation (HBGS) process for carbon dioxide pre-combustion capture." *Energy* 85: 261–279. https://doi.org/10.1016/j.energy.2015.03.103.

Bailera, M., P. Lisbona, L. M. Romeo, & S. Espatolero. 2017. "Power to gas projects review: Lab, pilot and demo plants for storing renewable energy and CO_2." *Renewable and Sustainable Energy Reviews* 69: 292–312. https://doi.org/10.1016/j.rser.2016.11.130.

Barco-Burgos, J., U. Eicker, N. Saldaña-Robles, A. L. Saldaña-Robles, & V. Alcántar-Camarena. 2020. "Thermal characterization of an alkaline electrolysis cell for hydrogen production at atmospheric pressure." *Fuel* 276: 117910. https://doi.org/10.1016/j.fuel.2020.117910.

Battaglia, P., G. Buffo, D. Ferrero, M. Santarelli, & A. Lanzini. 2021. "Methanol synthesis through CO_2 capture and hydrogenation: Thermal integration, energy performance and techno-economic assessment." *Journal of CO_2 Utilization* 44: 101407. https://doi.org/10.1016/j.jcou.2020.101407.

Bessarabov, D., H. Wang, H. Li, & N. Zhao. 2016. *PEM Electrolysis for Hydrogen Production: Principles and Applications*. Boca Raton, FL: CRC Press.

Bi, L., S. Boulfrad, & E. Traversa. 2014. "Steam electrolysis by solid oxide electrolysis cells (SOECs) with proton-conducting oxides." *Chemical Society Reviews* 43(24): 8255–8270. https://doi.org/10.1039/C4CS00194J.

Bockris, J. O., & T. N. Veziroglu. 2007. "Estimates of the price of hydrogen as a medium for wind and solar sources." *International Journal of Hydrogen Energy* 32(12): 1605–1610. https://doi.org/10.1016/j.ijhydene.2007.04.037.

Böhringer, C. 2003. "The Kyoto Protocol: A review and perspectives." *Oxford Review of Economic Policy* 19(3): 451–466. https://doi.org/10.1093/oxrep/19.3.451.

Bos, M. J., & D. W. F. Brilman. 2015. "A novel condensation reactor for efficient CO2 to methanol conversion for storage of renewable electric energy." *Chemical Engineering Journal* 278: 527–532. https://doi.org/10.1016/j.cej.2014.10.059.

Bos, M. J., S. R. A. Kersten, & D. W. F. Brilman. 2020. "Wind power to methanol: Renewable methanol production using electricity, electrolysis of water and CO_2 air capture." *Applied Energy* 264: 114672. https://doi.org/10.1016/j.apenergy.2020.114672.

Bossavy, A., R. Girard, & G. Kariniotakis. 2016. "Sensitivity analysis in the technical potential assessment of onshore wind and ground solar photovoltaic power resources at regional scale." *Applied Energy* 182: 145–153. https://doi.org/10.1016/j.apenergy.2016.08.075.

Bozzano, G., & F. Manenti. 2016. "Efficient methanol synthesis: Perspectives, technologies and optimization strategies." *Progress in Energy and Combustion Science* 56: 71–105. https://doi.org/10.1016/j.pecs.2016.06.001.

Buttler, A. 2018. "Technoökonomische Bewertung von Polygenerationskraftwerken und Power-to-X-Speichern in einem nachhaltigen Energiesystem." https://nbn-resolving.org/urn:nbn:de:101:1-2018122822384006956958.

Buttler, A., & H. Spliethoff. 2018. "Current status of water electrolysis for energy storage, grid balancing and sector coupling via power-to-gas and power-to-liquids: A review." *Renewable and Sustainable Energy Reviews* 82: 2440–2454. https://doi.org/10.1016/j.rser.2017.09.003.

Carmo, M., D. L. Fritz, J. Mergel, & D. Stolten. 2013. "A comprehensive review on PEM water electrolysis." *International Journal of Hydrogen Energy* 38(12): 4901–4934. https://doi.org/10.1016/j.ijhydene.2013.01.151.

Chaniago, Y. D., M. A. Qyyum, R. Andika, W. Ali, K. Qadeer, & M. Lee. 2019. "Self- recuperative high temperature co-electrolysis-based methanol production with vortex search-based exergy efficiency enhancement." *Journal of Cleaner Production* 239: 118029. https://doi.org/10.1016/j.jclepro.2019.118029.

Chen, L., F. Chen, & C. Xia. 2014. "Direct synthesis of methane from CO_2–H_2O co- electrolysis in tubular solid oxide electrolysis cells." *Energy & Environmental Science* 7(12): 4018–4022. https://doi.org/10.1039/C4EE02786H.

Cinti, G., D. Frattini, E. Jannelli, U. Desideri, & G. Bidini. 2017. "Coupling solid oxide electrolyser (SOE) and ammonia production plant." *Applied Energy, 192*, 466–476. https://doi.org/10.1016/j.apenergy.2016.09.026.

Costa, R., D. M. A. Dueñas, G. Futter, T. Jahnke, M. Riegraf, G. Schiller, & A. Surrey. 2019. "Solid oxide cells for power-to-X: Application & challenges." *ECS Transactions* 91(1): 2527. https://doi.org/10.1149/09101.2527ecst.

Dahl, P. J., T. S. Christensen, S. Winter-Madsen, & S. M. King. 2014. "Proven autothermal reforming technology for modern large-scale methanol plants." *Nitrogen+ Syngas International Proceedings Conference & Exhibition*, 2–3.

Deka, D. J., S. Gunduz, T. Fitzgerald, J. T. Miller, A. C. Co, & U. S. Ozkan. 2019. "Production of syngas with controllable H_2/CO ratio by high temperature co- electrolysis of CO_2 and H_2O over Ni and Co- doped lanthanum strontium ferrite perovskite cathodes." *Applied Catalysis B: Environmental* 248: 487–503. https://doi.org/10.1016/j.apcatb.2019.02.045.

Demirdelen, T., F. Ekinci, B. D. Mert, İ. Karasu, & M. Tümay. 2020. "Green touch for hydrogen production via alkaline electrolysis: The semi-flexible PV panels mounted wind turbine design, production and performance analysis." *International Journal of Hydrogen Energy* 45(18): 10680–10695. https://doi.org/10.1016/j.ijhydene.2020.02.007.

Dieterich, V., A. Buttler, A. Hanel, H. Spliethoff, & S. Fendt. 2020. "Power-to-liquid via synthesis of methanol, DME or Fischer–Tropsch-fuels: A review." *Energy & Environmental Science* 13(10): 3207–3252. https://doi.org/10.1039/D0EE01187H.

Dimitriou, P., & R. Javaid. 2020. "A review of ammonia as a compression ignition engine fuel." *International Journal of Hydrogen Energy* 45(11): 7098–7118. https://doi.org/10.1016/j.ijhydene.2019.12.209.

Ding, H., W. Wu, & D. Ding. 2019. "Advancement of proton-conducting solid oxide fuel cells and solid oxide electrolysis cells at Idaho National Laboratory (INL)." *ECS Transactions* 91(1): 1029. https://doi.org/10.1149/09101.1029ecst.

Duboviks, V., R. C. Maher, M. Kishimoto, L. F. Cohen, N. P. Brandon, & G. J. Offer. 2014. "A Raman spectroscopic study of the carbon deposition mechanism on Ni/CGO electrodes during CO/CO_2 electrolysis." *Physical Chemistry Chemical Physics* 16(26), 13063–13068. https://doi.org/10.1039/C4CP01503G.

Elsernagawy, O. Y. H., A. Hoadley, J. Patel, T. Bhatelia, S. Lim, N. Haque, & C. Li. 2020. "Thermo-economic analysis of reverse water-gas shift process with different temperatures for green methanol production as a hydrogen carrier." *Journal of CO2 Utilization* 41: 101280. https://doi.org/10.1016/j.jcou.2020.101280.

Elvers, B., & G. Bellussi (eds.). 2011. *Ullmann's Encyclopedia of Industrial Chemistry*, 7th edn. New York: Wiley-VCH.

EU. 2011. *Roadmap for Moving to a Low-Carbon Economy in 2050*. Brussels: European Commission. www.roadmap2050.eu/.

FAO. 2017. *World Fertilizer Trends and Outlook to 2020: Summary Report*. www.fao.org/publications/card/en/c/cfa19fbc-0008-466b-8cc6-0db6c6686f78/.

Fasihi, M., O. Efimova, & C. Breyer. 2019. "Techno-economic assessment of CO_2 direct air capture plants." *Journal of Cleaner Production* 224: 957–980. https://doi.org/10.1016/j.jclepro.2019.03.086.

Foster, S. L., S. I. P. Bakovic, R. D. Duda, S. Maheshwari, R. D. Milton, S. D. Minteer, M. J. Janik, J. N. Renner, & L. F. Greenlee. 2018. "Catalysts for nitrogen reduction to ammonia." *Nature Catalysis* 1(7): 490–500. https://doi.org/10.1038/s41929-018-0092-7.

Fu, Q., Y. Kansha, C. Song, Y. Liu, M. Ishizuka, & A. Tsutsumi. 2016. "An elevated- pressure cryogenic air separation unit based on self-heat recuperation technology for integrated gasification combined cycle systems." *Energy* 103: 440–446. https://doi.org/10.1016/j.energy.2015.09.095.

García, S., M. V. Gil, C. F. Martín, J. J. Pis, F. Rubiera, & C. Pevida. 2011. "Breakthrough adsorption study of a commercial activated carbon for pre-combustion CO_2 capture." *Chemical Engineering Journal* 171(2): 549–556. https://doi.org/10.1016/j.cej.2011.04.027.

Giglio, E., F. A. Deorsola, M. Gruber, S. R. Harth, E. A. Morosanu, D. Trimis, S. Bensaid, & R. Pirone. 2018. "Power-to-gas through high temperature electrolysis and carbon dioxide methanation: Reactor design and process modeling." *Industrial & Engineering Chemistry Research* 57(11): 4007–4018. https://doi.org/10.1021/acs.iecr.8b00477.

Grigoriev, S. A., V. N. Fateev, D. G. Bessarabov, & P. Millet. 2020. "Current status, research trends, and challenges in water electrolysis science and technology." *International Journal of Hydrogen Energy* 45(49): 26036–26058. https://doi.org/10.1016/j.ijhydene.2020.03.109.

Habib, M. A. et al. 2011. "A review of recent developments in carbon capture utilizing oxy-fuel combustion in conventional and ion transport membrane systems." *International Journal of Energy Research* 35(9): 741–764. https://doi.org/10.1002/er.1798.

Ham, Y. S., Y. S. Park, A. Jo, J. H. Jang, S.-K. Kim, & J. J. Kim. 2019. "Proton-exchange membrane CO_2 electrolyzer for CO production using Ag catalyst directly electrodeposited onto gas diffusion layer." *Journal of Power Sources* 437: 226898. https://doi.org/10.1016/j.jpowsour.2019.226898.

Hansen, M. K., D. Aili, E. Christensen, C. Pan, S. Eriksen, J. O. Jensen, J. H. von Barner, Q. Li, & N. J. Bjerrum. 2012. "PEM steam electrolysis at 130°C using a phosphoric acid doped short side chain

PFSA membrane." *International Journal of Hydrogen Energy* 37(15): 10992–11000. https://doi.org/10.1016/j.ijhydene.2012.04.125.

Hazarika, J., & M. S. Manna. 2019. "Electrochemical reduction of CO_2 to methanol with synthesized Cu_2O nanocatalyst: Study of the selectivity." *Electrochimica Acta* 328: 135053. https://doi.org/10.1016/j.electacta.2019.135053.

Hidalgo, D., & J. M. Martín-Marroquín. 2020. "Power-to-methane, coupling CO_2 capture with fuel production: An overview." *Renewable and Sustainable Energy Reviews* 132: 110057. https://doi.org/10.1016/j.rser.2020.110057.

Hochman, G. et al. 2020. "Potential economic feasibility of direct electrochemical nitrogen reduction as a route to ammonia." *ACS Sustainable Chemistry & Engineering* 8(24): 8938–8948. https://doi.org/10.1021/acssuschemeng.0c01206.

Hu, B., A. N. Aphale, M. Reisert, S. Belko, O. A. Marina, J. W. Stevenson, & P. Singh. 2018. "Solid oxide electrolysis for hydrogen production: From oxygen ion to proton conducting cells." *ECS Transactions* 85(10): 13. https://doi.org/10.1149/08510.0013ecst.

Huang, H., R. Can Samsun, R. Peters, & D. Stolten. 2021. "Greener production of dimethyl carbonate by the power-to-fuel concept: A comparative techno-economic analysis." *Green Chemistry* 23(4): 1734–1747. https://doi.org/10.1039/D0GC03865B.

Huang, Z., G. Grim, J. Schaidle, & L. Tao. 2020. "Using waste CO_2 to increase ethanol production from corn ethanol biorefineries: Techno-economic analysis." *Applied Energy* 280: 115964. https://doi.org/10.1016/j.apenergy.2020.115964.

Huang, W., & G. Yuan. 2020. "A composite heterogeneous catalyst C-Py-Sn-Zn for selective electrochemical reduction of CO_2 to methanol." *Electrochemistry Communications* 118: 106789. https://doi.org/10.1016/j.elecom.2020.106789.

Hübler, M., & A. Löschel. 2013. "The EU Decarbonisation Roadmap 2050: What way to walk?" *Energy Policy* 55: 190–207. https://doi.org/10.1016/j.enpol.2012.11.054.

Ikäheimo, J., J. Kiviluoma, R. Weiss, & H. Holttinen. 2018. "Power-to-ammonia in future North European 100% renewable power and heat system." *International Journal of Hydrogen Energy* 43(36): 17295–17308. https://doi.org/10.1016/j.ijhydene.2018.06.121.

IEA. 2020a. *Global Energy Review 2020*. Paris: IEA. www.iea.org/reports/global-energy-review-2020.

IEA. 2020b. *World Energy Outlook 2020*. www.iea.org/reports/world-energy-outlook-2020.

Ince, A. C., M. U. Karaoglan, A. Glüsen, C. O. Colpan, M. Müller, & D. Stolten. 2019. "Semiempirical thermodynamic modeling of a direct methanol fuel cell system." *International Journal of Energy Research* 43(8): 3601–3615. https://doi.org/10.1002/er.4508.

Iqbal Cheema, I., & U. Krewer. 2018. "Operating envelope of Haber–Bosch process design for power-to-ammonia." *RSC Advances* 8(61): 34926–34936. https://doi.org/10.1039/C8RA06821F.

Irfan Malik, M., Z. O. Malaibari, M. Atieh, & B. Abussaud. 2016. "Electrochemical reduction of CO_2 to methanol over MWCNTs impregnated with Cu_2O." *Chemical Engineering Science* 152: 468–477. https://doi.org/10.1016/j.ces.2016.06.035.

Jacobson, M. Z. et al. 2015. "100% clean and renewable wind, water, and sunlight (WWS) all-sector energy roadmaps for the 50 United States." *Energy & Environmental Science* 8(7): 2093–2117. https://doi.org/10.1039/C5EE01283J.

Jansen, D., M. Gazzani, G. Manzolini, E. van Dijk, & M. Carbo. 2015. "Pre-combustion CO_2 capture." *International Journal of Greenhouse Gas Control* 40: 167–187. https://doi.org/10.1016/j.ijggc.2015.05.028.

Jeerh, G., M. Zhang, & S. Tao. 2021. "Recent progress in ammonia fuel cells and their potential applications." *Journal of Materials Chemistry A* 9(2): 727–752. https://doi.org/10.1039/D0TA08810B.

Jørgensen, C., & S. Ropenus. 2008. "Production price of hydrogen from grid connected electrolysis in a power market with high wind penetration." *International Journal of Hydrogen Energy* 33(20): 5335–5344. https://doi.org/10.1016/j.ijhydene.2008.06.037.

Jouny, M., W. Luc, & F. Jiao. 2018. "General techno-economic analysis of CO_2 electrolysis systems." *Industrial & Engineering Chemistry Research* 57(6): 2165–2177. https://doi.org/10.1021/acs.iecr.7b03514.

Kaithal, A., C. Werlé, & W. Leitner. 2021. "Alcohol-assisted hydrogenation of carbon monoxide to methanol using molecular manganese catalysts." *JACS Au* 1(2): 130–136. https://doi.org/10.1021/jacsau.0c00091.

Kanniche, M., R. Gros-Bonnivard, P. Jaud, J. Valle-Marcos, J.-M. Amann, & C. Bouallou. 2010. "Pre-combustion, post-combustion and oxy-combustion in thermal power plant for CO_2 capture." *Applied Thermal Engineering* 30(1): 53–62. https://doi.org/10.1016/j.applthermaleng.2009.05.005.

Karaoglan, M. U., A. C. Ince, A. Glüsen, C. O. Colpan, M. Müller, D. Stolten, & N. S. Kuralay. 2021. "Comparison of single-cell testing, short-stack testing and mathematical modeling methods for a direct methanol fuel cell." *International Journal of Hydrogen Energy* 46(6): 4844–4856. https://doi.org/10.1016/j.ijhydene.2020.02.107.

Kotb, Y., S.-E. K. Fateen, J. Albo, & I. Ismail. 2017. "Modeling of a microfluidic electrochemical cell for the electro-reduction of CO_2 to CH_3OH." *Journal of the Electrochemical Society* 164(13): E391. https://doi.org/10.1149/2.0741713jes.

Kotowicz, J., D. Węcel, & M. Brzęczek. 2021. "Analysis of the work of a "renewable" methanol production installation based on H_2 from electrolysis and CO^2 from power plants." *Energy* 221: 119538. https://doi.org/10.1016/j.energy.2020.119538.

Koytsoumpa, E. I., C. Bergins, & E. Kakaras. 2018. "The CO_2 economy: Review of CO_2 capture and reuse technologies." *The Journal of Supercritical Fluids* 132: 3–16. https://doi.org/10.1016/j.supflu.2017.07.029.

Kuhl, K. P., E. R. Cave, D. N. Abram, & T. F. Jaramillo. 2012. "New insights into the electrochemical reduction of carbon dioxide on metallic copper surfaces." *Energy & Environmental Science* 5(5): 7050–7059. https://doi.org/10.1039/C2EE21234J.

Küngas, R. 2020. "Review: Electrochemical CO_2 reduction for CO production: Comparison of low- and high-temperature electrolysis technologies." *Journal of the Electrochemical Society* 167(4): 044508. https://doi.org/10.1149/1945-7111/ab7099.

Kyriakou, V., I. Garagounis, E. Vasileiou, A. Vourros, & M. Stoukides. 2017. "Progress in the electrochemical synthesis of ammonia." *Catalysis Today* 286: 2–13. https://doi.org/10.1016/j.cattod.2016.06.014.

Lau, L. C., K. T. Lee, & A. R. Mohamed. 2012. "Global warming mitigation and renewable energy policy development from the Kyoto Protocol to the Copenhagen Accord: A comment." *Renewable and Sustainable Energy Reviews* 16(7): 5280–5284. https://doi.org/10.1016/j.rser.2012.04.006.

Lee, M.-Y., K. T. Park, W. Lee, H. Lim, Y. Kwon, & S. Kang. 2020. "Current achievements and the future direction of electrochemical CO_2 reduction: A short review." *Critical Reviews in Environmental Science and Technology* 50(8): 769–815. https://doi.org/10.1080/10643389.2019.1631991.

Lei, L., T. Liu, S. Fang, J. P. Lemmon, & F. Chen. 2017. "The co-electrolysis of CO_2–H_2O to methane via a novel micro-tubular electrochemical reactor." *Journal of Materials Chemistry A* 5(6): 2904–2910. https://doi.org/10.1039/C6TA10252B.

Lei, L., J. Zhang, Z. Yuan, J. Liu, M. Ni, & F. Chen. 2019. "Progress report on proton conducting solid oxide electrolysis cells." *Advanced Functional Materials* 29(37): 1903805. https://doi.org/10.1002/adfm.201903805.

Lenser, C. et al. 2020. "Solid oxide fuel and electrolysis cells." In O. Guillon (ed.), *Advanced Ceramics for Energy Conversion and Storage*, pp. 387–547. New York: Elsevier. https://doi.org/10.1016/B978-0-08-102726-4.00009-0.

Leonzio, G., E. Zondervan, & P. U. Foscolo. 2019. "Methanol production by CO_2 hydrogenation: Analysis and simulation of reactor performance." *International Journal of Hydrogen Energy* 44(16): 7915–7933. https://doi.org/10.1016/j.ijhydene.2019.02.056.

Li, W., H. Wang, Y. Shi, & N. Cai. 2013. "Performance and methane production characteristics of H_2O–CO_2 co-electrolysis in solid oxide electrolysis cells." *International Journal of Hydrogen Energy* 38(25): 11104–11109. https://doi.org/10.1016/j.ijhydene.2013.01.008.

Liu, Z., H. Yang, R. Kutz, & R. I. Masel. 2018. "CO_2 electrolysis to CO and O_2 at high selectivity, stability and efficiency using sustainion membranes." *Journal of the Electrochemical Society* 165(15): J3371. https://doi.org/10.1149/2.0501815jes.

Low, Q. H., N. W. X. Loo, F. Calle-Vallejo, & B. S. Yeo. 2019. "Enhanced electroreduction of carbon dioxide to methanol using zinc dendrites pulse- deposited on silver foam." *Angewandte Chemie International Edition* 58(8): 2256– 2260. https://doi.org/10.1002/anie.201810991.

Luo, Y., Y. Shi, W. Li, & N. Cai. 2018. "Synchronous enhancement of H_2O/CO_2 co- electrolysis and methanation for efficient one-step power-to-methane." *Energy Conversion and Management* 165: 127–136. https://doi.org/10.1016/j.enconman.2018.03.028.

Lyngfelt, A. 2020. "Chemical looping combustion: Status and development challenges." *Energy & Fuels* 34(8): 9077–9093. https://doi.org/10.1021/acs.energyfuels.0c01454.

Manage, M. N., D. Hodgson, N. Milligan, S. J. R. Simons, & D. J. L. Brett. 2011. "A techno-economic appraisal of hydrogen generation and the case for solid oxide electrolyser cells." *International Journal of Hydrogen Energy* 36(10): 5782–5796. https://doi.org/10.1016/j.ijhydene.2011.01.075.

Mohn, K. 2020. "The gravity of status quo: A review of IEA's World Energy Outlook." *Economics of Energy & Environmental Policy* 9(1). https://doi.org/10.5547/2160-5890.8.2.kmoh.

Momeni, M., M. Soltani, M. Hosseinpour, & J. Nathwani. 2021. "A comprehensive analysis of a power-to-gas energy storage unit utilizing captured carbon dioxide as a raw material in a large-scale power plant." *Energy Conversion and Management* 227: 113613. https://doi.org/10.1016/j.enconman.2020.113613.

Mondal, M. K., H. K. Balsora, & P. Varshney. 2012. "Progress and trends in CO_2 capture/separation technologies: A review." *Energy* 46(1): 431–441. https://doi.org/10.1016/j.energy.2012.08.006.

Morales, J. M., A. J. Conejo, H. Madsen, P. Pinson, & M. Zugno. 2014. *Integrating Renewables in Electricity Markets*, Vol. 205. https://doi.org/10.1007/978-1-4614-9411-9.

Nazemi, M., S. R. Panikkanvalappil, & M. A. El-Sayed. 2018. "Enhancing the rate of electrochemical nitrogen reduction reaction for ammonia synthesis under ambient conditions using hollow gold nanocages." *Nano Energy* 49: 316–323. https://doi.org/10.1016/j.nanoen.2018.04.039.

Ni, M. 2012. "An electrochemical model for syngas production by co-electrolysis of H_2O and CO_2." *Journal of Power Sources* 202: 209–216. https://doi.org/10.1016/j.jpowsour.2011.11.080.

Ozturk, M., & I. Dincer. 2021. "An integrated system for ammonia production from renewable hydrogen: A case study." *International Journal of Hydrogen Energy* 46(8): 5918–5925. https://doi.org/10.1016/j.ijhydene.2019.12.127.

Peters, R., M. Baltruweit, T. Grube, R. C. Samsun, & D. Stolten. 2019. "A techno economic analysis of the power to gas route." *Journal of CO_2 Utilization* 34: 616–634. https://doi.org/10.1016/j.jcou.2019.07.009.

Qi, W., Y. Gan, D. Yin, Z. Li, G. Wu, K. Xie, & Y. Wu. 2014. "Remarkable chemical adsorption of manganese-doped titanate for direct carbon dioxide electrolysis." *Journal of Materials Chemistry A* 2(19): 6904–6915. https://doi.org/10.1039/C4TA00344F.

Ram, M. et al. 2019. *Global Energy System Based on 100% Renewable Energy*. www.energywatchgroup.org/wp-content/uploads/EWG_LUT_100RE_All_Sectors_Global_Report_2019.pdf.

Rivera-Tinoco, R., M. Farran, C. Bouallou, F. Auprêtre, S. Valentin, P. Millet, & J. R. Ngameni. 2016. "Investigation of power-to-methanol processes coupling electrolytic hydrogen production and catalytic CO_2 reduction." *International Journal of Hydrogen Energy* 41(8): 4546–4559. https://doi.org/10.1016/j.ijhydene.2016.01.059.

Samanta, A., A. Zhao, G. K. H. Shimizu, P. Sarkar, & R. Gupta. 2012. "Post-combustion CO_2 capture using solid sorbents: A review." *Industrial & Engineering Chemistry Research* 51(4): 1438–1463. https://doi.org/10.1021/ie200686q.

Savaresi, A. 2016. "The Paris Agreement: A new beginning?" *Journal of Energy & Natural Resources Law* 34(1): 16–26. https://doi.org/10.1080/02646811.2016.1133983.

Schemme, S., R. C. Samsun, R. Peters, & D. Stolten. 2017. "Power-to-fuel as a key to sustainable transport systems: An analysis of diesel fuels produced from CO_2 and renewable electricity." *Fuel* 205: 198–221. https://doi.org/10.1016/j.fuel.2017.05.061.

Schwabe, F., L. Schwarze, C. Partmann, W. Lippmann, & A. Hurtado. 2019. "Concept, design, and energy analysis of an integrated power-to-methanol process utilizing a tubular proton-conducting solid oxide electrolysis cell." *International Journal of Hydrogen Energy* 44(25): 12566–12575. https://doi.org/10.1016/j.ijhydene.2018.11.133.

Shipman, M. A., & M. D. Symes. 2017. "Recent progress towards the electrosynthesis of ammonia from sustainable resources." *Catalysis Today* 286: 57–68. https://doi.org/10.1016/j.cattod.2016.05.008.

Shiva Kumar, S., & V. Himabindu. 2019. "Hydrogen production by PEM water electrolysis: A review." *Materials Science for Energy Technologies* 2(3): 442–454. https://doi.org/10.1016/j.mset.2019.03.002.

Siddiqui, O., & I. Dincer. 2019. "Design and analysis of a novel solar-wind based integrated energy system utilizing ammonia for energy storage." *Energy Conversion and Management* 195: 866–884. https://doi.org/10.1016/j.enconman.2019.05.001.

Siddiqui, O., H. Ishaq, G. Chehade, & I. Dincer. 2020. "Experimental investigation of an integrated solar powered clean hydrogen to ammonia synthesis system." *Applied Thermal Engineering* 176: 115443. https://doi.org/10.1016/j.applthermaleng.2020.115443.

Skafte, T. L., P. Blennow, J. Hjelm, & C. Graves. 2018. "Carbon deposition and sulfur poisoning during CO_2 electrolysis in nickel-based solid oxide cell electrodes." *Journal of Power Sources* 373: 54–60. https://doi.org/10.1016/j.jpowsour.2017.10.097.

Smith, W. A., T. Burdyny, D. A. Vermaas, & H. Geerlings. 2019. "Pathways to industrial-scale fuel out of thin air from CO_2 electrolysis." *Joule* 3(8): 1822–1834. https://doi.org/10.1016/j.joule.2019.07.009.

Song, C., Q. Liu, N. Ji, S. Deng, J. Zhao, Y. Li, Y. Song, & H. Li. 2018. "Alternative pathways for efficient CO_2 capture by hybrid processes: A review." *Renewable and Sustainable Energy Reviews* 82: 215–231. https://doi.org/10.1016/j.rser.2017.09.040.

Song, Y., X. Zhang, K. Xie, G. Wang, & X. Bao. 2019. "High-temperature CO_2 electrolysis in solid oxide electrolysis cells: Developments, challenges, and prospects." *Advanced Materials* 31(50): 1902033. https://doi.org/10.1002/adma.201902033.

Suryanto, B. H. R., H.-L. Du, D. Wang, J. Chen, A. N. Simonov, & D. R. MacFarlane. 2019. "Challenges and prospects in the catalysis of electroreduction of nitrogen to ammonia." *Nature Catalysis* 2(4): 290–296. https://doi.org/10.1038/s41929-019-0252-4.

Toghyani, S., E. Afshari, E. Baniasadi, S. A. Atyabi, & G. F. Naterer. 2018. "Thermal and electrochemical performance assessment of a high temperature PEM electrolyzer." *Energy* 152: 237–246. https://doi.org/10.1016/j.energy.2018.03.140.

Tucker, M. C. 2020. "Progress in metal-supported solid oxide electrolysis cells: A review." *International Journal of Hydrogen Energy* 45(46): 24203–24218. https://doi.org/10.1016/j.ijhydene.2020.06.300.

Valera-Medina, A. et al. 2021. "Review on ammonia as a potential fuel: From synthesis to economics." *Energy & Fuels*. https://doi.org/10.1021/acs.energyfuels.0c03685.

Wan, Y., J. Xu, & R. Lv. 2019. "Heterogeneous electrocatalysts design for nitrogen reduction reaction under ambient conditions." *Materials Today* 27: 69–90. https://doi.org/10.1016/j.mattod.2019.03.002.

Wang, M. 2020. *Comparative Techno Economic Analysis of Ammonia Electrosynthesis*. https://prism.ucalgary.ca/handle/1880/111617.

Wang, J., S. Chen, Z. Li, G. Li, & X. Liu. 2020. "Recent advances in electrochemical synthesis of ammonia through nitrogen reduction under ambient conditions." *ChemElectroChem* 7(5): 1067–1079. https://doi.org/10.1002/celc.201901967.

Wang, X., Z. Feng, B. Xiao, J. Zhao, H. Ma, Y. Tian, H. Pang, & L. Tan. 2020. "Polyoxometalate-based metal–organic framework-derived bimetallic hybrid materials for upgraded electrochemical reduction of nitrogen." *Green Chemistry* 22(18): 6157–6169. https://doi.org/10.1039/D0GC01149E.

Wang, W. B., J. W. Liu, Y. D. Li, H. T. Wang, F. Zhang, & G. L. Ma. 2010. "Microstructures and proton conduction behaviors of Dy-doped BaCeO3 ceramics at intermediate temperature." *Solid State Ionics* 181(15): 667–671. https://doi.org/10.1016/j.ssi.2010.04.008.

Wang, L., M. Rao, S. Diethelm, T.-E. Lin, H. Zhang, A. Hagen, F. Maréchal, & J. Van Herle. 2019. "Power-to-methane via co-electrolysis of H_2O and CO_2: The effects of pressurized operation and internal methanation." *Applied Energy* 250: 1432–1445. https://doi.org/10.1016/j.apenergy.2019.05.098.

Wang, L., Y. Xu, T. Chen, D. Wei, X. Guo, L. Peng, N. Xue, Y. Zhu, M. Ding, & W. Ding. 2021. "Ternary heterostructural CoO/CN/Ni catalyst for promoted CO_2 electroreduction to methanol." *Journal of Catalysis* 393: 83–91. https://doi.org/10.1016/j.jcat.2020.11.012.

Wang, Y., L. Zhao, A. Otto, M. Robinius, & D. Stolten. 2017. "A review of post- combustion CO_2 capture technologies from coal-fired power plants." *Energy Procedia* 114: 650–665. https://doi.org/10.1016/j.egypro.2017.03.1209.

Wijayanta, A. T., & M. Aziz. 2019. "Ammonia production from algae via integrated hydrothermal gasification, chemical looping, N_2 production, and NH_3 synthesis." *Energy* 174: 331–338. https://doi.org/10.1016/j.energy.2019.02.190.

Wilcox, J., P. C. Psarras, & S. Liguori. 2017. "Assessment of reasonable opportunities for direct air capture." *Environmental Research Letters* 12(6): 065001. https://doi.org/10.1088/1748-9326/aa6de5.

Wu, F., M. D. Argyle, P. A. Dellenback, & M. Fan. 2018. "Progress in O_2 separation for oxy-fuel combustion: A promising way for cost-effective CO_2 capture: A review." *Progress in Energy and Combustion Science* 67: 188–205. https://doi.org/10.1016/j.pecs.2018.01.004.

Xie, K., Y. Zhang, G. Meng, & J. T. S. Irvine. 2011. "Direct synthesis of methane from CO_2/H_2O in an oxygen-ion conducting solid oxide electrolyser." *Energy & Environmental Science* 4(6): 2218–2222. https://doi.org/10.1039/C1EE01035B.

Xu, H., B. Chen, J. Irvine, & M. Ni. 2016. "Modeling of CH4-assisted SOEC for H_2O/CO_2 co-electrolysis." *International Journal of Hydrogen Energy* 41(47): 21839–21849. https://doi.org/10.1016/j.ijhydene.2016.10.026.

Yang, D., T. Chen, & Z. Wang. 2017. "Electrochemical reduction of aqueous nitrogen (N 2) at a low overpotential on (110)-oriented Mo nanofilm." *Journal of Materials Chemistry A* 5(36): 18967–18971. https://doi.org/10.1039/C7TA06139K.

Yang, L., T. Wu, R. Zhang, H. Zhou, L. Xia, X. Shi, H. Zheng, Y. Zhang, & X. Sun. 2019. "Insights into defective TiO_2 in electrocatalytic N_2 reduction: Combining theoretical and experimental studies." *Nanoscale* 11(4): 1555–1562. https://doi.org/10.1039/C8NR09564G.

Yapicioglu, A., & I. Dincer. 2019. "A review on clean ammonia as a potential fuel for power generators." *Renewable and Sustainable Energy Reviews* 103: 96–108. https://doi.org/10.1016/j.rser.2018.12.023.

Yuan, J., M.-P. Yang, Q.-L. Hu, S.-M. Li, H. Wang, & J.-X. Lu. 2018. "Cu/TiO_2 nanoparticles modified nitrogen-doped graphene as a highly efficient catalyst for the selective electroreduction of CO_2 to different alcohols." *Journal of CO2 Utilization* 24: 334–340. https://doi.org/10.1016/j.jcou.2018.01.021.

Zappa, W., M. Junginger, & M. van den Broek. 2019. "Is a 100% renewable European power system feasible by 2050?" *Applied Energy* 233–234: 1027–1050. https://doi.org/10.1016/j.apenergy.2018.08.109.

Zhang, H., & U. Desideri. 2020. "Techno-economic optimization of power-to-methanol with co-electrolysis of CO_2 and H_2O in solid-oxide electrolyzers." *Energy* 199: 117498. https://doi.org/10.1016/j.energy.2020.117498.

Zhang, X., Q. Liu, X. Shi, A. M. Asiri, Y. Luo, X. Sun, & T. Li. 2018. "TiO_2 nanoparticles–reduced graphene oxide hybrid: An efficient and durable electrocatalyst toward artificial N_2 fixation to NH_3 under ambient conditions." *Journal of Materials Chemistry A* 6(36): 17303–17306. https://doi.org/10.1039/C8TA05627G.

Zhang, H., L. Wang, J. Van Herle, F. Maréchal, & U. Desideri. 2020. "Techno-economic comparison of green ammonia production processes." *Applied Energy* 259: 114135. https://doi.org/10.1016/j.apenergy.2019.114135.

Zhang, D., & K. Zeng. 2016. "Status and prospects of alkaline electrolysis." In D. Stolten & B. Emonts (eds.), *Hydrogen Science and Engineering: Materials, Processes, Systems and Technology*, pp. 283–308. New York: John Wiley & Sons. https://doi.org/10.1002/9783527674268.ch13.

Zhao, R. et al. 2019. "Recent progress in the electrochemical ammonia synthesis under ambient conditions." *EnergyChem* 1(2): 100011. https://doi.org/10.1016/j.enchem.2019.100011.

Zhou, Y., Z. Zhou, Y. Song, X. Zhang, F. Guan, H. Lv, Q. Liu, S. Miao, G. Wang, & X. Bao. 2018. "Enhancing CO_2 electrolysis performance with vanadium-doped perovskite cathode in solid oxide electrolysis cell." *Nano Energy* 50: 43–51. https://doi.org/10.1016/j.nanoen.2018.04.054.

Index

Absorber 29, 82, 194
Absorption Chiller 2, 83, 194, 262
Ash Content 257

Battery 119
Biomass 49
Blade Coolant 150, 151
Boiler and Evaporators 252
Bottoming Cycle 5

CHP Plants 94, 169
Co-Gasification 35
Coal Gasifier 36
Coal Mass Ratio 38, 39
Combined Cycle 3, 4
Combustion Chamber 101, 149
Compression Ratio 143
Compressor Inlet Temperature 57, 58
Concentrated Solar Power 3, 56, 260
Condenser 250
Cost Index 29, 102
Cryogenic Fuel 49

Diathermic Oil 258
Dry Working Fluid 245–46

Economic Analysis 198
Economic Impacts 129
Energy Analysis 222
Energy Security 129
Engineering Equation Software 15
Environmental Analysis 198
Environmental Impacts 193
Evaporative Condenser 250
Evaporator 17
Exergoeconomic 18, 185
Exergoeconomic Analysis 225
Exergy Analysis 224
Exergy Destruction Rate 178, 183
Exergy Efficiency 176, 178, 282

Fuel Cells 48
Fuel Utilization 138
Fuel Utilization Ratio 142, 153, 154

Gasification 3, 39, 76
Gasifier 35
Genetic Algorithm 19
Geothermal Energy 48
GWP 241

HRSG 126, 139
Hybrid Efficiency 153

Industrial Waste Heat 48
Intercooler 143
Internal Combustion Engine 48
IRGT Cycle 230, 231
Isentropic Working Fluid 46

Kalina Cycle 136, 168

Levelized Cost 20, 168, 198
LHV 36

Molten Carbonate Fuel Cell 7, 138
Multi-Criteria Optimization 13, 19, 24, 171, 178, 186, 187
Multi-Objective Optimization 13, 138, 200, 207

Nuclear Reactors 47

ODP 241

P-T-F Pathways 270, 275
Pareto Frontier 24, 26, 27, 30
Performance Map 157, 159
Photovoltaic Cell 2, 117
Power-To-Ammonia 284
Power-To-Fuel 269
Power-To-Methanol 279
Pressure Ratio 139, 143

Proton Exchange Membrane Fuel Cell 73, 131

Recirculation Ratio 147, 153, 154, 160
Recuperator 144
Reforming 274

Sensitivity Analysis 151, 152, 153, 160
Shifting 99
Simulation 35, 104, 141
Solar Thermal 259
Solar Evacuated Thermal Collector 195
Solar Panel 117
Solid Oxide Fuel Cells 5, 73, 93, 97
Solid Waste Biomass 36
Specific CO_2 Emission 229, 232, 234, 237
Strong Solution 223
Subcritical 45, 56
Supercritical 45
Sustainability Index 13, 16, 19, 31
Syngas 36, 37, 38

Thermoelectric Generator 168, 169, 171
Thermophysical Properties 49, 51, 58, 65
Total Cost Rate 13, 18, 19, 20, 21, 23, 27, 28, 29, 138, 176, 177, 178, 179, 180, 181, 183, 184, 185, 186, 229, 232, 234, 236, 237
Total Energy Efficiency 228
Total Exergy Efficiency 229
Total Revenue Requirement 104
Transcritical 45, 46, 66
Trigeneration 74, 81, 219
Turbine Inlet Temperature 140, 151, 156

Waste Gasifier 172
Waste Heat Recovery 48, 219
Water Electrolyzer 272
Wet Working Fluid 46